U. Lindemann (Ed.)

Human Behaviour in Design

Springer

Berlin
Heidelberg
New York
Hong Kong
London
Milan
Paris
Tokyo

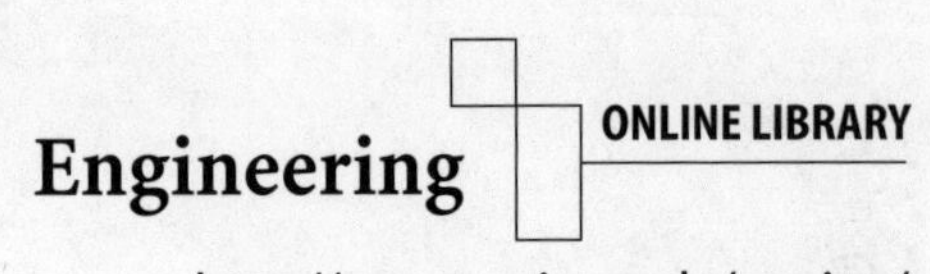

Udo Lindemann (Ed.)

Human Behaviour in Design

Individuals, Teams, Tools

With 95 Figures

Springer

Professor Dr.-Ing. Udo Lindemann

Technische Universität München
Institute for Product Development
Boltzmannstr. 15
85748 Garching
e-mail: sekr.@pe.mw.tum.de

ISBN 3-540-40632-8 Springer-Verlag Berlin Heidelberg New York

Library of Congress Cataloging-in-Publication-Data applied for

A catalog record for this book is available from the Library of Congress.

Bibliographic information published by Die Deutsche Bibliothek. Die Deutsche Bibliothek lists this publication in the Deutsche Nationalbibliographie; detailed bibliographic data is available in the Internet at http://dnb.ddb.de

Springer-Verlag Berlin Heidelberg New York
a member of BertelsmannSpringer Science+Business Media GmbH
http://www.springer.de

Printed in Germany

Typesetting: camera-ready by author
Cover design: medio Technologies AG, Berlin
Printed on acid free paper 62/3020/M - 5 4 3 2 1 0

Contents

Preface

Udo Lindemann

"Human Behaviour in Design" addresses some important aspects of creative engineering design. An informal group of German scientists in engineering design and cognitive psychology have been co-operating for some years and discussing the basic issues of design thinking. The main topics are the role of the interaction between two complementary modalities – image (*„Bild"*) and concept (*„Begriff"*), internal and external components of design thinking, and design strategies – both for individual designers and design teams. One of the goals is to improve and evaluate tools and methods that support design.
This group was initiated by Klaus Ehrlenspiel during a set of workshops called the "Ladenburger Diskurs" in 1992 and '93. We have to thank Klaus Ehrlenspiel for his initiative and his vision of future research.
After ten years of regular meetings and a number of bilateral research projects between members of this group, a number of papers had been published in journals and at conferences (both in the fields of psychology as well as engineering). It then seemed that a strong link to other international activities was required. Because of this impression the group organised a conference and invited a number of experts in the addressed area of research. "Human Behaviour in Design" was the title of this conference, which took place in March 2003 in the castle of Hohenkammer in Germany.
All of the participants prepared a paper and a poster from their field of research. Based on these documents and additional modifications following the joint discussion, we were able to create this book. In addition the stream chairs tried to summarise the outcome – results and questions - of the discussions within the streams as well as in the plenary session.
I want to say thank you to all participants for all their efforts before, during, and after the conference.
Conferences as well as books have to be prepared and organised. This was done by the "Bild und Begriff" group during a set of meetings and with additional individual work. Key persons within this team were Hans Stricker and Martin Pache, who had to carry the main organisational load of the conference along with this book. Thank you to the whole "Bild und Begriff" group, the chairpersons as well as the organising team.
Financial support was given by the Deutsche Forschungsgemeinschaft DFG (German Science Foundation) which carried most of the cost of the conference. With this basis we were able to invite key persons from around the globe to participate at the conference and prepare this book. The participants came from 6 European countries, from Australia, India, Israel and the USA. Thank you to all of the participants at the conference and the authors of the papers and the conclusion/discussion sections prepared for this book.

The venue of the conference was Schloss Hohenkammer, an old water castle located in the Bavarian countryside in the south of Germany. This location was an excellent site for intensive and deep discussions amongst the participants.
We have to thank the publisher Springer for supporting this book. From the very beginning the collaboration was target oriented and fruitful.

Introduction

"Human Behaviour in Design" is supporting the discussion concerning the role of human beings and their strategies in engineering design.
What is the origin of this book?
In 1992 and 1993 scientists of engineering design and cognitive psychology met under guidance of Gerhard Pahl and Johannes Müller near Mannheim in Germany for extensive workshops concerning common questions and problems around design and research on design. The results were published as the "Ladenburger Diskurs" (Pahl 1994). Because of an intensive discussion about the importance and the influence of images (in German "Bild") and concepts / words (in German "Begriff") between Dietrich Dörner (psychologist from Bamberg) and Klaus Ehrlenspiel (engineering designer from Munich) during one of these workshops the idea of a special workshop came up. Klaus Ehrlenspiel invited a number of scientists to discuss intensively the questions around images and concepts ("Bild und Begriff") in Munich in February 1994. This was the beginning of the workshops "Bild und Begriff". There have been ten annual meetings to date.
Questions of the first workshop in Munich were addressing items like these:

- What is the importance of generating sketches and drawings when teaching and learning design?
- Under which circumstances are concepts more important than images?
- Are there talents preferring images or others preferring concepts?
- Is creativity more related to images than to concepts?
- Is it a disadvantage that technical universities have the main focus on analytical, discursive and mathematical training neglecting the creative and image oriented aspects?
- Is there an overlap of art and engineering design?

During the years a number of topics have been discussed within the "Bild und Begriff" workshops. Some examples:

- Image and concept (word) are important representations during design.
    - What is the relation between internal representations within our mind compared to external representations like sketches?
    - What is the importance of physical models in this context?
- External representation in the engineering design process
    - Is the role of the various kinds of representation different in distinct phases of design e.g. conceptual design compared to embodiment design?
    - What is the specific content of the different forms of representations?
    - Which kind of representation best supports creativity?
- Acting in engineering design
    - Which items are influencing the steps and the strategies of acting in design?

  - How can we support a target-oriented process?
- Analysis of problem solving in design
  - How can we measure the external as well as the internal effort of specific processes when acting as a problem solver?
  - What are the characteristics of individual and team oriented design processes?
- Cognitive outsourcing
  - What is the role of computers within design processes?
  - Do we lose creativity when working with computer systems?
  - What are the characteristics of computer systems to support creativity?

The members of the "Bild und Begriff" group are:

- Bamberg, Institute of Theoretical Psychology, Dietrich Dörner and Petra-Badke-Schaub
- Berlin, Institute of Engineering Design and Methodology, Lucienne Blessing
- Braunschweig, Institute of Design Technology, Hans-Joachim Franke
- Darmstadt, Institute of Product Development and Machine Elements, Herbert Birkhofer
- Dresden, Research group Knowledge, Thinking and Acting, Winfried Hacker
- Ilmenau, Institute of Design Technology, Günter Höhne
- Jena, Institute of General Psychology, Werner Krause
- Munich, Institute of Product Development, Udo Lindemann
- Stuttgart, University of Applied Science, Rüdiger von der Weth

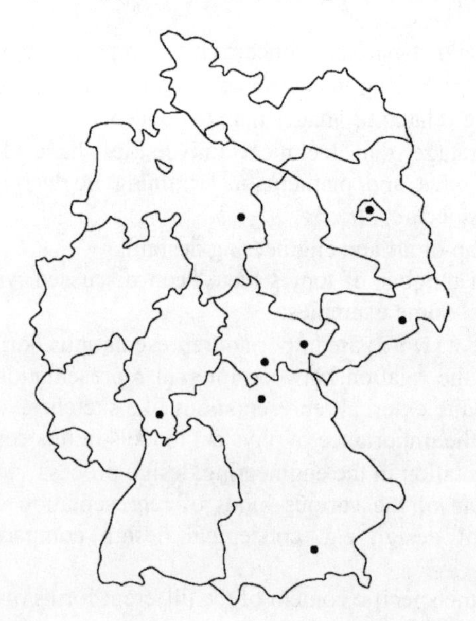

Fig. 1. Network Germany

In addition to the actual group of „Bild und Begriff", a number of scientists have participated at the workshops:
Peter Auer, Christina Bartl, Beate Bender, Bernd Bender, Rainer Bernard, Torsten Brix, Norbert Dylla, Renate Eisentraut, Eckart Frankenberger, Gerd Fricke, Mehmet Göker, Joachim Günther, Judith Jänsch, Sven Lippardt, Robert Lüdcke, Franz Müller, Johannes Müller, Martin Pache, Gerhard Pahl, Frank Pietzcker, Udo Pulm, Ralph Reimann, Simone Riemer, Anne Römer, Carsten Rückert, Pierre Sachse, Frauke Schroda / Jahn, Martina Schütze, Erdmunde Sommerfeld, Horst Sperlich, Joachim Stempfle, Ralf Stetter, Hans Stricker, Stefan Wallmeier, Guido Weißhahn, and others.
The German Science Foundation (DFG – Deutsche Forschungsgemeinschaft) supported a number of bilateral projects between scientists of cognitive psychology and engineering design. All the results were published at international conferences as well as in respected journals in the UK and the US. These projects addressed for example the design and problem solving strategies of individuals, critical situations in team work or the role and effect of sketches and physical models on the design process.
In 1997 some members of the "Bild und Begriff" group organised a conference with the title "Designers – the Key to Successful Product Development", which addressed a similar topic, the outcome was published as a book with the same title (Frankenberger et al. 1998).
In between a number of further research projects have been finalised. Based on our joint discussions and additional, new findings from our research, the "Bild und Begriff" group decided to initiate a small conference to deepen the discussion and the transfer of knowledge in this specific field.
The aim of this conference was:

- to start the discussion on a broad international basis
- to build up a global network
- to initiate new ideas for research projects
- to define the well established findings which can be integrated into university teaching courses
- to find aspects that can be transferred to become industrial practice
- to show the importance of our findings for industry

Aims connected to the content:

- the internal and external representation of design elements
- the individual design process
- the design process within teams / groups
- the boundary conditions of the enterprise
- the impact of using computers in design processes
- methods to analyse design processes
- teaching and training of design

In preparing for the conference we structured the field into three subtopics:

1. Individual thinking and acting
2. Interaction between individuals
3. Methods, tools and prerequisites

The conference was supposed to initiate a face-to-face discussion within a limited circle of international experts.

The Topic "Individual Thinking and Acting in Design" was prepared and moderated by Winfried Hacker, who was supported by Lucienne Blessing. The participants in this topic:

- Winfried Hacker (psychology, Dresden – Germany)
- Lucienne Blessing (engineering, Berlin – Germany)
- Bernd Bender (engineering, Berlin – Germany)
- Klaus Ehrlenspiel (engineering, Munich – Germany)
- John Gero (architecture, Sidney – Australia)
- Gabi Goldschmidt (architecture, Haifa – Israel)
- Pertti Saariluoma (psychology, Jyväskylä – Finland)
- Barbara Tversky (psychology, Stanford - USA)
- Willemien Visser (psychology , Le Chesnay Cedex – France)
- Hans Stricker (engineering, Munich – Germany)
- Constanze Winkelmann (psychology, Dresden – Germany)

The Topic "Interaction between Individuals in Design" was prepared and moderated by Herbert Birkhofer, who was supported by Petra Badke-Schaub. The participants in this stream:

- Herbert Birkhofer (engineering, Darmstadt – Germany)
- Petra Badke-Schaub (psychology, Bamberg – Germany)
- Uday Athavankar (engineering, Bombay - India)
- Amaresh Chakrabarti (engineering, Bangalore - India)
- Özgür Eris (engineering, Stanford - USA)
- Hans-Joachim Franke (engineering, Braunschweig – Germany)
- Stig Ottosson (engineering, Halmstadt – Sweden)
- Stephen Scrivener (computer science, Coventry – UK)
- Ken Wallace (engineering, Cambridge - UK)
- Marc Ernzer (engineering, Darmstadt – Germany)
- Judith Jänsch (engineering, Darmstadt – Germany)
- Franz Müller (engineering, Munich – Germany)

The Topic "Methods and Tools for Design Activity and their Effect on Individual Behaviour" was prepared and moderated by Günter Höhne, who was supported by Mogens Andreasen. The participants in this topic:

- Günter Höhne (engineering, Ilmenau – Germany)
- Mogens Andreasen (engineering, Copenhagen - Danmark)
- Balakrishnan Chandrasekaran (computer science, Ohio – USA)
- Udo Lindemann (engineering, Munich – Germany)

- Martin Pache (engineering, Munich – Germany)
- Ralph Schönfelder (computer science, Ulm – Germany)
- Tetsou Tomiyama (engineering, Delft – Netherlands)
- Remco van der Lugt (engineering, Delft – Netherlands)
- Rüdiger von der Weth (psychology, Stuttgart – Germany)
- Torsten Brix (engineering, Ilmenau – Germany)
- Tim Jarratt (engineering, Cambridge - UK)

Because of the importance of the common discussion, there were only a few scheduled lectures for introductory purposes. The papers had been made available before the conference and were presented with help of posters. In a first section the posters were presented briefly within the streams followed by an overall discussion. Within the next section all participants were given the opportunity to look at the posters of the two other streams with some additional brief discussion. A second stream discussion and a final plenary discussion were the content of the second day of this conference.

This book presents the contents and the results of the conference "Human Behaviour in Design". Based on the various discussions most of the papers have been revised by the authors.
Following this general introduction the three streams are presented starting with a brief introductory section by the stream chairs followed by the revised papers of the participants. The condensed results from the stream discussions are to be found at the end of each stream written also by the stream chairs or their supporters. The result of the plenary discussion was written by Lucienne Blessing, who had also moderated this session of the conference.

References

Frankenberger E, Badke-Schaub P, Birkhofer H (eds) Designers, The key to successful product development. Springer, Berlin Heidelberg New York, 1998.

Pahl, G. (Ed): Psychologische und pädagogische Fragen beim methodischen Konstruieren. Ergebnisse des Ladenburger Diskurses vom Mai 1992 - Oktober 1993. Köln: Verlag TÜV Rheinland, 1994.

Conclusion and Outlook

It was of high value, to have just a few speakers and to spend some time with the posters of all the participants and to have a lot of time for discussion. By the excellent posters all the authors have concentrated their key messages for the presentation and the following discussion. All the posters are published in the web [www.pe.mw.tum.de/hbid/].

The discussion in the streams, the plenary as well as in coffee breaks or during lunch or dinner was open and constructive. Though the differences between the specific fields of research were enormous, all of us had benefits of learning from each other and also supplying inputs to others.

Fig. 2. The participants of the symposium

We did an important step to establish a more global network in this important area of research.

Fig. 3. Global network

It was decided to use the platform of the Design Society [www.thedesignsociety.org] to start a Special Interest Group (SIG Human Behaviour in Design") to bundle activities for the future. This will give an important support in transferring results to teachers and students as well as to practitioners in industry.

References

Human Behaviour in Design – Posters: www.pe.mw.tum.de/hbid/ (at least until 2005)
The Design Society: www.thedesignsociety.org

Topic I: Individual thinking and acting

Winfried Hacker, Technical University Dresden

Nine contributions were assigned to the topic of individual design activities. Activities and their components, the actions, include mental proceses and mental representations as well as psychomotor proceses and external representations.

The mental processes discussed in the papers are mainly cognitive ones, for example perception, thinking or remembering. Thinking again includes several types of proceses, e.g. reasoning, problem solving or decision making. Since design activities are knowledge rich, several kinds of knowledge,i.e. of memory representations, are inevitable. An important part of them is uncodified, unarticulate "tacit knowledge". The reason is that along with tought knowledge action regulation is based on own experience from learning by doing and on psychomotor and mental skills.

Since design activities are working activities they belong to the category of goal-directed activities. Goals as anticipations of intended results are the key component of design activities (Hacker 2001).

There are several kinds of the mentioned psychomotor components and external representations involved in designing. Though the papers are mostly concerned with sketching and sketches, one should be aware of speaking, gesticulating or writing, and of notes as well.

Finally an a working activity designing follows the rules of division of labour. Therefore individual design activities are actually dependent components of cooperative design activities — by no means only of "teamwork" —, whether this cooperation is a real or virtual one.

Following an alphabetical order Bender and Blessing present an empirically based model interpreting conceptualization as a feedback–driven iterative process with alternations between different levels of abstraction and between design and evaluation.

Ehrlenspiel discusses the impact of subconscious, automatic and tacit mental operations and representations in design in terms of mental economy.

Goldschmidt presents a system for the analysis of reasoning in design, based on the links between small design moves, the linkography, and applies it in the comprehension of the effectiveness of the design process.

Hacker and co-workers suggest to interpret the design process as a working process which may be analyzed in terms of Action Theory and illustrate this proposal by empirical results for the issues of (i) procedural characteristics of successful designing, (ii) the interaction of mental and psychomotor operations, and (iii) the reflective type of action control.

Kavakli and Gero present protocol studies applying coding schemes of operations which reveal differences between novice and expert designers and discuss them with respect to strategic knowledge thought of in terms of different chunking of operations.

Krause and co-workers contribute neurophysiological evidence for the cooperation of computation and visuo-mental imagery in solving mathematical problems especially in gifted subjects stressing, thus, the Dual Representation Hypothesis.

Saariluoma discusses theoretically the contribution of content-based design research and offers two examples of explanatory contents, the functional relations and risky thought models.

Tversky and her co-workers show in a first project that experts deliberately use perceptual reorganization strategies to produce design ideas, and in a second project that in visualizations, people produce spontaneously, they systemetically omit some information and highlight some other information.

Visser stresses the reuse-based and the opportunistic organization of design activities and discusses consequences for design assistence.

Reference

Hacker, W. (2001). Activity Theory in Psychology. In N.J. Smeser and P.B. Baltes (eds.), International Encyclopedia of the Social and Behavioral Sciexces. Oxfort: Elsevier.

Re-Interpretation of Conceptualisation – A Contribution to the Advance of Design Theory

Bernd Bender and Lucienne TM Blessing, Berlin University of Technology

Introduction and Objectives

To identify successful design strategies for early stages of the design process, individual design procedures have been investigated with the aim to determine the applicability of Design Methodology in these phases of the product development process. First results have been published at ICED'01 (Bender et al. 2001b). This paper provides results of a detailed analysis of the observed design procedures of N=71 participants (83 evaluable cases) and their effects on design performance. Based on the results of this empirical study a re-interpretation of conceptualisation is proposed as a contribution to the advancement of design theory.

Theory

A very common approach to design and design activity is the recourse to considerations of cognitive psychology, with the main focus on design problem solving. We try to show that an integration of concepts from action regulation theory (cf. Hacker 1999) can enhance our understanding not only of what design activity is but also how it could be improved.

Design Methodology and Problem Solving

The tradition of Design Methodology in Germany is based on the fundamental assumption that from a cognitive point of view, design is a complex problem solving activity: "Designers are often confronted with tasks containing problems they cannot solve immediately. Problem solving in novel areas of application and at different levels of concretisation is a characteristic of their work." (Pahl and Beitz 1996) To support designers and to rationalise design processes a prescriptive methodology has been developed which is nowadays basis of design education. This Design Methodology "[…] is a concrete course of action for the design of technical systems that derives its knowledge from design science and cognitive psychology, and from practical experience in different domains. It includes: plans of action to link working steps and design phases according to content and organisation; strategies, rules and principles to achieve general and specific goals, and methods to solve individual design problems or partial tasks." (Pahl and Beitz 1996) This approach led to Guideline VDI 2221 which prescribes a "Systematic

Approach to the Design of Technical Systems and Products". This guideline has had a remarkable impact not only in Germany but in many other countries and can be seen as typical for prescriptive approaches developed in middle Europe to support engineering design processes (Blessing 1995).

The core idea of problem solving in this approach is to reduce complexity by a hierarchical top-down decomposition of design problems (Fig. 1) and the application of appropriate methods on each level of abstraction: "A further, generally valid and established approach to solving problem is to proceed through several phases in increasingly concrete terms – the strategy being: *from the abstract or general to the concrete or specific*. Thus, for example, fundamental relationships are first determined in principle, and embodiment details not established until later. The following strategy proceeds in the same direction: *from the most important to the less important, or from the main problems to the sub-problems*." (VDI 2221 1987)

However, from the beginning a distinct discomfort with the strict hierarchical nature of this approach has been articulated in particular by professional designers. As a consequence it has been outlined that iterations are needed when applying this approach to real design processes: "[...] such strategies, and procedures based on them, should not necessarily be pursued in a strictly linear fashion. [...] It is therefore important for a plan of approach to include iteration, the repetition of phases or steps, with those of little importance being passed over rapidly or even skipped." (VDI 2221 1987)

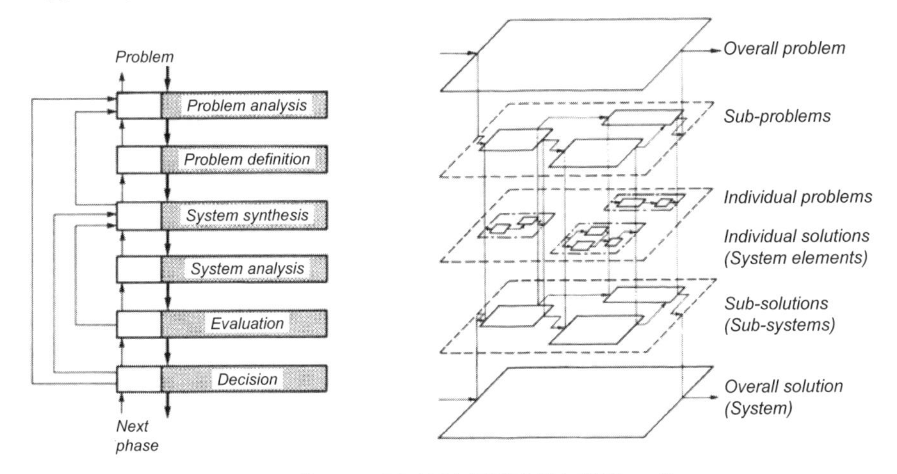

Fig. 1. Problem solving according to VDI 2221 (VDI 2221 1987, p. 4)

After all, this has been added to the linear approach afterwards and almost no guidance or criteria exist for how to manage this in practice: how to decide when to deviate from the top-down approach and when not, which steps and phases are to be skipped and which not. As a result, this methodology is often seen as being artificial, time-consuming and restricting and, therefore, is hardly accepted in an industrial engineering context. In addition, empirical research showed that strictly following the rules and plans of actions recommended by VDI 2221 does not necessarily lead to better design performance. But still, several of the core ideas, such as paying more attention to the early phases, have become established.

Empirical Design Research

Much empirical design research has been undertaken in the past 15 years, changing the way we look at individual design activity. Especially the prescriptive German tradition of Design Methodology is challenged by a number of results of this research.

Rutz investigated the design heuristics of engineering design students within a number of case studies and came to the conclusion that *problem formulation* and *problem decomposition*, the *change of levels of abstraction*, *search for and adaptation of solutions*, *trial-and-error behaviour*, and *iterative and recursive work* are characteristics of observed individual design processes. His main consideration was that "[…] the design process is not […] a linear process from 'abstract' to 'concrete' […]" but rather a "[…] continuous alternation between *'abstract' and 'concrete'* and between the *whole and its details*." (Rutz 1985)

Dylla found that two fundamentally different individual design strategies can be observed in early design stages, which he called *productive solution generation* and *corrective solution generation* (Dylla 1991). Productive solution generation is characterised by the synthesis of several independent solution variants in the beginning of the process followed by an evaluation process leading to the selection of a promising solution to be further refined. Corrective solution generation starts off with one more or less concrete solution idea followed by a stepwise refinement until a proper solution quality is reached.[1] Dylla pointed out that corrective solution generation prevails significantly (81% of all observed generations had been corrective) although a generative solution generation strategy is strongly recommended by Design Methodology guidelines.

Fricke distinguished between *stepwise process-oriented* and *function-oriented* design strategies (Fricke 1996). The stepwise process-oriented strategy follows a hierarchical and sequential plan of action, executing basic design operations step by step for all sub-systems more or less in parallel. Within a function-oriented procedure design operations are carried out for one initial problem area until a satisfying level of concretion is reached before the next problem area will be addressed in the same way. Fricke comes to the conclusion that the best design performance can be achieved by following a "*flexible-methodical proceeding*" (Fricke 1996), aiming at a "*balanced search*" for solutions (Fricke 1996).

Günther identified personal characteristics of designers supporting design performance. These are in particular *professional design experience*, high *heuristic competencies* and a *low disposition for emotional stress*, *regression* (i.e. behaviour of escape and avoidance when facing difficult problems) and

[1] This is very similar to the 'breadth first' vs. 'depth first' concepts discussed in the anglo-american context.

resignation (Günther 1998). He also identified process characteristics that had a negative influence on design performance: *incomplete clarification* and *bad memorisation of requirements*, *wrong weighting of sub-problems*, *too early moving from concept to embodiment* without an elaborated concept and a *largely unstructured procedure* (Günther 1998).

Rückert showed in a field study into design strategies and performance of graduate engineering design students that characteristics of the learning process are important for the efficacy of a Design Methodology Education (DME). He observed 96 participants within a period of 10 month in three groups with different levels of Design Methodology guidance: (a) without any methodological or systematic recommendations, (b) with a strict plan of action after VDI 2221 and (c) with a Design Methodology training but a self-determined application of methods within the design teams. He found out that a self-determined application of Design Methodology strategies led to better design results compared to the strict application of VDI 2221 (Rückert 1997). In addition, the acceptance of Design Methodology was significantly higher in the self-determined group (Rückert 1997).

One central conclusion can be drawn from these results: The concept of strict hierarchical top-down decomposition seems not to be an appropriate model of design problem solving, nor is it an appropriate model to be recommended to achieve a better design performance and to be accepted in practice. More or less all authors come up with proposals of a more "flexible-methodical" and "balanced" (Fricke 1996) or "flexible solution-oriented" (Rückert 1997), adaptation of Design Methodology strategies, rules and guidelines. This leads to the interesting question what 'flexible' and 'balanced' adaptation actually means and how the appropriate level of flexibility and the 'right balance' can be determined. To answer this question, considerations from the field of action regulation theory and in particular considerations of cognitive planning and goal-directed activity seem to be promising to extend the understanding of individual design behaviour and what it makes successful (cf. Hacker 1999). Some relevant findings are discussed in the following paragraphs.

Hayes-Roth and Hayes-Roth investigated individual planning processes and introduced the concept of an 'opportunistic model of planning' to describe planning processes. They presumed that the cognitive activity of planning comprises a set of basic activities (in their terminology 'cognitive specialists') which have to be co-ordinated to succeed in planning. They found out that "the activities of the various specialists are not coordinated in any systematic way. Instead, the specialists operate opportunistically, suggesting decisions whenever promising opportunities arise." (Hayes-Roth 1979)

Guindon in her research on individual strategies of software designers transferred these results to the design area and described the concept of *opportunistic decomposition* instead of top-down decomposition for dealing with complex and ill-structured problems. She observed individual design processes, deviating greatly from top-down approaches without having negative consequences for design performance. "The analyses show that these deviations are not special cases due to bad design habits or performance breakdowns but are, rather, a natural consequence of the ill-structuredness of problems in the early stages of design." (Guindon 1990) She comes to the conclusion that "[…] the early stages of the design process are best characterized as opportunistic, interspersed with top-down decomposition." (Guindon 1990) In detail, "opportunistic design is characterized by on-line changes in high-level goals and plans as a result of inferences and additions of new requirements. In particular, designers try to make the most effective use of newly inferred requirements, or the sudden discovery of partial solutions, and modify their goals and plans accordingly. […] Opportunistic planning is in fact a more general type of planning than hierarchical planning." (Guindon 1990)

Visser underlines these findings in her research. From an analysis of a number of empirical studies and from own studies she concluded that "[…] even if designers possess a pre-existing solution plan for a design problem, and if they can and do retrieve this plan to solve their problem (which is often possible for experts confronted with routine design), *yet* if other possibilities for action ('opportunities') are also perceived (which is often the case in real design) and if the designers evaluate the cost of all possible actions ('cognitive' and other costs), as they will do in real design, *then* the action selected for execution will often be an action other than the one proposed by the plan (it will be a selected opportunity)." (Visser 1994) In result "[…] even for experts involved in routine design, retrieval of a pre-existing plan does not characterise the organisation of their actual activity appropriately. Such a plan which, if it is followed, may lead to systematically organised activities, is supposed to be only one of various action-proposing knowledge structures." (Visser 1994) She identified *cognitive economy* as the most important category motivating designers to deviate from top-down strategies to opportunistic design.

One can discuss whether this descriptive research only identified the *most common strategies* or if it also can teach us the *best strategy* with respect to design performance. From a prescriptive point of view, one may argue that opportunistic behaviour might be very common under everydays' constraints of professional design practice but that a deliberate step-by-step strategy nevertheless would lead more efficiently to better or at least more reliable results when applied consequently. Clarifying this question has been the main motivation of our research.[2]

Questions and Hypotheses

To further investigate the question of 'best strategies' for design and based on indications of the many empirical results, some of which were mentioned above, the following research questions and hypotheses were formulated:

---

[2] Collaborative research project "Applicability of Design Methodology in Early Phases of the Product Development Process" together with W Hacker's group of the Technische Universität Dresden, funded by the Deutsche Forschungsgemeinschaft DFG

1. What differences in design performance can be observed between engineering design students with and without a Design Methodology Education (DME) when dealing with test design tasks from the conceptual and early embodiment stage of design?

We expected that participants with 'fresh' DME would not achieve better design performance compared to participants without any DME (control groups). Due to the fact, that the deliberate application of freshly learnt methods and procedures causes additional cognitive load they should need more time to solve the same task resp. deliver similar or even worse design performance within the same time limit compared to the control group. An advanced level of DME then should lead to better design performance because an internalisation and proceduralisation of methods and procedures should enable this trained group to apply pre-defined plans for dealing with design problems at least episodically.

2. What differences in individual design procedures can be observed between engineering design students with and without DME when dealing with test design tasks from the conceptual and early embodiment stage of design?

We expected to observe different individual design procedures and, following the theoretical concepts of opportunistic vs. hierarchical approaches, and relying on the findings of the empirical research mentioned above, we assumed four individual types of 'design styles':

a. *Hierarchically phase-oriented*: a procedure strictly following VDI 2221 which can be characterised as organised sequentially and hierarchically at the same time (cf. Fricke 1996). The process thus involves a combination of hierarchical decomposition and sequential goal accomplishment. Therefore a design activity is executed for all subsystems before the next activity takes place, which is again executed for all subsystems, and so on.
b. *Hierarchically object-oriented*: An overall task can also be decomposed in terms of systems and sub-systems to be designed. Such a design style would be characterised by the execution of all design activities – or subsets of activities within 'hierarchical episodes' – for one sub-system before they are executed for the next sub-system, and so on.
c. *Opportunistic and associative*: This is a knowledge-driven resp. data-driven procedure following the concepts of opportunistic planning with hierarchical episodes. Instead of starting with a hierarchical decomposition of the task, the designer may begin with a preliminary decomposition and then start elaborating a part of the system. A possible reason is that the designer remembered a suitable solution for this part and took this opportunity as a starting point. This is a mixture of a hierarchical decomposition with an opportunistic, local and bottom-up, proceeding. It is less guided by a total goal but more by the association between the step that has just been finished and the opportunity to do the same step again on a different (sub-)problem.
d. *'Muddling through' trial and error*: From the point of view of action regulation, a trial and error-like procedure, hardly following any observable systematic or methodical guideline, should be identified. This involves a more

or less unsystematic trying to cope with different parts of the system, different design activities and (sub-)problems instantly every time they occur or come to conscious consideration. This procedure also can be classified as opportunistic but misses any kind of hierarchical episodes.

1. Are there any correlations between observed individual design strategies and the achieved design performance?

We expected that the different design styles mentioned above would have a significant influence on design performance. The main question was if we could find a 'best strategy' based on design performance. In addition we expected that participants with 'fresh' DME would follow the recommendations of Design Methodology rather strictly, especially in the early concept stage. We also expected that this would lead to a more time consuming procedure or to similar or even worse design performance within the same time limit compared to the control group. Participants on an advanced DME-level should be able to adapt Design Methodology rules and guidelines more flexibly and in a 'solution-oriented' way, which in result should lead to better design performance.

Methods

A quasi-experimental design of experiment was developed to answer the research questions and to verify the hypotheses (Fig. 2). Mechanical engineering students with three different levels of DME – identified by examining educational and professional background – each were confronted in a laboratory environment with a *conceptual design task*, *a concept evaluation task* and an *embodiment design task* at each of the three predetermined stages of their studies. The conceptual design task and the concept evaluation task consisted of a verbal description of a design problem without any visualisation, and had to be finished within one hour. The embodiment design task comprised a verbal description, a sketch of the working structure and technical data. This task had to be finished within 3.5 hours.

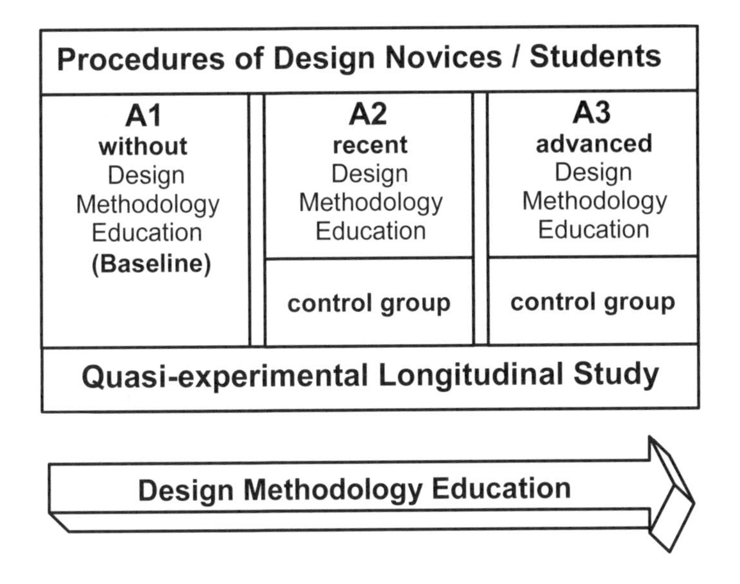

Fig. 2. Overview of the quasi-experimental empirical study[3]

The A1 group consisted of engineering design students at the end of their basic machine elements course without any DME so far. This group can be seen as the baseline of the study. The A2 group consisted of students having taken part in a one-year Design Methodology lecture and the students of the A3 group had in addition taken part in a 6 month Design Methodology project. In parallel, on the A2 and A3 level we investigated the procedures and the design performance of control groups consisting of mechanical engineering students from the same semesters not having taken part in the Design Methodology courses. This led to in total 71 participants with different levels of DME – including control groups without DME – who were confronted with standardised design tasks out of the area described above. The homogeneity of all tasks with respect to the criteria *conflicting aims*, *complexity*, *transparency*, *degrees of freedom* and *required knowledge* has been validated by design experts using a method for systematic analysis of design problems and tasks (for details of the task design see Bender et al. 2002).

Data of the design activity was observed using photo-documentation, protocols and different pencil colours according to different test periods. Mental representations of the design activity were investigated by using a card sorting technique and questionnaire. For the assessment of the participants' individual design performance, a formalised method based on strategies of value analysis has

[3] Due to the fact that there have not been enough participants following the Design Methodology course of study continuously we were not able to follow a genuine longitudinal approach. In the A2 and A3 stages of the study therefore students took part who did not attend the earlier stages.

been developed to achieve maximum evaluation reliability and reproducibility. Design quality was measured in three different sub-categories: (a) the quality of the documented *concept variants*, (b) the quality of the documented *evaluation of concept variants*, and (c) the quality of the documented *embodiment designs*. For each of these sub-categories a multi-criteria set of objectives had been formulated containing those features that describe design success in the specific sub-category. Multi-criteria evaluation charts allowed the evaluation of design performance in every sub-category using these objectives and criteria, their relative weights, and ordinal value scales. To further increase evaluation reliability, a team of three design experts carried out this assessment.

For the observation, analysis and categorisation of the individual design heuristics of the participants we followed a systematic approach described in detail in Bender et al. 2001a. Each observed design activity was classified following a scheme of pre-defined basic design operations, and the deviation of the observed sequences of activities from the VDI 2221 process was determined. In particular, the number of visits to the design stages defined in VDI 2221was counted, as well as the number of transitions from one basic design operation to another and the 'size of these transitions'[4] – which one might also call 'jumps' – compared to VDI 2221 (similar to Fricke 1996). In addition, we recorded for every transition if this transition at the same time involved a change of sub-system or not. The mean values of the transition sizes in both cases were determined for every participant and entered into a portfolio diagram along different axes. This allowed a valid distinction of the postulated different types of design procedures (see results in Fig. 5).

Based on experience from the teaching area, we assumed that below a distinct level of design performance the influence of individual procedures on design performance is biased by a lack of basic design competencies. Therefore, for further analysis of correlations between design procedures and design performance we left out those participants with embodiment design performance lower than 40% of the achievable maximum of credit points.[5] To analyse group differences, non-parametric testing methods, especially the H-Test after Kruskal and Wallis and the Median-Test were used[6].

Results

Different DME Levels and Design Performance

A quantitative analysis of design performance and its correlation with different levels of DME (for details see Bender et al. 2001b) showed that, as expected,

[4] transition size = number of operations that were jumped - forward or backward - according to strict hierarchical approach in VDI 2221

[5] We proved the validity of the 40% limit by a sensitivity analysis.

[6] The α-values for both tests are given in the diagrams. 'H' indicates the H-Test, 'M' the Median-Test.

recent DME (A2) had no significant positive influence on embodiment design quality. A positive influence of advanced and practiced DME (A3) could be found (Fig. 3). This matches our hypotheses very well.

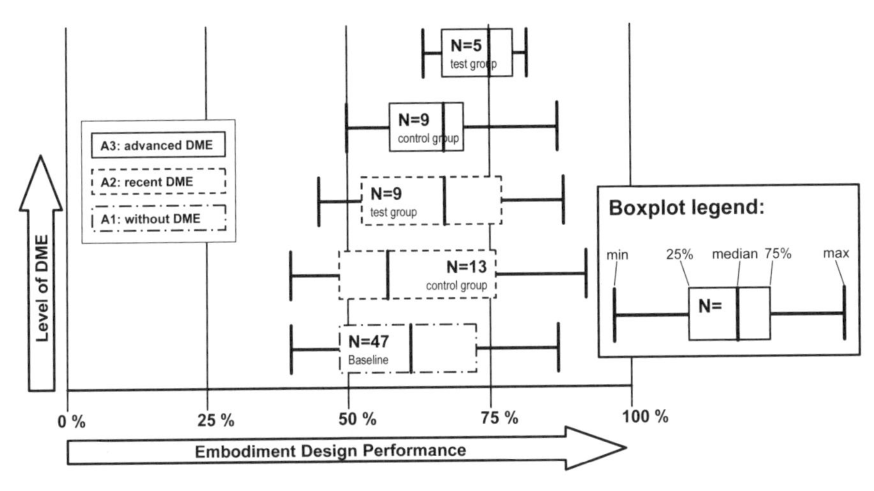

Fig. 3. Embodiment design performance under the influence of DME

For the conceptual design task, neither the recent (A2) nor the advanced DME-level (A3) was found to have a positive influence on conceptual design performance. To the contrary, a slight tendency towards a lower performance within the DME-group in comparison to the group without DME could be observed. For the concept evaluation task, the evaluation quality of all participants on the baseline level (A1) is low. The evaluation performance on both DME levels increases compared to the baseline level, including the control groups without DME, but the DME groups show a tendency towards a more positive effect on evaluation competency. This is at least partly in contrast to our hypotheses, where we expected that on the A3-level the performance should have been significantly better compared to the baseline level (A1) and the control group.

Interestingly DME had only a distinct positive influence on the embodiment design performance. For the conceptual and the concept evaluation task even the group with advanced DME did not perform significantly better than the control groups. However, a tendency to a narrower bandwidth of performance and a shift of the minimum value to a higher level can be identified at least at the advanced DME level. Nevertheless, the influence of DME as we have practiced over the years is not as convincing as we might have wished – and not for the phases we expected it to be effective.

Individual Design Procedures and Design Performance

To further determine the influence of specific design heuristics on design performance, the correlation of the observed individual design styles and procedures with design performance has been investigated in more detail. One interesting result is illustrated in Fig. 4, showing how often the participants visited pre-defined design stages by carrying out the typical basic design operations of these stages. We observed that compared to the group with medium design performance, excellent design students as well as poor ones both visit the early design stages significantly more often. These results corroborate our hypothesis that excellent designers analyse a task in more detail, plan their procedure thoroughly and elaborate rough concepts before starting with detailed design. In contrast, poor designers spend more time in simply clarifying and understanding the problem. The hypotheses that 'poor systematic designers' and novices often stick to a very formal pseudo-systematic resp. "unreasonably methodical" (Fricke 1996) plan of work to avoid making decisions instead of starting to solve the problem is confirmed by these results.

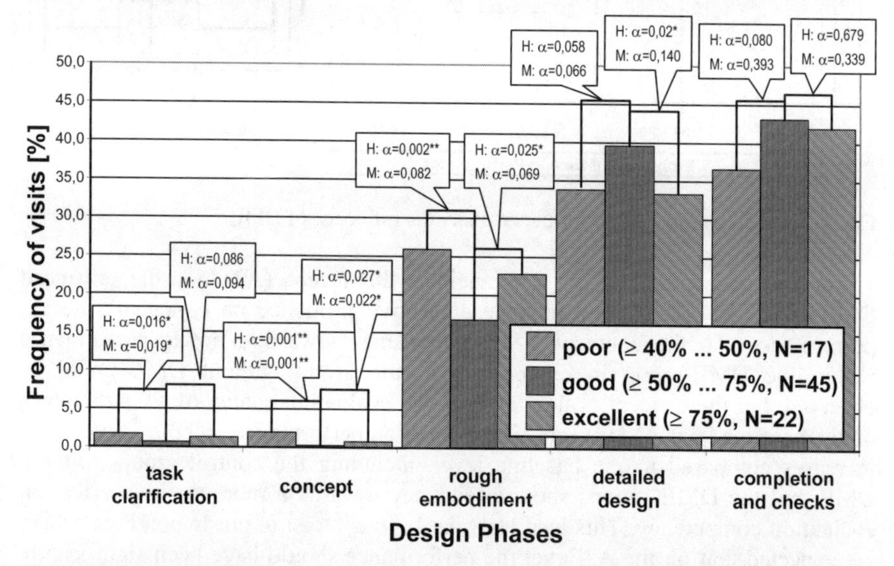

Fig. 4. Frequencies of visits of design phases in relation to design performance (in %)

We assume that the group with medium design performance follows a more opportunistic strategy, starting with early concretisation and continuing with an iterative and corrective procedure (Dylla 1991). But in this stage this is still speculation to be investigated further.

Combining the methods described above, we were indeed able to identify the four postulated individual design styles and a significant influence of these individual procedures on design performance in the early embodiment stage. We determined the medium size of activity transitions for every participant (a) involving and (b) not involving a sub-system transition, entered these data into a portfolio diagram and dichotomised the data along both axes at the 25%

percentile, where the most significant differences were found.[7] Then we looked at the embodiment design performance of these groups of participants. The result is documented in Fig. 5. The group with the worst design performance is the one following strictly the hierarchically phase-oriented procedure. These participants reached only about 50% (median) of the achievable maximum of 60 credit points. The hierarchically object-oriented and the opportunistic and associative group reached significantly better performance (about 70% of maximum performance). Even the group with the trial and error design style does not achieve a worse embodiment design performance compared to the hierarchically phase-oriented group, there is even a tendency to a better performance (about 60% of max. performance).

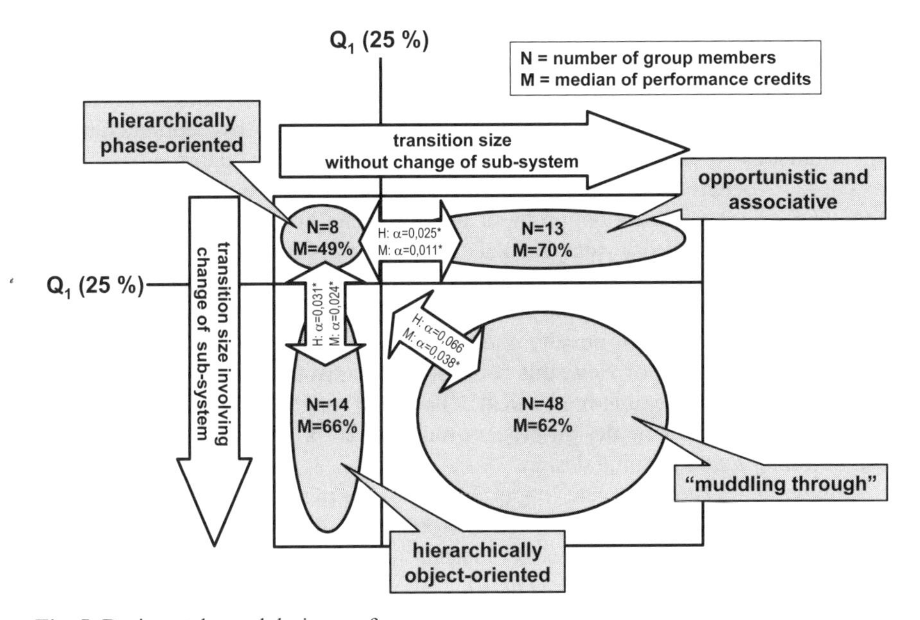

Fig. 5. Design styles and design performance

According to these results we are able to answer our main question mentioned above: Flexible and opportunistic design strategies are obviously not only more common in real design processes but as a matter of fact lead to better design performance.

[7] We assumed that it would be a rather small medium transition size to distinguish hierarchical from other procedures but we did not have a hypothesis for the concrete position of this transition size limit. We therefore chose a data-driven approach and the 25% percentile turned out to be a stable dichotomisation point. Its validity was proven in a sensitivity analysis.

Conclusion

As a main conclusion, the systematic and hierarchical approach of established Design Methodology still seems to support design activities that take advantage of hierarchical strategies. This seems to hold for the embodiment design stage which is focussing on refinement and concretisation rather than on creation. Here we found a positive influence of an advanced DME. For ill-structured design problems involving at least a medium level of novelty and complexity – like they often do in the conceptual design stage in practice – strictly hierarchical strategies are artificial and inappropriate.

However, irrespective of the DME background strictly hierarchically phase-oriented procedures lead to poor design performance, while a flexible adaptation of a systematic approach helps to achieve good performance. Flexibility can be seen in two main directions. Cognitive load might be reduced by (a) approaching the design problem through decomposition along the design objects (or sub-systems or 'problem-areas') or (b) by approaching it opportunistically, taking advantage of obvious and promising opportunities.

Thus, to better understand successful strategies for early design stages, the cognitive concept of design problem solving benefits from considerations of action regulation theory (cf. Hacker 1999). Hayes-Roth's & Hayes-Roth's, Guindon's, and Visser's concept of 'opportunism' as a natural consequence of the ill-structured character of complex and knowledge-rich design problems involving at least a medium level of novelty could be proved as valid for successful design heuristics. In our point of view, this concept is the most promising for determining in concrete what a 'flexible-methodical', 'balanced' or 'flexible solution-oriented' approach to a systematic design process might be in professional design practice and in teaching and learning design.

Combining systematic and opportunistic strategies into a prescriptive but flexible model leads us to a re-interpretation of conceptualisation as a feedback-driven learning process instead of a strictly goal-directed and hierarchical top-down procedure (Fig. 6).

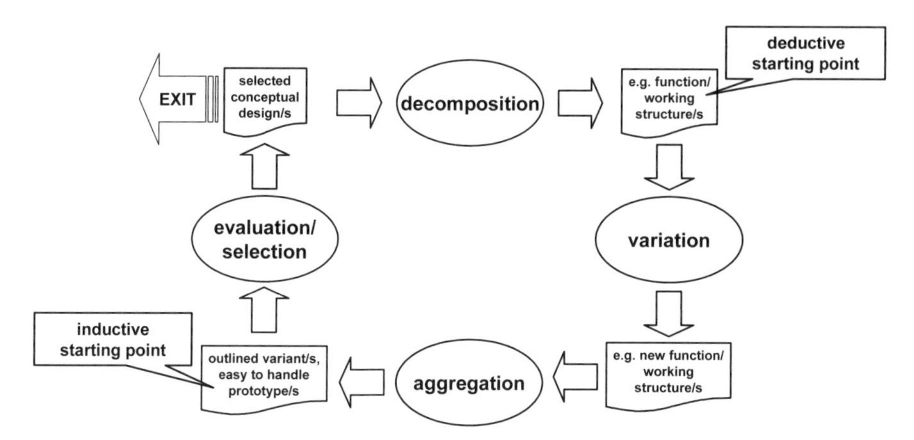

Fig. 6. Re-interpretation of conceptual design as a feedback driven learning process

In this approach decomposition and aggregation refer to a logical and hierarchical breakdown using different levels of abstraction and representation. Conceptualisation can start at an inductive starting point with one or more concrete, but probably ill-defined, concept variants, based on 'opportunities' like e.g. existing solutions or the results of an intuitive search for solutions. The resulting variants can then be refined by following the other steps in the loop. Alternatively one may start on a higher level of abstraction at a deductive starting point as proposed in classical Design Methodology. But no matter which starting point is chosen, the whole loop has to be carried out according to the requirements of the actual task, to prevent early fixation and neglect of innovative solutions.

The creation of new solutions should not only rely upon variation and experimentation at the abstract levels of function and working principle. The aggregation phase is crucial for solution finding. Outlining a solution by sketching or impromptu-modelling does not only illustrate an idea, but is an important creation technique in its own right, supporting the mental interaction of internal and external activity.

The core idea of this approach is to re-interpret conceptualisation as a cyclical process of learning. Strategies to support this early design stage therefore in our opinion should be based on this process model and should integrate considerations of the support of effective processes of (experiential) learning such as techniques of deliberate reflection.

Acknowledgements

The ideas and results presented here are based on collaborative research together with W Hacker and F Pietzcker (TU Dresden), and U Kammerer (TU Berlin) into the „Applicability of Design Methodology in Early Phases of the Product Development Process". The project was funded by the Deutsche Forschungsgemeinschaft.

References

Bender et al. 2001a

Bender B, Kammerer U, Pietzcker F, Blessing LTM, Hacker W (2001) A Systematic Approach to Observation, Analysis, and Categorisation of Design Heuristics. In: Culley S et al. (eds) Proceedings of the International conference on Engineering Design ICED'01: Design Research – Theories, Methodologies, and Product Modelling. Professional Engineering Publishing, Bury St Edmonds & London, pp197–204

Bender et al. 2001b

Bender B, Kammerer U, Pietzcker F, Blessing LTM, Hacker W (2001) Successful Strategies for Complex Design in Early Phases of the Product Development Process – An Empirical Study. In: Culley S et al. (eds) Proceedings of the International conference on Engineering Design ICED'01: Design Research – Theories, Methodologies, and Product Modelling. Professional Engineering Publishing, Bury St Edmonds & London, pp173–180

Bender et al. 2002

Bender B, Kammerer U, Pietzcker F, Blessing LTM, Hacker W (2002) Task Design and Task Analysis for Empirical Studies into Design Activity. In: Marjanovic D (ed): Proceedings of the 7th International Design Conference DESIGN2002, Dubrovnik, vol 1, pp 119–124

Blessing 1995

Blessing LTM (1995) Comparison of design models proposed in prescriptive literature. In: Proceedings of the COST A3 / COST A4 International Research Workshop on "The role of design in the shaping of technology", Lyon, 2–3 February 1995, published by the European Committee, 1996, pp 187–212

Dylla 1991

Dylla N (1991) Denk– und Handlungsabläufe beim Konstruieren. Hanser, München, Wien

Fricke 1996

Fricke G (1996) Successful Individual Approaches in Engineering Design. Research in Engineering Design 8: pp 151–165

Günther 1998

Günther J (1998) Individuelle Einflüsse auf den Konstruktionsprozess: Eine empirische Untersuchung unter besonderer Berücksichtigung von Konstrukteuren aus der Praxis. Shaker, Aachen

Guindon 1990

Guindon R (1990) Designing the Design Process. Human-Computer Interaction 5: pp 305–344

Hacker 1999

Hacker W (1999) Konstruktives Entwickeln als Tätigkeit – Versuch einer Reinterpretation des Entwurfsdenkens (design problem solving). Zeitschrift für Sprache & Kognition18(3/4): pp 88–97

Hayes-Roth 1979

Hayes-Roth B, Hayes-Roth F (1979) A Cognitive Model of Planning. Cognitive Science 3: pp 275–310

Pahl & Beitz 1996
Pahl G, Beitz W (1996) Engineering Design. A Systematic Approach. 2nd edn. Springer, London Berlin Heidelberg New York
Rückert 1997
Rückert C (1997) Untersuchungen zur Konstruktionsmethodik – Ausbildung und Anwendung. VDI-Verlag, Düsseldorf
Rutz 1985
Rutz A (1985) Konstruieren als gedanklicher Prozess. Ph.D. thesis, TU München
VDI 2221 1987
VDI 2221 (1987) Systematic Approach to the Design of Technical Systems and Products. VDI-Guideline. VDI-Verlag, Düsseldorf (latest German edn 1993)
Visser 1994
Visser W (1994) Organisation of Design Activities: Opportunistic with Hierarchical Episodes. Interacting with Computers 6 (no 3): pp 235–274

On the Importance of the Unconscious and the Cognitive Economy in Design

Klaus Ehrlenspiel, Technical University München

1. Something about information processing in our brain.

Initially a quotation of Prof. G. Roth (Institute for Brain-Research, University of Bremen): "Will, thinking and behaviour of man are governed to a great extent by the limbic centres, which work fundamentally unconscious. Our rational thinking has a very limited access to these processes". (Roth 2000).

In every second our senses transmit about 11 millions of Bits of "information" to the brain. Especially the eye. In the same time our conscious experience is processing only 40 Bits, which is nothing by comparison. The lion's share will be processed unconscious. Besides this our conscious experience lags about 0,5 seconds behind the real event (Noretranders 1994). Vice versa a lot of our "free" decisions will be anticipated by about 0,5 seconds due to our nervous excitation (Noretranders 1994). ("The brain has made the decision meanwhile we think about it": J. Paulus). For instance it is processed unconsciously the whole complex conversion from thinking to syntax, semantic and articulation to speaking and backward to thinking at the brain of the hearer (Thompson 1994). Apart from this, we are "designing and simulating" the world around us without becoming aware of this. (For instance: a paper seems white even in an orange sunset light).

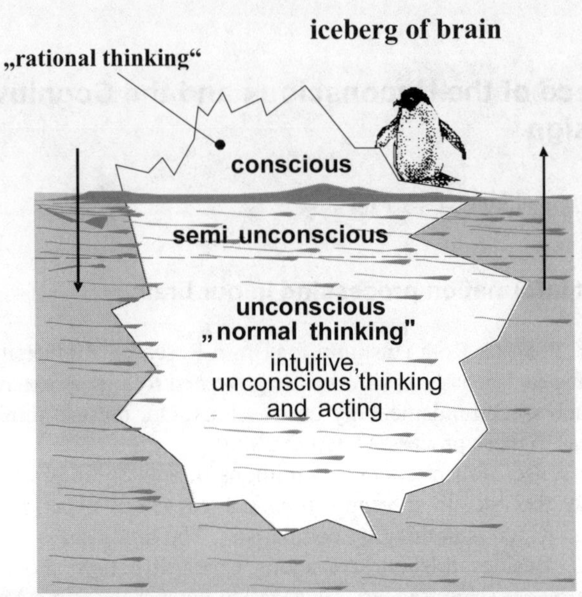

Fig. 1. Iceberg analogy: The main part of our thinking and acting is running more or less unconsciously in a *"normal operation"*: Essential mental processes run as routine in the unconscious ("under water"). The *"rational operation"* is based over this ("over the water") The conscious and the unconscious are in continuous exchange (Arrows!)

It's similar to a computer: We don't see the complex calculations inside, but only the results on the screen. – The centre of our thinking and acting is not situated in the consciousness, but in the unconscious! Ernst Pöppel (1999) writes: "According to the architecture of our brain, perception is happening with simultaneous recollection and evaluation."

The brain is not making a new design of the perceived world in every instant, but takes according to the cognitive economy (!) existing hypothesis (pre-judices!) as a basis for confirmation or rejection. "This is an automatism, which nobody can evade". – Our thinking and acting is similar to an iceberg, of which the main size (6/7 under water) and shape is hidden to conscious awareness (Origin of this model comes from Sigmund Freud. *Fig.1*). Conscious and unconscious are like the two sides of a coin. It depends which one is up (Müller J. personal remark 2003). This means that even if we are in a clear consciousness, the main part of our thinking and acting is running more or less unconsciously in *"normal operation"* (Müller 1991). – Finally, if we are blocked-up mentally, we change over to *"rational operation"*, with its methodical, discursive and rational processing and steering. This lasts longer, but often it is more safe.

We may assume that conscious thinking and acting is prepared by unconscious motivations and standards of value (Wulf 2001). The conscious thinking is so to say supported by the unconscious (see Fig.1).

Thinking underlies the economic principle (principle of simplicity; cognitive economy). Therefore for an engineer the typical demand is that he should use methodical planning in the “rational operation” as little as possible and only if necessary from the point of view of product quality, but working in the “normal operation” with routine and knowledge of experience as much as possible; (Müller 1991; Ehrlenspiel 2002). (See the two rules at the end).

A student having solved a design problem in his diploma thesis expressed this understanding in the following words. “Most of the time I did not consciously work methodically, but intuitively. I drove with my car, so to say, using two wheel drive and the normal working gear. Only when I came onto difficult terrain did I switch over to four wheel drive, i.e. design methodology. And in most cases I was successful”.

The unconscious is important not only for the individual, but also for communication with others too. Facial play and gestures give an unconscious exchange even in advance to verbal communication. Estimates exist, that these can be dominant. According to Watzlawick, it is not possible not to communicate. G. Roth writes: Feelings dominate relative to reason. This is biological reasonable, because they reflect experience of life in a concentrated manner. This could not be represented consciously with all relevant details. Frankenberger (1997) and Wulf (2001) have found, that communication with its unconsciousness and its feelings is one main factor for the quality of the design. This especially is valid to team work.

2. Expressions used in this paper

For me as an engineer, not knowing the different schools of psychology, the expression of unconscious is not the unconscious according to Freud's psychoanalysis. My orientation is the modern brain research (physiology of the brain with biological, chemical, and physical methods). According to Pöppel (2000) unconscious is all that, what cannot communicated to others e.g. by language, by text, by gestures. As unconscious mental processes are meant such, which cannot be communicated rationally, because they are not processed in the cerebral cortex (Thompson 1994). Synonymous expressions are intuitive, implicit, tacit knowledge, unintentional.

Nevertheless, in the actual state of science an absolute separation of conscious and unconscious is not possible (Perrig et al. 1993). Therefore most statements in this paper have the character of hypothesis and are not statistically safe. They are meant to give impulses for the analysis of practise and for future research.

3. What importance for design has the distinction between conscious and un-conscious mental processes?

1. Our *design processes* are by far more *unconscious* than *conscious*. This is particularly valid for designers, who are not trained methodically. Most of designers belong to this group. Unconscious per “trial and error” learned

methods therefore mostly run unconsciously. Among these there are effective and non-effective ones, even such which cause errors.

2. *Unconscious methods run much faster than conscious ones.* This is common in work sciences (Hacker 1992). It's clear: Routine processes are quicker than processes, in which every step must be reflected on or taken from printed prescriptions. (Example: driving a car unconsciously is much quicker than planning actual processes or looking in the handbook before pressing the clutch, switching a gear and so on). P-designers, as practitioners in Fig.3 need only 60% of the time of consciously working methodical m-designers.
3. Our *knowledge how to teach and how to learn* to recognise quickly "critical situations" (Frankenberger 1997), and essential functions, parts, possibilities of errors, is poor. This, up to now, happens more or less unconsciously. Research must be done.
4. *Research in Design Methodology* should try to clarify unconscious processes too. That means design research should not be limited to purely technical influences (such as physical effects, shape, material, kinds of contact and fixing of parts), but should take into account human influences and human interaction between individuals, groups, and within the company. These often influence quality, cost and lead-time of a future product to a high degree. (Frankenberger 1997; Badke-Schaub 2001).
5. *Research itself is determined by unconscious influences* developing in the society. According to Heymann (2002), German research in design methodology can be divided into two periods. The first from about 1960 to 1985 is the "strict" one, which was oriented analogous to the success of physics, chemistry and computer technology. It was unconscious for the researchers of that time, that their design science idea was too strict! – The second period for more "flexible" and intuitive processes started in West-Germany about 1984/1986 in the "Bamberg Circle" with the psychologist Dörner as a change of paradigm.
6. *Creative ideas come out of the unconscious.* They arise in problems, for which initially no solution can be seen. The right idea results in a pattern-recognition and combination of former experience (Dörner 1995; Broelmann 2002). The phases of that process are "elaboration" (intensive working at the problem), "incubation" (leave the problem alone for a unconscious ripening), "illumination" (the sudden idea in a relaxed situation) and the "verification" (rational evaluation and realisation). – I often have observed these phases in my life and the same has happened to other people to whom I told my observation: After the study of a problem in the evening (e.g. a mechanical problem, a software problem or structuring a future paper), I awake several times in the night and every time I write down what comes to my consciousness. In the morning I look at the notes: some are interesting, some duplicated, some useless or silly. But as a whole I have an idea for a solution. I thank my little fairies. – The spark of intuition results probably from a "bisociation", i.e. a holistic merge of (at least) two separate areas in the brain (Broelmann 2002).

The *consequence* of the above is, that we cannot derive all our thinking and acting from free, rational, and logical decisions. This gives a feeling of modesty, of tolerance for our own errors and these of others. This leads to a willingness for the acceptance of the meaning and behaviour of others. And this may be positive for a good group climate in design.

4. Proof of above statements by empirical research and from the literature

4.1 Unconscious processes of practitioners in design

In a cooperation with the experimental psychologist Dörner, University of Bamberg, Günther's thesis (1998) had the aim of comparing the different behaviour in design between graduate engineers and pure practitioners working on the same task, a device (Günther and Ehrlenspiel 1999). All test persons had to make a layout-drawing on a conventional drawing-board. Time was unlimited. (They worked between about 5 and 15 hours). The basis for the evaluation were videos and recordings of what the test-persons said. (Loud speaking of their thinking was recommended). The evaluation of the design results was made by two independent teams at the TU München and TU Darmstadt.

Test persons where five individual working practitioners, which included draftsmen, technicians and master-craftsmen, without any higher education and two graduate designers. The first are nominated *"p-designers"* (from "practice"), the latter *"m-designers"*, because they were trained in design-methodology. Guenther took the two out of 20 tested people, because their design processes were most similar to the German-guide-line for design methodology VDI 2221 (1993) and further they had a certain amount of experience with industrial companies.

The task in Fig.2 shows a mechanical device, but redrawn in 3D from the solution of one test person. The main function is to adjust an optical device in several directions for rectifying pictures. This optical system is guided up and down on a column and should allow an inclination in the α- and β-direction. The 8 partial functions PF where not given to the test persons.

The m-designers were acting methodically, mainly conscious in contrast to the p-designers. The p-designers where asked after the end of their work, if they knew how and why they had chosen their procedure as they did. But they where astonished over this question and said. “Why you ask? It's clear, that's just the usual procedure. No problem”. Nevertheless it seems to me, that parts of their process are conscious, others unconscious with routine or even creative acting. The exact separation of conscious and unconscious is difficult, as mentioned before.

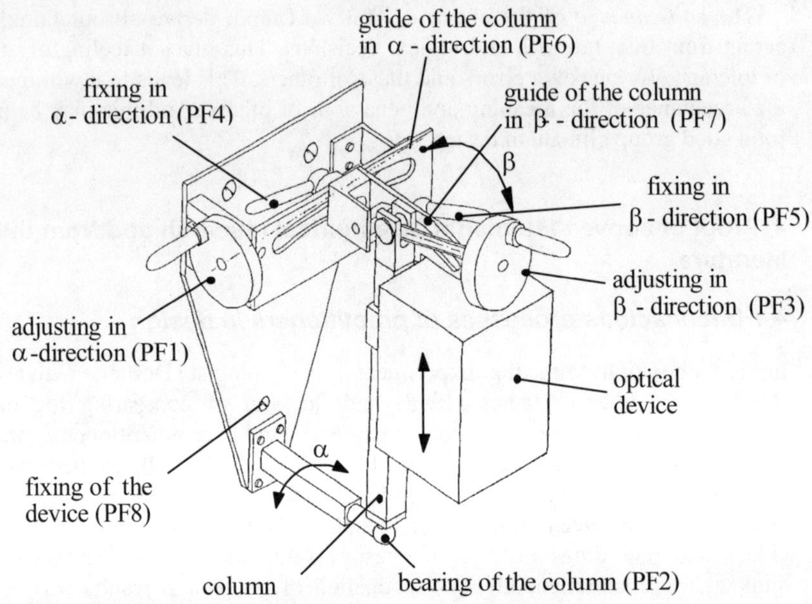

Fig. 2. Example for the mechanical device "wall mounting" with 8 partial (or sub-) functions to be designed (Ehrlenspiel 2002)

Hans as an example for a *m-designer* works as follows:

- In a *phase-oriented procedure* he tries to solve all 8 partial functions according to the VDI 2221. He is going stepwise through all phases of design: Task clarification, conceptual design, embodiment design (rough and final). (The phase 'detail design' was not demanded here).
- After the initial clarification of the task, he searches several solutions for every function, which he than combines into concept-variants. The best will then be selected from these. This is the *"generative" procedure*, as used in the VDI 2221 (1993).
- This procedure is similar to multitasking on the PC, a kind of *"open" design,* which takes a large amount of working memory instead of a *"quick conclusion"* design of the p-designers. Therefore the m-designers have to note and sketch their ideas and document them a lot more than the p-designers. This takes time.
- Hans needed 12 hours by contrast to the p-designer Rolf who needed only 5.3 hours. Although there was no given time-limit, p-designers seem to have internalised the time-pressure of practice and look for short processes. M-designers follow a *quality-oriented* strategy. Time for them is secondary. In contrast to these, p-designers follow a *time-oriented strategy* with a fairly good, an *"acceptable" solution.* In Fig.3 about 87% the quality of m-designers.
  To reach this solution, the procedure of *p-designers* (e.g. *Rolf*) is as follows:

- With a *sub problem-oriented procedure*, Rolf first solves the question of the bearing as the essential sub-problem (or partial problem; partial function) and works it through all phases until the concrete drawing in the final embodiment design phase. He even omits one phase: the rough embodiment design. All other solutions for the other partial functions or sub problems were found one after the other and were adapted to this first solution: This may be named the *"corrective" procedure* (Fig.4).

- Typical for this procedure is: A short clarification of task; very few variants for solutions, choosing the best with very short, intuitive evaluation; very poor documentation of the process. Main effort is the documentation of the design of the product. This, and the mostly unconscious process, saves time!

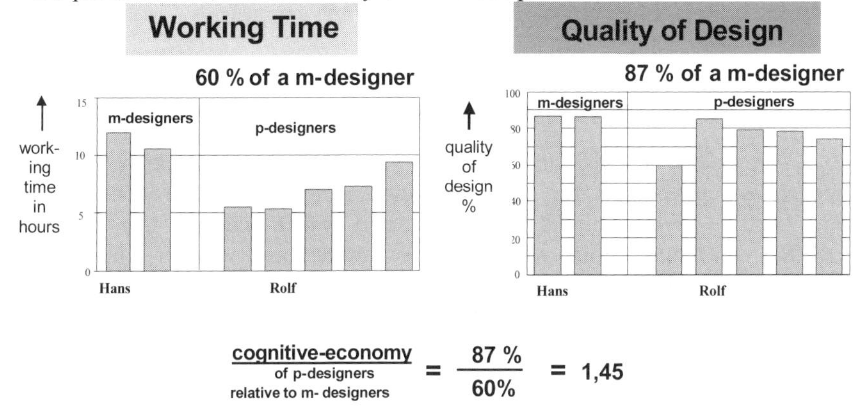

Fig. 3. Working time and quality of design of the m- and p-designers. P-designers need in average 60% of the time of the m-designers and realize 87% of their quality. (For cognitive economy see Fig.7).

In Fig. 3 you can see that the working time of the five p-designers is on average about 60% of that of the m-designers. The relative quality of their design is 87%. The two designers Hans and Rolf are indicated. – It seems that the strategy of the *m-designers* is better for *new and high quality products*, whereas that of the *p-designers* is better for *variants of known products if time is tight*. It is clear that most of design tasks in industry are concerned with the latter.

It is obvious that these statements have a poor basis, seen from the number of test people, but this is a first research and the evaluation of the experiments took several weeks per person. Besides this it was difficult to get comparable test-people. Nevertheless it gives us an idea about what happens in practice and puts forward questions for future research. It is clear too, that we need more information on the behaviour of methodical trained high school engineers, which have gathered experience a longer time in industrial practice, as the p-designers have. Perhaps they are much better than the p-designers and comparable short in time. For complex tasks this likely would happen (see explanations to Fig.9).

In Fig.4 the difference between m- and p-designers are given. Due to restriction of length it is not possible to explain all expressions (Ehrlenspiel 2002). The (mostly unconscious) methods on the right were new and not thought by design-methodology in BRD before.

m-designer (e.g.Hans)	p-designer (e.g. Rolf)
phase-oriented procedure	subproblem-oriented procedure
generative procedure	corrective procedure
open design	quick concluding
quality-oriented strategy	time-oriented strategy

Fig.4. The difference between m- and p-designers. Both use methods. The latter more or less unconscious.

4.2 From the "creative feasibility study" to a computer program

Tropschuh (1988) has shown in his Dr.-thesis that it is possible for creative, unconscious procedures for the configuration of complex marine gearing components to be transmitted first in an expert system and after that in a usual FORTRAN-program. This was done in cooperation with a known gear manufacturer. The produced designs connect Diesel engines, gas turbines, propeller thrust bearings, free wheel clutches, electro generators and spur gears as well as planetary gears. (Example from Tropschuh's work in Fig.5)

Fig.5. Marine gear for two Diesel engines with 6180 KWeach. Speed 530/135 rpm. With two multiple-disc clutches.
Further two generators for electric power switched by two other multiple-disc clutches.

Initially the project engineers could not believe, that their "creative and quick, mostly unconscious decisions, resulting from experience over years" (their statement!), could ever have been formalized. – But with a systematic analysis of the possible configurations as well as with a careful observation and questioning of their decisions the tacit knowledge of the engineers was lifted to consciousness.

Finally the engineers made their decision-program out of Tropschuh's logic and used it to shorten their processes.

4.3 The influence of external, psychomotoric activities on design (sketching, gesticulating, speaking, writing)

Hacker differentiates in the process of design between *internal*, mental steps (e.g. concluding, decision making) and *external*, psychomotoric steps (e.g. sketching, gesticulating, speaking, writing) (Hacker 2002), which latter influence the quality and the timing of design too. These external activities force us to concentrate on the essentials, because it is – due to the restrictions of our brain – not possible to speak about everything or to sketch everything.

Pache investigated the conceptual solution for a device to laser weld sheet metals in the automotive industry. (See the paper Pache and Lindemann in this book: Fig.4). Shape and dimensions of the sheet were given to the student, as well as the laser and a linear cylinder for actuation. One of the main problems was to guide the laser along the complex shape of the sheet metal.– The process of design was investigated in a similar manner to 4.1.

Pache's Fig.4, state 1 shows the first sketch of the laser guide by two tracks. From this the student generated a new solution principle, shown in state 3. The laser now is driven by a belt, running over pulleys. – Probably this resulted from a "bisociation": an analogy of the bent track to the curve of a belt. Sudden creative processes will be released by the abstraction of the sketched object. This may even be triggered off by gestures.

This in all probability was an intuitive idea, because looking at an image silently combined with gestures is often unconscious. As we know the use of gestures during speaking is mostly unconscious and not planned.

4.4 The history of inventions or findings

Broelmann (2002), analysing the history of the origin of the gyro-compass, asks if the usual explanation of the origin of many new principles may be right, as emerging from the physical and technical sciences. This may be valid for some, e.g. atomic power stations emerged from nuclear physics. – But for most inventions this is not valid, as Broelmann assumes, and demonstrates using the gyro-compass as an example. The inventors were the psychologist Narziss Ach, the „art historian" Hermann Anschütz-Kaempfe, the engineer Otto Schlick, the mechanic John Serson (p.26) and the industrialist Werner von Siemens (p.13). None of them were scientists! – The beginning was the study of a toy gyroscope, whose horizon was independent of the inclination of its support. Just this was needed on ships to determine the height of the sun from a non-discernible horizon. Additionally the spinning axis is stable and not influenced by the earth's rotation. – To summarize the whole development over years, involving several persons: The beginning was an intuitive, at least partly unconscious handling of simple models, the making of better models and the study of their characteristics. Just such a presentation of facts started established scientists (i.e. the professors Grammel; Felix Klein; August Föppl; Sommerfeld; Einstein; Magnus) to look for theoretical models and systematisation.

But there are *direct verbalisations* too from the initially unconscious generation of even scientific knowledge. *Albert Einstein* has written: "For my kind of thinking the verbs or the language, as it is written or spoken, do not seem to play any role. The mental units, which serve as elements of my thinking, are certain signs and pictures of more or less lucidity" (p.258). – Known too is how the idea arose for the molecular structure of the Benzol C_6 H_6 by *August Kekulé* in 1865. He and some colleagues had a long search without success before. One evening, in front of his fire he had a dream: a serpent biting her own tail. This brought up the idea of the Benzol ring structure.

Conclusion: The usual view, that scientific or technical rational mental processes plays a major part in new invention or discovery is wrong. Intuitive, unconscious mental processes are often dominant initially (Müller and Franz 1998; Ehrlenspiel 2002). In addition, for invention, playing with and experimental handling of concrete models can be decisive. – Scientific methods are then essential for the further development and optimising of born inventions!

4.5 The "sound design" or the result of the "design feeling"

Any skilled designer sees straight away (unconscious), which parts in a drawing are sound, which not. From his long experience with similar products, he knows if the proportions are right. If the walls are too thick or too thin, if the size or the number of nuts or bolts is ok and so on. – But as Dörner (1995) writes, the corresponding feeling is principally conservative, because it results from a pattern matching of hundreds of similar drawings in the past. If there are new conditions and problems the "intuitive feeling" goes wrong. It is similar to a skier, who is used to skiing with dreamlike safety on the steepest slope. But when the weather has changed, the conditions altered, he probably initiates an avalanche. – But as most designers work on similar products all the time, they follow their design feelings and don't calculate, since it is neither affordable nor necessary. It's typical, that only about 3 to 5 % of the total design time in mechanical industry is spent on calculations. – It is not astonishing that this feeling is not only valid for the shaft ends of Fig.6, but for drawn faces too. The proportions in the middle are sound for shafts.

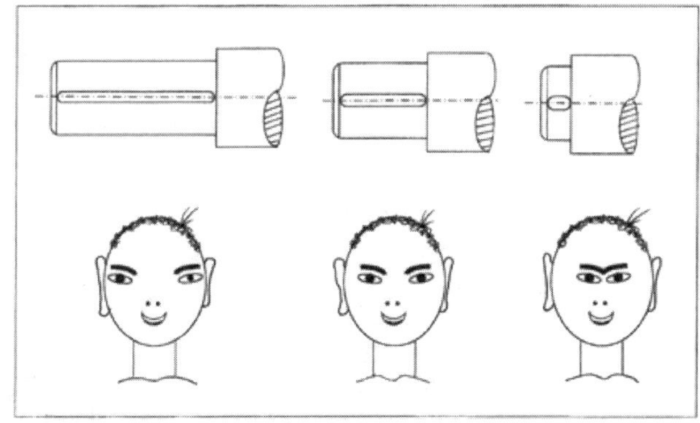

Fig.6. The unconscious design feeling is evaluating the shaft in the middle as sound, because the relation key length to shaft diameter is technically optimal. This is not the case right and left. – The portrait in the middle is sound too, because the eye distance is correct.

In the middle the bearing key length l, divided by the diameter d, is l/d = 1,1, and this corresponds to the Niemann (1975) rule l/d = 1,1...1,4 for steel. Left the shaft is too long (l/d = 1,9) and will break under the torque because the key has not share the load over its whole length. On the right, the shaft is too short (l/d = 0,18). To take same torque it should have several keys, which will not share the torque equally. – This is just a good feeling, not a calculated one. A portrait painter has the same feeling: only the spacing of the middle eyes is sound. (Rule: between the two eyes there should be space for a third eye).

Conclusion: We have seen above with the p-designers (4.1), as well as here with "design feeling", that the experience, accumulated in the unconscious, gives a "feeling for the sound", a common sense for design, which plays an essential role in practical design. – This includes a sound relation between the result (quality) and the effort (time) of an activity, which will be regarded below as "cognitive economy". Assuming that most of our actions are those of information processing, to which belongs more than just thinking, this is valid for the "information economy" too. (See Fig.7).

5. Cognitive economy as a main influence on human behaviour

Cognitive economy originally is a philosophical expression for the evaluation of different alternative theories (Müller and Franz 1998). A hypothetical theory, which is clearer and needs less mental effort, is nearer to truth. In this book the papers of Badke-Schaub and Stempfle, Gero, Goldschmidt, Tversky et al. and Visser have a certain background to cognitive economy. In our context, it is defined as the fraction of benefit and effort of a mental operation.

Fig. 7 shows this fraction, in which the numerator stands for the result of mental or information-processing and the denominator for the effort or time for processing.

The expression has two roots

- The limited capacity of the *working memory*. This is the bottleneck of conscious capacity for complex mental processes (Hacker 1989; Hacker 1989a).
- Probably the *evolution strategy of nature*: To reach an advantageous result nature in general is looking for minimal effort.

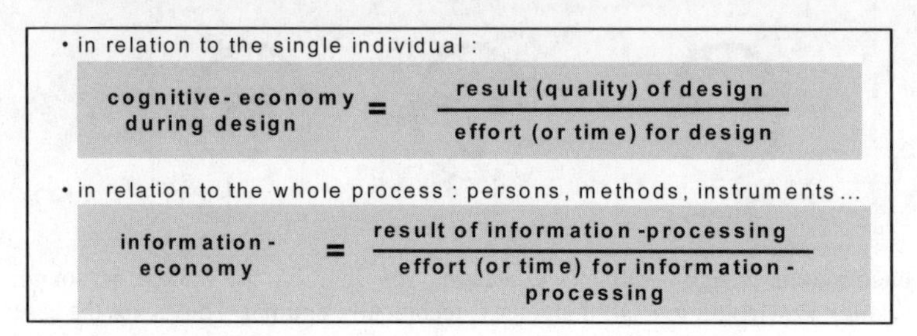

Fig.7. *What is the meaning of cognitive- and information- economy?* – Information economy is not discussed here. It means the whole process of information treatment. This includes persons, technical and “human-centred” methods (e.g. for organisation, project planning, communication), means and instruments (e.g. computers, experimental equipment...) (Ehrlenspiel 2002)

The *effort* is easy to measure, e.g. over time if all other influences on the effort are constant or negligible. – The *result* or quality, on the other hand, is very difficult to determine. In our case, with the "wall-mounting device" (4.1) it was possible and done by the judgement of two independent groups in München and Darmstadt.

From *Fig.3* we have seen that the working time of the five p-designers is on average about 60% of that of the m-designers. The relative quality of their design is 87%. The relative cognitive economy therefore is 87/60 = 1,45, that means 45% higher than for the m-designers. They have as a motto “An 80 % solution is sufficient; time is what matters most. If quality is not sufficient, we rework the design”. Design as continuous error correction! (Hacker 2002)

How to reach high cognitive economy?

Fig. 8 shows some types of behaviour necessary according to the restrictions of our brain. In a very simple model our brain will be divided in three parts: An ultra-short-time memory (a filter for the sensorial input), the working (respectively the short-time -) memory and the long-time memory. The working /short-time memory has a very limited capacity of 7 +/- 2 chunks[1]. These are units for information-processing. To simplify even more: In a PC the working / short-time

[1] According to Hacker and Osterland (1995) and Hacker at al. (1997) the short-time memory is just remembering items, meanwhile the working memory additionally carries out processes. Both is limited. It is distinguished between a numerical and a textual span. For better understanding an example: We can multiply 24 x 230 = 5520 direct in our brain, but normally we can not multiply 241 x 231 = 55671 without an external memory: we have to write the numbers!

memory is comparable to the PC working memory, the long-time memory to the disc. – A higher cognitive economy is reached usually unconsciously by reduction of the effort or time for the process holding an acceptable quality of output.

The *first 5 points* in Fig.8 are typical for problem-solving in general. Because of the limited capacity of our working memory we try to realise small and "quickly concluding" processes: working in steps. The same is valid for iterative working. This means, after having found a provisional solution, this will be documented and therefore it is out of the brain. Later on it will be retrieved for further improvement, and so on. This takes time, but the cognitive effort before documentation is not too high. Swinging between… means, that we are not able to store e.g. the principle of function of a design together with all its details. – A lot of problem-solving methods and the design methodology are based at least on points 1 to 5, and were developed without knowing this background of our restricted brain.

Point 6 to 10 were partly explained in section 4.1 according to the p-designers. In our experience the tasks should not be too complex and within the designers experience: "well defined problems"!

But it is a question, which tasks are specially suitable for the corrective or sub-problem oriented design. Short-time design could have lower design quality! – Point 8 and 9 are supported by the paper of Badke-Schaub and Stempflc in this book. As well point 10 corresponds to "process 1" of this paper. Avoiding to analyse reduces mental energy and time. In this case evaluation results from an intuitive feeling. The cited statement of Pöppel (1999) is experimentally confirmed: "Perception is happening with simultaneous recollection and evaluation". The points 8 to 10 were observed too from Wulf (2001) and described as typically for (natural) human problem solving.

Behaviour due to restrictions of the working - memory

1. working in steps
2. iterative working
3. swinging between the whole and the detail
4. swinging between the abstract and the concrete
5. doing the essential first
6. sub problem-oriented design
7. corrective design
8. applying known solutions, if possible out of own experience
9. reduction of alternative solutions
10. evaluation of solutions without analysis

Behaviour due to restrictions of the long - time memory

11. storing information in external memories (e.g. paper, computer)
12. working by division of labour: integrating the knowledge of others
13. asking competent, experienced and motivated colleagues: use of the brain of others

Fig.8. Behaviour *due* to the restrictions of our brain. Point 1 to 5 is a background of methodical acting and usually is unconscious. Point 6 to 10 reduces design time.

Point 11 to 13 result mainly from the restrictions of the long-time memory. We know only a small part of what we should know. Therefore we have to store and to search. Not only in sheets, books and data-stores but also in the "brains of other persons": division of labour! This is mainly to be observed in complex organisations. Therefore it is necessary to look for good cooperation and motivation.– Point 13 results from Frankenberger (1997), Badke-Schaub (2001), Badke-Schaub et al. (2001a). It is much more effective for quality as for time to ask experienced people about a certain problem, than to search in electronic stores; except were it is just the question of numerical standard data. (See too the paper of Ken Wallace in this book).

I like to emphasize the *research methodology* of Blessing et al. (1998): In Fig.8 a provisional <u>description</u> of methods, practitioners use, is given by the first 10 points (more or less unconsciously and in an opportunistic manner: see the paper of W. Visser in this book). But our knowledge about these methods is rather poor. Therefore we should intensify the research how methodical untrained people work. Than we should analyse and evaluate this behaviour to come to a <u>prescription</u> for an improved behaviour (e.g. for higher quality in a restricted time). The <u>training</u> of this and later <u>efficiency-control</u> and further improvement could lead to a new design methodology, which is nearer to a "natural design behaviour". This again could improve the acceptance, the motivation and the teaching of design methodology. Probably a *two stage methodology* comes out: For simple problems and just an acceptable solution the "standard thinking" according to Fig.1. And for higher quality and complex problems the "rational thinking" as sketched below.

Fig. 9. gives an overview on *how to realise high cognitive- and information-economy* for rather complex and innovative tasks or for "ill-defined problems". The same is valid, if high quality of the product is required. In this case it is necessary to solve the task methodically right from the very beginning to avoid errors and therefore to reduce long iterative cycles. By this a reduction of effort and overall time is possible compared to the usual "muddling through". This is not contradictory to point 2 in Fig.8, that our restricted brain has the tendency to quick final processes, trial and error and therefore to (short) iterative cycles. This kind of behaviour is probably valid just for simple problems or parts of complex problems. In contrast to this, for complex products and processes, where the cycles last much longer and could affect a lot of persons, a planned, methodical process and team work is necessary, because our specialisation is increasing.

How to reach high cognitive and information economy for complex products / processes and high quality of products? (Added up for the whole process)
1. reduce long lasting iterative cycles! How?
•emphasize the clarification of the task in a team
•in the team should be persons from other departments and
persons, who have to do with that job (task) in the future.
Further: the client (user), the supplier, at least temporary.
•plan time of process and cost of product and process
•motivate the team; give responsibility
2. use methodical design (tools) flexibly
•emphasize early phases of the process
•use phase-oriented design
•use generative design
•use of creativity techniques to overcome barriers
•evaluate solutions after analysis
3. use computer-tools for information-search, modelling (CAD, Virtual Reality, Rapid Prototyping, TRIZ...), project-planning, data storage

Fig.9. Behaviour for complex and innovative tasks – "ill defined problems" – and for products of high quality. This results in high cognitive and information economy added up for the whole process. (Last point of 2 corresponds to "process 2" of the paper Badke-Schaub / Stempfle in this book)

Summary

It should be emphasized that unconscious processes play an enormous role in design. This was shown for the finding of new concepts, products, ideas (section 4.1 to 4.4), as well as for the intuitive feeling on the soundness of known products (4.5). The cognitive economy resulting from the limitations of our brain is a central thread and one basis for design methodology (Fig.8).

Finally two rules for design:

- In "standard design areas"[2] work with quick unconscious search for an acceptable *solution* in the area of "standard thinking" with minimum *effort* (Fig.1, Fig.7, Fig.8).
- For "innovative design" look for the best *solution* with "rational thinking" (Fig.1, Fig.9) using design methodology at a higher, but acceptable *level of effort.*

[2] standard design area = known, optimized products with general known area of solutions for the designer

References

Badke-Schaub, P (2001) Training zum Erkennen und Bewältigen kritischer Situationen in Projektgruppen. In Fisch R, Badke-Schaub P, Stempfle J, Wallmeier, S (2001a) Transfer of Experience in Critical Design Situations. In Cully S, Duffy A, Mc Mahon C, Wallace K (eds) Design Management – Process and Information Issues, pp 251-258, London: Professional Engineering Publishing

Blessing L, Chakrabarti A, Wallace K (1998) An overview of Descriptive Studies in Relation to a General Design Research Methodology. In Frankenberger E, Badke-schaub P, Birkhofer H (eds): Designers - the Key to Successful Product Development; London: Springer

Broelmann J (2002) Intuition und Wissenschaft in der Kreiseltechnik 1750 bis 1930. München: Deutsches Museum; eingereicht als Dissertation TU München

Dörner D (1995) Konstruktion und Intuition. Tagung „Effizienter Entwickeln und Konstruieren" 23./24.3.1995; Düsseldorf: VDI Berichte 1169; p 1-10

Ehrlenspiel K (2002).: Integrierte Produktentwicklung. München: 2. edn Hanser

Frankenberger E (1997) Arbeitsteilige Produktentwicklung. Diss. TH Darmstadt 1997; Düsseldorf: VDI-Verlag; VDI-Fortschrittsberichte, Reihe 1; Nr. 291

Günther J (1998) Individuelle Einflüsse auf den Konstruktionsprozeß. Aachen: Shaker-Verlag; Diss.TU München

Günther J, Ehrlenspiel K (1999) Comparing designers from practice and designers with systematic design education. Design Studies Vol. 20, No. 5; Elsevier Science Ltd. p. 439-451

Hacker W (1989) On the utility of procedural rules. Ergonomics; 32, 7, p. 717-732

Hacker W (1989a) How to feed the computer quickly? Pros and cons of hierarchical data organisation. In. Klix F, Streitz N.A., Waern Y and Wandtke H (eds) Man-Computer interaction research; pp 253-271; Amsterdam: North Holland

Hacker W (1992) Expertenkönnen – Erkennen und Vermitteln. Göttingen: Hogrefe

Hacker W (2002) (ed) Denken in der Produktentwicklung. R. Hampp Verlag

Hacker W, Osterland D (1995) Mentale Koordinationskapazität – Einfluß von Text- und Arbeitsgedächtnismerkmalen auf das Verstehen von Instruktionstexten. Zeitschrift für experimentelle Psychologie, 42 (XLII), p 646-671

Hacker, W, Handrick S, Heimann I, Oehm D, Richter F, Sachse P, Schneider M (1997) Individuelle Unterschiede im Arbeitsgedächtnis für numerisches Material: - Rechenspanne – Differenzierungsfähigkeit und Differenzierungsquellen. TU Dresden, Inst f. Allg. Psychologie, Forschungsberichte, vol 48

Heymann M (2002) Kunst und Wissenschaft in der Technik des 20.Jahrhunderts. Zur Bedeutung von „tacit knowledge" bei der Gestaltung von Technik. Manuscript for a planned book

Müller J (1991) Akzeptanzbarrieren als berechtigte und ernstzunehmende Notwehr kreativer Konstrukteure. Nicht immer nur böser Wille, Denkträgheit oder alter Zopf. In: Hubka V (ed) Proceedings of ICED 1991, Zürich. Zürich: Edition Heurista p. 769–776. (Schriftenreihe WDK 20)

Müller J, Franz 1 (1998) Anforderungen des Konstrukteurs an die Entwicklung von Wissenssystemen für das Konstruieren. VDI-Berichte 775; Düsseldorf; VDI-Verlag

Niemann G (1975).: Maschinenelemente. Bd.1 (2.edn) Berlin: Springer (see 3. edn 2001)

Noretranders T (1994) Spüre die Welt. Die Wissenschaft des Bewußtseins. Reinbeck: Rowohlt

Perrig J, Wippich W, Perrig-Chiello P (1993) Unbewußte Informationsverarbeitung. Bern: Huber

Pöppel E (1999) Informationsverarbeitung im Gehirn. Der GMD Spiegel 3 / 4; pp 48-51

Pöppel E (2000) Grenzen des Bewußtseins. Wie kommen wir zur Zeit, und wie entsteht Wirklichkeit? Frankfurt: Insel-Verlag

Roth G (2000) Geist ohne Gehirn? Bonn: Forschung und Lehre pp 249-251; Heft 5

Thompson R.F (1994) Das Gehirn.(2. edn), Heidelberg: Spektrum Akad.Verlag

Tropschuh P.F (1989) Rechnerunterstützung für das Projektieren mit Hilfe eines wissensbasierten Systems. München: Hanser; equal: Dissertation TU München 1988

VDI-Richtlinie 2221(1993) Methodik zum Entwickeln und Konstruieren technischer Systeme und Produkte. Düsseldorf: VDI-Verlag

Wulf J. (2001) Elementarmethoden zur Lösungssuche. Diss. TU München, Reihe Produktentwicklung München, München: Verlag Dr. Hut

I am grateful to Prof. Dr. J. Müller, Chemnitz and Ms. Dr. P. Badke-Schaub, Bamberg, for correction, D. Welbourn, Cambridge for improving the English.

Strategic knowledge differences between an expert and a novice designer

Manolya Kavakli, Charles Sturt University, Australia
John S Gero, University of Sydney, Australia

1. Introduction

In this chapter, we explore the effect of strategic knowledge in conceptual design process by examing the differences in cognitive processes and groupings between a novice and an expert. In our previous studies, we found that the expert's cognitive activity and productivity (in terms of image generation) were three times as high as the novice's in the overall design process (Kavakli et al. 1999). We investigated the structure of cognitive actions in the design protocols, and found that there is evidence for the coexistence of the cognitive actions (Kavakli and Gero 2001). Certain groups of cognitive actions increase and decrease in parallel with each other in the protocols of the novice and expert designers. We suggested that the differences in the performance of designers could be attributed to the differences in the structure of those concurrent cognitive actions. Investigating the concurrent cognitive actions, we found that the expert's cognitive actions are well organized and clearly structured, while the novice's cognitive performance has been divided into many groups of concurrent actions. (Kavakli and Gero 2002). In this chapter, we focus on explaining the difference in performance between the expert and novice in terms of their respective strategic knowledge.

2. Strategic Knowledge, Chunks, and Cognitive Segments

Beginning in the early 1950s a number of researchers (Miller 1953; Attneave 1954) noted that if an individual's response tendencies change with experience then observing his behaviour yields information as to what experiences he has had. This approach suggests the value of analyzing the detailed sequence of both observable and the inferred operations performed by an individual engaged in a cognitive task, that is, engaged in the process of taking in, storing, transforming, and retrieving information (Estes 1976). This kind of information processing theory necessitates close attention to the information processing requirements of a task and thus is often useful in uncovering constraints on the learner that would not otherwise be apparent. We take sketching in conceptual design as a form of mental imagery processing (Kavakli and Gero 2001). Mental imagery processing (Kosslyn et al. 1984) consists of image generation (drawing production), inspection (attention), transformation (reinterpretation), and information retrieval from a case base in long term memory. Eventually, all of these processes affect the rate of cognitive activity due to the limit of human short term memory.

Broadbent (1971) maintained that information is recorded in only two forms: transient excitations and long-term records. He compared short-term memory to address registers in a computer. The registers themselves do not store data; instead they point to data held in another storage medium. Short term memory consists of a limited number of *working registers*, each of which excites or activates some record in a long-term storage. If short-term memory is maintained by working registers, a basic question is how many registers are available. Using a variety of evidence, Miller (1956) showed that short-term memory can hold about seven *chunks* of information, where a chunk is the amount of information in a schema – a single bit, a decimal digit, a word, or a phrase. The basic property of a chunk is not its size, but its unity as a well-learned, familiar pattern (Sowa 1984). Broadbent (1975) argued that a better estimate is three working registers rather than seven, because only three or four items can be recalled with a high degree of accuracy, although the average span of short-term memory is about seven items. Extra items beyond three or four are remembered accurately only if they have associations to other items in the list. When people recall items from a familiar category, they tend to group their responses in bursts of two or three.

The design protocols used herewere collected as a retrospective report after the design session. We used the content-oriented protocol analysis method to investigate concurrent cognitive actions of designers. These protocols were divided into segments, indexed and coded according to the information categories classified by Suwa and Tversky (1997). Information on procedures of protocol parsing and coding can be found in Kavakli and Gero (2001 and 2002). A *cognitive segment* consists of cognitive actions that appear to occur simultaneously. We assume these cognitive actions as chunks in Miller's definition. We found that the design protocol of the expert includes 2,916 actions (chunks) and 348 segments, while the novice's protocol includes 1,027 actions and 122 segments. In both protocols, each segment includes 8 cognitive actions on average. However, considering that the same amount of time was given to both participants, the expert's design protocol is 2.8 times as rich as the novice's in terms of actions. There were also 2.8 times as many segments in the expert designer's session as in the novice's.

3. Concurrent Cognitive Processing

The cognitive activity of designers appears to be parallel to the drawing production on pages in both design protocols. Cognitive actions including looking, perceptual and functional actions, as well as certain types of goals, increase and decrease in parallel with each other in both protocols. If the cognitive activities slow down at some stage, this may be because of not only one activity, but also other activities having different roles that occur together. Therefore, we could look for the reason for the drop in the performance in concurrent cognitive actions, rather than only within a certain group of cognitive actions (Kavakli and Gero

2002). We investigated the concurrent cognitive actions in three levels of a tree-like structure: root, primary branching, secondary branching.

3.1. Concurrent Cognitive Processing

First, we investigated the major groups of cognitive actions (in terms of pages produced by the designers) that indicate strong correlations in both design protocols. As can be seen in Table 1 there are strong correlations between major categories of cognitive actions in the root in both design protocols. There are differences here between the expert and novice that will impact on our later analysis.

Table 1. Correlation coefficients of cognitive actions in pages

expert-page	Drawing	Looking	Perceptua	Functional	Goals	Moves
Drawing	1.000					
Looking	0.864	1.000				
Perceptual	0.998	0.909	1.000			
Functional	0.998	0.951	0.998	1.000		
Goals	0.995	0.829	0.996	0.996	1.000	
Moves	0.975	0.635	0.968	0.978	0.975	1.000

expert-page	Drawing	Looking	Perceptua	Functional	Goals	Moves
Drawing	1.000					
Looking	0.968	1.000				
Perceptual	0.786	0.898	1.000			
Functional	0.744	0.828	0.670	1.000		
Goals	0.655	0.806	0.981	0.617	1.000	
Moves	0.951	0.862	0.680	0.504	0.529	1.000

3.2. Primary Branching of Concurrent Cognitive Processing

Second, we investigate the correlations between subcategories of cognitive actions to find the structure in concurrent cognitive processing. Our purpose here is to explore the top-down correlations between a specific group of drawing actions (Dc: Depicting drawings) and others. We will narrow down our exploration and focus on only the concurrent actions highly correlated with depicting drawings, Table 2. Table 2 gives the full list of the correlation values of cognitive actions, while Table 3 only lists the codes and definitions of highly correlated concurrent actions with Depicting Drawings in both design protocols. The full list of the codes with their definitions can be found in Kavakli and Gero (2001 and 2002).

Table 2. . Correlations with Depicting Drawings

Action code	novice	expert	Action code	novice	expert
Drf	0.34	0.03	Frei	0.21	0.20
Dts	0.98	0.58	Fo	0.51	0.83
Dtd	-0.75	0.25	Fnp	0.60	0.31
Dsy	0.74	0.35	Fop	0.21	0.68
Dwo	0.75	0.32	Fi	0.26	0.24
L	0.99	0.81	G1-1	-0.29	0.45
Psg	0.71	-0.17	G1-2	0.73	0.67
Posg	0.64	0.27	G1-3	0.21	0.44
Pfn	0.66	0.45	G1-4	0.85	0.14
Pfp	0.90	0.15	G2	0.38	0.34
Pof	-0.27	0.53	G3	0.71	0.21
Prp	0.98	0.74	G4	0.58	0.19
Prn	0.28	0.70	Ma	-0.29	0.31
Por	0.92	0.57	Mod	0.60	0.07
Fn	0.86	0.75	Moa	0.89	0.69

We categorize the concurrent cognitive processing into two groups: primary branching and secondary branching. Primary branching lists the cognitive actions that directly correlate with depicting drawings. Secondary branching lists the cognitive actions that highly correlate with each action in primary branching. Table 4 lists the primary concurrent actions highly correlated with depicting drawings. In this and the following tables, "-" refers to negative strong correlations, "~" refers to substantial correlations and blank line refers to the cognitive actions which do not correlate. Otherwise, action code refers to positive strong correlations.

Table 3. Concurrent Action Codes

Cognitive Action	Action Code
Looking at old depictions	L
Overtracing	Dts
Mention of a relation	Por
Discovery of a spatial or an organizational relation	Prp
Creation of a new relation	Prn
Continual or revisited thought of a function	Fo
Association of a new depiction with a function	Fn
Motion over an area	Moa

Goals directed by the use of explicit knowledge or past cases	G1-2
Writing	Dwo
Depicting symbols	Dsy
Tracing over the sketch on a different sheet	Dtd
Discovery of a new space as a ground	Psg
Discovery of a new feature of a new depiction	Pfp
Goals not supported by knowledge, requirements or goals	G1-4
Goals to apply introduced functions in the current context	G3

Table 4. Primary Concurrent Actions Correlated with Depicting Drawings (Dc)

Root Action Code	Primary Action Code: Novice	Primary Action Code: Expert
Dc	L	L
	Dts	~Dts
	Por	~Por
	Prp	Prp
		Prn
	~Fo	Fo
	Fn	Fn
	Moa	~Moa
	G1-2	~G1-2
	Dwo	
	Dsy	
	-Dtd	
	Psg	
	Pfp	
	G1-4	
	G3	

( ) positive strong correlation; (-) negative strong correlation; (~) substantial correlation.

As we can see in Table 4, strong correlations in both design protocols are seen between depicting drawings (Dc) and:

- looking actions (L),
- discovery of a relation (Prp), and
- association of a new depiction with a function (Fn).

In addition to these, in the expert's design protocol, there are also strong correlations between depicting drawings (Dc) and:

- creation of a new relation (Prn)
- revisited thought of a function (Fo).

There are weak correlations in these categories in the novice's. However, except for these two (Prn and Fo), there are many actions that occur together in the novice's protocol in parallel to depicting drawings. Concurrent actions in the novice's design protocol indicates a long list of correlations.

Besides these, tracing over the sketch on a different sheet (Dtd) is also strongly negatively correlated with depicting drawings (Dc) for the novice. On the contrary, discovery of a new space as a ground (Psg) is, surprisingly, negatively (though weakly) correlated for the expert.

3.3. Secondary Branching

Table 5 lists secondary concurrent actions (that occur parallel to the primary concurrent actions). The first column indicates the primary concurrent action code, which is parallel to depicting drawings. The second column indicates its correlation value with depicting drawings in the novice's design protocol, while the third indicates the same in the expert's. Secondary concurrent actions listed in a row are the ones that strongly correlate with the primary concurrent action in the first column.

Table 5. Secondary Concurrent Actions Correlated with Depicting Drawings (Dc)

Primary Action Code	Secondary Action Codes: Novice	Secondary Action Codes: Expert
L	Dc, Dts, -Dtd, Dwo, Psg, Posg, Pfp, Prp, Por, Fn, G1-2, G1-4, G3, Moa	Dc, Prp, Por, Fo
Dts	Dc, Pfn, -Prn, Fi, G1-1, Ma	Dtd
Por	Dc, Dts, -Dtd, Dwo, L, Posg, Prp, Fo, G1.2, G1.4, G2, G3	L, Prp, Fo
Prp	Dc, Dts, -Dtd, Dwo, L, Psg, Posg, Pfp, Por, Fn, G1-2, G1-4, G3, Moa	Dc, L, Pof, Por, Fo
Prn		Dc
Fo	-Dtd, Pfn, Por, Frei, Fop, G1-3, G1-4, G2, G3	Dc, L, Prp, Por
Fn	Dc, Dsy, L, Psg, Pfp, Prp, -Pof	Dc
Moa	Dc, Dts, Dsy, L, Psg, Pfp, Prp, Fn, Fnp, Mod	Dc, Fn, Fop, G1-2
G1-2	Dc, Dts, Dwo, L, Psg, Posg, Prp, Prn, Por, -G1.1, G1.4, G4, -Ma	Moa
Dwo	Dc, Dts, L, Posg, Prp, Prn, Por, G1-2, G1-4, G2, G3	
Dsy	Dc, Psg, Pfp, -Pof, Fn, Fnp, Mod, Moa	
Dtd	-Dc, -Dts, -L, -Pfn, -Prp, -Por, -Fo, -Fi, -G1-4, -G3	
Psg	Dc, Dts, Dsy, L, Pfp, Prp, Fn, Fnp, -G1.1, G1-2, G4,-Ma, Mod, Moa	
Pfp	Dc, Dts, Dsy, L, Psg, Fo, Fi, G3	
G1-4	Dc, Dts, -Dtd, Dwo, L, Posg, Prp, Por, Fo, G1-2, G2, G3	
G3	Dc, Dts, -Dtd, Dwo, L, Posg, Pfn, Prp, Por, Frei, Fo, Fop, G1-3, G1-4, G2	

( ) positive strong correlation
(-) negative strong correlation
(~) substantial correlation

We can see in Table 5 that many concurrent cognitive actions coexist in the novice's design protocol, while only a small group of cognitive actions occurs in parallel in the expert's. Table 5 indicates that the expert's cognitive activity is based on the coexistence of a limited number of actions (5 at most) for each primary concurrent action code. However, in the novice's protocol secondary concurrent actions range from 7 to 16, which is more than the human short term memory can manage at one time (Miller 1956). Whereas the expert's working registers stay in the limits. In the context of chunks defined by Sowa (1984), as we remarked in the beginning of this chapter, the main property of a working register is its unity as a well-learned, familiar pattern. Do the structure of working registers, as shown in this case study, have some reference to well-learned design strategies through experience?

4. Tree Structure of Concurrent Cognitive Processing

Taking it one step further, now we will group the secondary concurrent actions to see the associated groups of working registers suggested by Broadbent (1975) who argued that a better estimate is three working registers rather than seven, because only three or four items can be recalled with a high degree of accuracy, although the average span of short-term memory is about seven items. Our purpose in this chapter is neither to test nor verify the number of working registers in design cognition, but to use it as a basis for structural differences in concurrent cognitive processing. Table 6 is the classification of the secondary concurrent actions into associated groups.

Table 6. Structures of Concurrent Actions Correlated with Depicting Drawings (Dc)

NOVICE		EXPERT	
Primary Action	Secondary Action	Primary Action	Secondary Action
L	A-1 {Fo}, B, C, D	L	A
Dts	G, -J-1{G4}, Dc	~Dts	Dtd
Por	A, B, D-2{Psg, Pfp}, E	~Por	A-1{Dc}
Prp	A-1 {Fo}, B, C, D	Prp	A, Pof
		Prn	Dc
~Fo	B-1{Dwo}, H, Por	Fo	A
Fn	A-2{Por, Fo}, D-2{Dts, Posg}, F	Fn	Dc
Moa	A-2{Por, Fo}, D-1{Posg}, F-1{-Pof}, K	~Moa	Dc, Fop, C
G1-2	A-1 {Fo}, D-1{Pfp}, J, Dwo, G1.4	~G1-2	C-1{Fn}
Dwo	A-1 {Fo}, B-1{-Dtd}, D-2{Psg, Pfp}, E, Prn		
Dsy	D-2{Dts, Posg}, Dc, F, K, Moa		
-Dtd	-A, -Dts, -G, -G1-4, -G3		
Psg	A-2{Por, Fo}, C, D-1{Posg}, J-1{Prn}, K, Dsy		
Pfp	A-2{Por, Prp}, D-1{Posg}, Dsy, Fi, G3		
G1-4	A, B, D-2{Psg, Pfp}, E		
G3	A, B, D-2{Psg, Pfp}, H		

Group A = {Dc, L, Prp, Por, Fo}	Group F = {Dsy, -Pof, Fn}
Group B = {-Dtd, Dwo, G1-4, G3}	Group G = {Pfn, Fi}
Group C = {Fn, G1-2, Moa}	Group H = {Pfn, Frei, Fop, G1-3, G2}
Group D = {Dts, Posg, Psg, Pfp}	Group J = {Prn, -G1-1, G4, -Ma}
Group E = {G1-2, G2}	Group K = {Fnp, Mod}
X-n {w, z}	
X: group code	- strong negative correlation
-n: number of missing group members	~ substantial correlation
{w, z}: missing members	

In the expert's design protocol in Table 6, as implied by the fourth column, looking actions (L) highly correlate with a group of actions including depicting drawings (Dc), discovery of a relation (Prp), mention of a relation (Por), and revisited thought of a function (Fo). The same group can be seen in the novice's design protocol, as shown in the fourth column, but there is a member of this particular group missing in the novice's design protocol. Since this list of concurrent actions appears more than once in both the expert's ad novice's design protocols, we call this list of actions (including the action code in the first column) a group and label it Group A in Table 6. Group C is another example of this type of grouping. It consists of a group of actions taking place as a full list in both design protocols. It includes the association of a new depiction with a function (Fn), goals directed by the use of explicit knowledge or past cases (G1-2), and motion over an area (Moa). Similar to Group A, Group C also appears with a missing member in some other categories in both design protocols. The other groups (B, D, E, F, H, J, K) are produced with the same criteria of appearing at least once as a full group among concurrent actions. Groups in which all the members are negative are represented by a - prefix. We represent the missing members in a group with a - followed by the number missing and the group member itself in parenthesis {}, for example, A-1{Fo} means group A less one member, Fo.

As we can see in Table 6, in the expert's protocol, strong correlations can be seen in the coexistence of only one group of secondary concurrent actions. It is either A including depicting drawings (Dc), looking actions (L), discovery of a relation (Prp), mention of a relation (Por) and revisited functions (Fo), or C including association of a new depiction with a function (Fn), goals directed by the use of explicit knowledge or past cases (G1-2), and motion over an area (Moa). Whereas, in the novice's protocol, cognitive performance has been divided into many groups of actions, B, C, D, E, F, G, H, J, K, in addition to A. The novice's secondary concurrent actions appear to be combinations of these groups of actions. For each action code, the associated groups of concurrent actions range from 1 to 3 in the expert's cognitive processing, while they range from 3 to 5 in the novice's.

5. Conclusion

This case study shows the effect of strategic knowledge on performance. We may categorize the differences between a novice and an expert in terms of: productivity; rate of cognitive activity; structure of concurrent action; number of cognitive processes and groupings and strategic knowledge.

The expert in our experiment governs his performance in a more efficient way than the novice, because of the clear organization and the structure of his cognitive actions. We have provided evidence that the expert's cognitive activity is based on a tree structure including a small group of concurrent actions in each branch (up to 5 in the primary and up to 6 in the secondary branching of cognitive processing). However, in the novice's protocol, cognitive performance has been divided into many groups of concurrent actions with a tree structure including many concurrent actions in each branch with up to 14 in the primary and up to 16 in the secondary levels. The novice deals with 2.8 times as many concurrent actions as the expert (14 compared to 5 associated groups). The expert's design protocol is 2.8 times as rich as the novice's in terms of actions. There were also 2.8 times as many segments in the expert designer's session as in the novice's. Is this coincidental or an indicative of the value of his strategic knowledge to govern his performance?

We have used Miller's magical number seven (Miller 1956) to support our hypothesis. We have applied the 7 +/- 2 test to both protocols and found that the groupings for the novice fail the test whilst those of the expert pass it. Thus, the number of concurrent cognitive actions of the expert is between the limits of human short term memory, whereas, it is beyond the capacity of human short term memory for the novice. We have also used Broadbent's associated grouping test (Broadbent 1975) based on maximum 3 to 4 members. We have found that for each action code, the associated groups of concurrent actions range from 1 to 3 in the expert's cognitive processing, while they range from 3 to 5 in the novice's. The number of associated groups for the novice fail the Broadbent's test, whilst those of the expert pass it. These tests provide empirical evidence for a different strategic knowledge that structure their respective cognitive actions, and that may be the cause of the difference in performance between the novice and the expert.

These tests also provide evidence for a systematic expansion in primary and secondary branching in the expert's design protocol, and an exhaustive search in the novice's. Adelson and Soloway (1985) also report evidence on a systematic expansion in the experts' design protocols in their experimental findings. Granovskaya et al. (1987) stated that the process of amalgamating a new basis for classification (allowing one to reduce exhaustive search when choosing a strategy for solving a problem) involves changes in an alphabet of motion components. The changes result in the complete disappearance of external movements and formation of structures, which replace motions in analysis. Once these structures are present, the recognition process rate is increased so much that it gives the impression of being an insight. Our experimental results highlight the nature of this insight and suggest that strategic knowledge governs both cognitive activity and professional performance. Even with this limited evidence, our case study raises a question: to become experts, do designers have to learn strategies that would serve as an aid for perceptual and cognitive grouping?

References

Adelson, B., Soloway, E.: 1985, The role of domain experience in software design, *in* B. Curtis (ed.), *Human Factors in Software Development*, IEEE Computer Society Press, Washington, DC, pp. 233-242.

Attneave, F: 1954, Informational aspects of visual perception, *Psychological Review,* 61(3): 183-193.

Broadbent, D.E: 1971, *Decision and Stress,* Academic Press, London.

Broadbent, D.E.: 1975, The magic number seven after fifteen years., *in* A. Kennedy & A. Wilkes (eds.), *Studies in Long-Term Memory,* Wiley, Chichester, pp. 3-18.

Estes, W.K.: 1976, Structural aspects of associative models for memory, *in* Cofer, C.N. (ed.), *The Structure of Human Memory*, Freeman, California, pp. 31-53.

Granovskaya, R.M., Bereznaya, I.Y., Grigorieva, A.N: 1987, *Perception of Form & Forms of Perception*, Lawrence Erlbaum, London.

Kavakli, M., Suwa, M, Gero, J.S., Purcell, T: 1999, Sketching interpretation in novice and expert designers, *in*: Gero, J.S., Tversky, B (eds), *Visual and Spatial Reasoning in*

Design, Key Centre of Design Computing and Cognition, University of Sydney, Sydney, pp. 209-219.

Kavakli, M., Gero, J.S: 2001, Sketching as mental imagery processing, *Design Studies,* 22(4): 347-364.

Kavakli, M., Gero, J.S: 2002, The structure of concurrent cognitive actions: a case study on novice and expert designers, *Design Studies,* 23(1): 25-40.

Kosslyn, S.M., Brunn, J., Cave, K.R., Wallach, R.W: 1984, Individual differences in mental imagery ability: A computational analysis, *Cognition*, 18: 195-243.

Miller, G.A: 1953, What is information measurement?, *American Psychologist*, 8: 3-11.

Miller, G.A: 1956, The magical number seven, plus or minus two: Some limits on our capacity for processing information, *Psychological Review*, 63: 81-97.

Sowa, J.F: 1984, *Conceptual Structures: Information Processing in Mind and Machine*, Addison Wesley, Reading, Massachusetts.

Suwa, M. and Tversky, B.: 1997, What do architects and students perceive in their design sketches?: A protocol analysis, *Design Studies,* 18(4): 385-403.

This is based on a paper presented at *SKCF01*.

Cognitive economy in design reasoning

Gabriela Goldschmidt, Technion – Israel Institute of Technology

Introduction

Whether we define designing as systematic problem solving or, conversely, as free-wheeling artistic creation, we postulate that there is agreement about the following:

During the process of designing, designers continually reason about prospective features of the designed entity and the rationale for accepting or rejecting them. This is taken to be true whether the designer is an individual working by him or herself or a team working on a design assignment jointly.

If we accept the notion of the primacy of reasoning in designing, we must ask ourselves: what characterizes design reasoning, and what can be considered "good" (i.e. effective) reasoning, as opposed to "bad" (i.e. ineffective) reasoning in the particular case of designing? The effectiveness of designing can be concluded from evaluating its outcome. We can evaluate a proposed design product (complete or partial, fully developed or conceptual) by judging its merits against explicit and implicit criteria, both pre- established and constructed on the fly. Clients, teachers and peers habitually evaluate design proposals, often using comparisons between alternatives as a means to calibrate their assessments.

We propose that it is also possible to evaluate design reasoning. This can be done once we identify salient characteristics of design reasoning and develop a method to measure them in a straight-forward quantitative manner. This paper presents one such characteristic and a method to measure it: we propose that designers link the moves they generate to previous moves in order to ensure that the best possible fit is achieved between the emerging design and criteria for its evaluation (given and self-generated).

We shall claim that the effectiveness of the process of designing is commensurate with the density of links among design moves. This proposition is based on the belief that the cognitive system is 'economy minded' and therefore prioritizes a process in which a solution (design proposal) can be achieved, inclusive of its rationale, with as little 'waste' as possible. In our case, 'waste' means design moves that display little or no linkage to other design moves, thereby not contributing to the effort to construct a network of interconnections between the diverse facets of a design configuration and the reasons for opting for them. In other words, a "good" process is economical in that the moves generated while it unfolds maintain a high level of inter-connectivity such that the resultant design configuration satisfies requirements and desires, and is coherent, comprehensive, and free of contradictions.

There are many reasons for wishing to study the relationship between a design outcome and the process that brought it into being. First, a better understanding of the differences between effective and ineffective design processes may help in developing design support tools and interventions aimed at ameliorating the results. Second, we may gain insights relevant to design education. Finally, what is true for design reasoning may have parallels in other processes of creative endeavor about which we know at present painfully little.

The building block: the design move

Design tasks are complex because even in the case of limited assignments the problems are usually ill-defined and/or ill-structured. Ill-defined, because many potential outcomes may be acceptable and a precise definition of the task or problem must be developed along with the solution. Ill-structured, because the information provided at the outset is insufficient and more information must be 'imported' into the task in order to reach a satisficing (to use Simon's terminology) result. The ill-defined and ill-structured nature of design problem-solving is especially evident in the early, conceptual phase, in which a search is conducted: ideas are generated and deliberated until one concept is chosen for development in subsequent stages. Because of its special significance, we shall concentrate on the search during the conceptual phase of the design process. Since we want to capture the designer's reasoning at a cognitive level, we must work with 'on-line' information which is obtained by recording live design sessions: a team at work or an individual who is asked to 'think out loud', and transcribing the recorded data (and any graphic output) into protocols.

A design proposal, final or interim, is a solution – an actual or potential physical configuration[1] – that is always presented along with its rationale, so as to enable its assessment and comparison to alternative proposals. The rationale is not added to the solution once it has been finalized. Rather, the design solution and its rationale are developed concurrently in the process of designing, which is composed of design moves. A design move, analogous to a chess move, is a step, an act, an operation that transforms the design situation in some measure relative to the state in which it was prior to that move. A typical move comprises one sentence, sometimes two (see example in the next section). To understand the process of designing we must understand how designers reason on the fly as they generate design moves. Therefore, protocols of design sessions are parsed into design moves. The design move, then, is our basic unit of analysis. Moves are achieved in a few seconds (in one of our studies an individual industrial designer took an average of 10.9 seconds to generate a move, and a team of three industrial designers made a move every 4.2 seconds. See Goldschmidt 1995,199, Table 1). Therefore, even a brief design session is composed of a large number of moves. Given the limitations of the analysis system described below we limit our

---

[1] In this paper we address design leading to tangible products such as are normally envisaged in architecture, mechanical engineering, industrial design, etc.

inspections to relatively short vignettes. These, however, when wisely selected, are enough to allow comparisons of segments of interest: by different designers, e.g., experts and neophytes, or ordinary episodes and ones where breakthroughs are observed. Such comparisons are a major target of the cognitive inquiry we focus on, for which brief segments are perfectly suitable.

Links among moves

In this study we are not interested in the contents of moves, which we take to be task and context dependent, whereas our interest rests with the designer or designers' cognitive effectiveness (taken to be content and context free). We choose to focus on interconnectivity among moves, or the network of links among them. We therefore engage in establishing the existence, or lack thereof, of links between every pair of moves within a designated segment (of up to approximately 100 moves). The number and location of these links, we claim, tells us a lot about the nature and structure of reasoning. In particular, we would like to advance the following proposition:

The effectiveness of design reasoning, assessed by how successful the resultant design solution is judged to be, is commensurate with the ratio of linkage among design moves in the process of designing.

To investigate this proposition we developed *Linkography*, a system for the notation and analysis of links among design moves. First we establish backlinks: based on contents of moves and using common sense, we ask whether a link exists (yes/no) between every move and all preceding moves in a sequence under scrutiny (i.e., for n moves the question is asked n(n-1):2 times). For example, move 64 in a segment taken from a protocol of a team session designing a bicycle rack for a backpack was found to link with move 35, as both address joinery features of a so-called "tray" (rack):

64 you have a the tray would zip clip

35 maybe the tray could have plastic snap features in it so you just like kkkkk snap your backpack down in it

The backlinks are notated in a *Linkograph* – basically a matrix of links that can also be displayed as in Fig. 1 below. The backlinks of a move are all the links it maintains with preceding moves, as determined by assessors. Once all backlinks are notated, we can derive forelinks that moves protract. A move's forelinks are the links subsequent moves maintain with it, and they can be established only after the fact. In our Linkographs forelinks are mechanically derived, not determined by judgment as is the case with backlinks.

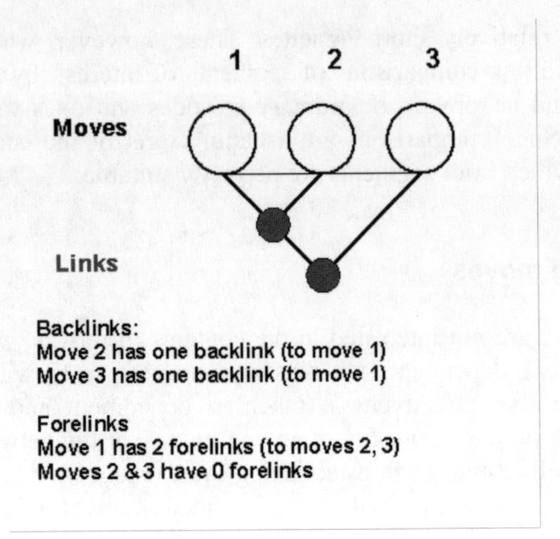

Fig. 1. Linkograph of 3 moves. Backlinks are stringed on / lines, Forelinks are stringed on \ lines

Once we have notated all links, we are no longer concerned with the contents of moves (in the present study). We want to reveal the 'link economy': according to our proposition effective design reasoning typically yields a high proportion of interlinked moves. If this is true, then the reverse should also be true: A high number of links is indicative of a fruitful process of design reasoning. To begin with we can simply calculate the ratio of links to moves in a given segment; this ratio is called Link Index (L.I.[2]). We are likely to find a higher L.I. in segments where design progress is observed, as compared to segments in which little or no progress can be reported. However, the picture afforded by Link Indices is limited; we therefore proceed to refine our analyses in the hope of gaining insights into design reasoning through an exploration of the 'economy' of links.

Critical moves

Moves differ in the amount of links they generate, both backward and forward. We can count the number of backlinks and forelinks that moves are credited with, and single out those moves which have a particularly high number of links in either direction (or rarely, in both directions). What is a high number of links? The answer to this question is that we must establish a threshold level for each investigation, depending on its 'granularity' and the purpose of the study (for segments of up to 100 moves, the threshold typically varies between 3 and 8. See

[2] van der Lugt (2001) calls this "link density."

[3] It is useful to indicate the threshold level; sometimes we use more than one such level in the same study.

Goldschmidt 1995). If we are right in attributing value to links as indicators of effective design reasoning, then there is reason to pay special attention to those moves that are especially link-intensive (potentially we can also get interested in their contents. We do that elsewhere, e.g., Goldschmidt and Weil 1998). Moves that generate a high number of links (< backward or forward >) are called Critical Moves (CMs). The notation indicates the direction of links that awarded the move its criticality, and the threshold number of links used to define criticality[3]. For example, <CM^6 indicates a critical move at the threshold level of 6 backlinks. As we shall see, Critical Moves are most helpful in understanding what we term as the *cognitive economy of design reasoning*. Fig.2 demonstrates the notation of CMs in a Linkograph.

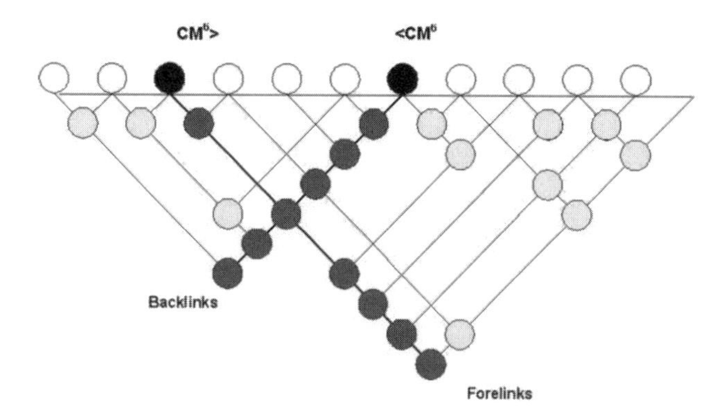

Fig. 2. Linkograph with one CM^6> and one <CM^6

To instantiate the significance of CMs, let us look at a critical path study. *Critical path*: the sequence of CMs in a given segment may be referred to as critical path. If a critical path determined this way resembled a critical path established using a different method to denote high significance, this would support the hypothesis that links among moves are noteworthy, and link-intensive moves have a special standing in the design process. Wang and Habraken (1982) conducted a protocol study of decisions taken by a designer in a conceptual design exercise. They listed 35 steps made by the designer and, based on their own content-centered expert assessment, established a critical path (we borrow the term, in the present context, from their study) which includes 7 of these steps (one of which is in fact a combination of two steps), as follows:

Wang and Habraken, critical path: 0 1+2 9 11 13 18 19

On the basis of a graph (network representation) they drew we are able to plot a Linkograph of the steps, which we treat as moves (although they are not moves in the cognitive sense). Using a level of three links as the threshold for criticality, we have established a critical path of CM^{3}s:

Critical path based on Linkography: 0 4 9 13 18 19 32

We notice that the two paths: a) have the same number of members (in fact step 4 summarizes and largely repeats a combination of steps 1 and 2, which is why we have consolidated them); b) maintain a similarity of over 70% (5 out of 7 members are identical). This is a high degree of correspondence that lends support to the magnitude we ascribe to Critical Moves and their use in exploring design reasoning. Our assumption that linkage among moves is related to the effectiveness of design reasoning appears to be supported by the fact that key moves are also the ones with the highest number of links. Fig. 3 shows a Linkograph pertaining to the afore-mentioned bicycle rack design, by a team. The critical path in this segment is comprised of the moves: 2 16 30 31 32 35 41 42 44 50 51 54 64.

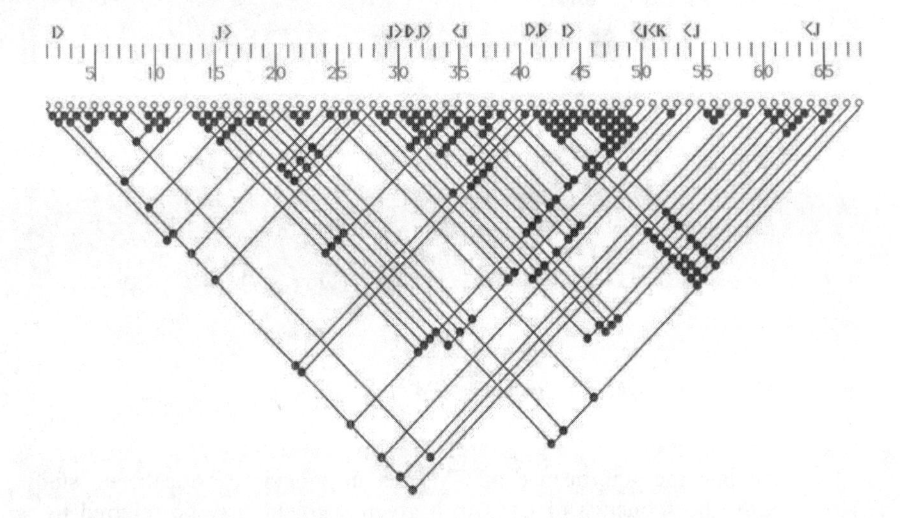

Fig. 3. Linkograph of 68 moves, team design (Delft Protocol Workshop).

Structure: Having established the significance of Critical Moves, let us inspect their position within the sequence of moves in a segment. In Fig. 3 above, Critical Moves are notated with the symbol < or > on top, indicative of a high number of backlinks or forelinks, respectively (the adjacent letters are initials of team-members who made those moves; see discussion in the next section). We can see that for the most part, forelinks are registered in the first part of the segment and backlinks are recorded in the second part (move 35 is a sole exception. In other cases the sequence of <CMs and CMs> can sometimes be less regular than in this segment). CMs> between moves 2 and 44 have mating <CMs; each pair of CM> and <CM describes a triangle in the Linkograph. Triangles may have overlaps and sometimes a small triangle is embeded in a larger one. Each defines design deliberations that are closely interrelated. For example, between moves 30 and 35 a first inspection of a leading idea takes place. Between moves 41 and 51 other ideas are looked at and then related to the idea expressed in 30-35. Each such

group of moves begins with a CM> and ends with a <CM; both are included in two larger triangles: 2-50 and 16-50 (move 51 "sticks out": an irregularity that is not uncommon in a non-linear process, where repetitions and lack of continuity are frequently observed). Likewise, they are also part of the triangle 30-64.

Besides some evidence regarding clear structuring of reasoning in the process of design reasoning (clarity of structure varies; the example above does not display a very sharp structure), the typical positioning of CMs> and <CMs at beginnings and endingss of small 'chunks' of reasoning, leads us to contemplate their different nature and different role in the process. A move with many forelinks is normally one that advances a design proposal, at least a tentative one, or a move that sets a goal. In contrast a move rich in backlinks summarizes, confirms or assesses a proposal. In between various attributes of the issue in question are being brought up and inspected. This is logical and not surprising; the novelty here is the expression of this logic in terms of the quantity of links generated by the moves that partake in the process. We shall return to the roles of Fore- and backlinks of moves later.

Comparing performance

A major goal of the study of design reasoning, and of human behavior in design in general, is gaining a better understanding of successful vs. unsuccessful performance. To this end we must define comparable parameters of performance that can be measured quantitatively. At the cognitive level we wish to use the ratio of linkage among design moves as an example of a yardstick for comparison. We shall compare: a) between an individual designer and a team; b) the contribution of team-members to the design process in terms of the generation of key moves. Both comparisons use the bicycle rack design exercise devised for the *Delft Protocol Workshop* (Cross et al. 1996). A detailed account of these comparisons can be found elsewhere (e.g. Goldschmidt 1995).

Team vs. individual: Teamwork is the order of the day in design, not only because tasks are too vast for an individual to deal with and required expertise is too diverse, but also because we tend to believe that several minds put together produce better work than a single mind working alone. We ask whether this is always the case; the question is relevant in cases where neither the scope of the task nor the formal types of knowledge required is at stake. We compare a single industrial designer with a team of three industrial designers (John, Ivan and Kerry), all of fairly similar background and experience and working on the same relatively small assignment under controlled (and similar) conditions. The two-hour experimental sessions were videotaped and transcribed. The comparison pertains to portions of 21 and 25 minutes (Team and individual respectively) in the middle of the session where similar issues were dealt with.

We base the comparison on the total number of links among design moves and on Critical Moves: their total number and the division into the two types of CMs by the back or fore orientation of the links. Since the team has generated more than twice the number of moves than the individual in the segments under scrutiny, we compare only ratios and percentages. Table 1 presents the findings regarding the Link Index, the percentage of CMs, and the ratio of <CM^7 to CM^7>.

Table 1. Links and Critical Moves, team vs. individual designer (Delft Protocols)

	Team	Individual
Link Index	2.75	2.67
% CM^7 of all moves	12.42	10.14
% <CM^7 : %CM^7>	43 : 57	43 : 57

The team reached slightly higher values for both the L.I. and the percentage of Critical Moves; however, the differences are statistically insignificant[4] ($p \geq 0.6$ for the L.I.s, and $p \geq 0.1$ for the CM percentage). In both processes more CMs based on forelinks were made, and the ratio between them and CMs based on backlinks is precisely the same. We conclude that in this particular case the team had no advantage over the individual designer in terms of the cognitive efficacy of design reasoning. This does not mean that no differences exist between the processes, or that teamwork is never advantageous. It does mean, however, that under certain circumstances an individual's reasoning in the process of designing can match that of a team (or possibly surpass it?). These findings have clear implications for the workplace in terms of human resources.

Contribution by team members: Among others, the success of teamwork depends on its membership. In a good team all members contribute to the shared effort. When no roles are assigned relative to expertise or otherwise, team members tend to carve out their own roles, explicitly or implicitly. As mentioned above, in the Delft team there were no background or expertise differences among the members, nor were roles assigned to them by an exterior agent. Ivan had agreed, when asked by his teammates, to act as 'time keeper'. According to some analyses (Cross and Clayburn-Cross 1996) Kerry, the only woman in the team, had been intimidated by her male partners, possibly leading to performance impediments. We are not in a position to undertake an analysis of personality or social factors and their impact on team performance, but we can measure the contribution of Kerry, Ivan and John to the process by looking at the number of Critical Moves that each of them generated (See Fig. 3 for a portion of the analysis in question). We do that at two levels: at 6 and 7 links per Critical Move. As we shall see, results differ between the two levels, and we can learn something

about work in a team from the differences. Table 2 below displays the percentage of CMs contributed by the three designers for both levels of analysis.

[4] At the level of CM^6, the team generated significantly more CMs than the individual.

Table 2. . Percentage of Critical Moves contributed by team members (Delft Team Protocol

	% CM		% <CM		% CM>	
Designer	CM^{6}	CM^{7}	<CM^{6}	<CM^{7}	CM^{6}>	CM^{7}>
Kerry	27.4	32.4	34.3	35.3	22.0	28.6
Ivan	37.0	24.3	28.6	17.6	43.9	33.3
John	35.6	43.2	37.1	47.1	34.1	38.1

The figures confirm an informal observation: John was the highest "scorer" at almost all levels (he also generated the largest number of moves overall: 38.3%). His performance at level 7 was better than at level 6. The same is true for Kerry, whose global contribution was lower than John's. In contrast Ivan's performance at level 6 was much higher than at level 7. How should we understand these level differences? We would like to propose that Ivan was content with the expression of new ideas, whereas John always felt the need to sum up and provide the final 'punch-line.' Kerry, on the other hand, was mainly concerned with very specific aspects of the design, and was less assertive when the discussion dealt with other issues. In addition, Ivan provides a cogent example of a designer who is much better at proposing ideas (CMs>) than at evaluating them (<CMs). Like in a sports team, team members excel in different roles (here implicitly designated). Through Linkography we can inspect how balanced the team membership is.

Cognitive economy and the non-linearity of design processes

Because they involve a search, design processes are not linear. More or less trial and error may be involved, but we almost always observe sidetracking, backtracking, etc. Is that also true at the cognitive level? Linkography makes it possible to explore this question. Let us once again enlist Critical Moves for a brief quantitative exploration of this question.

Although criticality is designated due to a high number of links in one direction (< or >) almost all CMs also generate links in the opposite direction. In the segment of the Delft team session described earlier we counted the links, in each direction, generated by 35 <CM^{6}s and 41 CM^{6}>s (including three <CM>). Table 3 below has the results.

Table 3. Back- and Forelinks formed by CM[6]s (Delft Team Protocol)

<CMs		CMs>	
< links	>	< links	>
74%	26%	27%	73%

We notice that interestingly, 3/4 of the links of CMs are made in the designated direction, but the remaining 1/4 are reserved for the inverse direction. This means that design reasoning is not a linear process; instead, it progresses via a back-and-forth dance of small steps. The designer's mind appears to attempt to make sure at all times that energy is not wasted. Every (critical) move is double-speared: it moves forward but also makes sure it is congruous with what has already been achieved; it validates what has been done thus far, with an eye on ways to proceed from that point. The pattern of back-and-fore- interlinked moves represents a cognitive strategy that controls the efficiency and effectiveness of reasoning in designing. Back and fore links ensure continuity, while also guaranteeing that progress is made. The strategy serves the need of sustaining a coherent design rationale for the emerging entity that is being designed.

Producing highly interconnected design moves entails considerable, if unselfconscious skill; the acquisition of this skill is, to a large degree, the goal of the practicum that is part of every design education curriculum.

Acknowledgment

Research for this paper was supported by a grant from the fund for the promotion of research at the Technion, hereby gratefully acknowledged.

References

Cross N, Christiaans H, Dorst K (eds.) (1996) Analyzing design activity. Wiley, Chichester

Cross N and Clayburn-Cross A (1996) Observations of teamwork and social processes in design. In: Cross N, Christiaans H, Dorst K (eds.) (1996) Analyzing design activity. Wiley, Chichester, pp 291-317

Goldschmidt G (1995) The designer as a team of one. Design Studies 16(2):189-210

Goldschmidt G, Weil M (1998) Contents and structure in design reasoning. Design Issues 14(3):85-100

Wang MH and Habraken NJ (1982) Notation of the design process: The six operations. Unpublished manuscript, MIT

van der Lugt R (2001) Sketching in design idea generation meetings. Ph.D. dissertation, Delft University of Technology

Entropy reduction in mathematical giftedness

Werner Krause, Gundula Seidel University of Jena, Germany
Frank Heinrich, University of Bamberg, Germany

A new measurement for mental performance

This paper deals with the elementary analysis of cognitive processes in mathematical problem solving for better diagnostics. The present experiment was designed to analyse the internal process and to localize the neural substrates involved in solving mathematical tasks, using EEG-coherence. The internal process is revealed by a sequence of microstates. Microstates are defined here as coherence maps which remain stable over time. The difference in performance between gifted and normal subjects in solving mathematical problems is reflected by the difference between the sequential and topographical properties of microstate sequences. Sequential properties were measured by means of entropy reduction. A higher entropy reduction was found in mathematically gifted subjects in contrast to the normal subjects. Topographical properties were measured by means of difference maps of microstates between mathematically gifted and normal subjects. Gifted subjects exhibit a higher coherence left and right parietal and left frontal. This result corroborates the double representation hypothesis.
This new method may be extended to measure mental performance more objectively.

The so-called double-representation hypothesis plays an important role in mathematical (and technical) giftedness. Obviously, particular mathematical problems can be solved by means of computation (e.g. using the equation of Pythagoras) or based on a visual mental imagery resolution strategy (e.g. using the imagery of triangles) (see Fig. 1). Mathematically gifted subjects may be able to activate both the computational and the imagery resolution strategy and to switch from one to the other (Klix 1992). In order to determine the differences in ability between gifted and normal subjects, we used mathematical tasks solvable with both modality strategies: the computation component of numerical mental calculation and the imagery component of figural mental calculation (Heinrich 1997). The strategies differ in the mental effort required.

Gifted subjects exhibit high-level cognitive performance. As an example, one of our mathematically gifted subject solved the following problem: `What is the number of diagonals of a 23-polygon?` in 28 seconds. To decide whether this phenomenon is caused by better cognitive process, an EEG-coherence analysis was used to subdivide the thinking process into a sequence of microstates with sequential and topographical properties. Sequential properties: Cognitive process is reflected by means of entropy reduction, as a kind of concatenation of microstates. Topographical properties: EEG-coherence maps were used to determine the activated cortical areas.

Method

Subjects. Twelve right handed normal (16 to 18 years old; mean, 17.5; 6 females) and twelve right handed highly mathematical gifted subjects (16 to 24 years old; mean, 17.9; 3 females) participated in the experiment. The normal subjects were enrolled at an ordinary highschool. The highly mathematical gifted subjects were enrolled at the special mathematical highschool. In order to exclude covariables, further criteria were proved. Firstly, they were tested for their performance on the verbal and non-verbal IQ test (Table 1). The two samples exhibited a significant difference on the non-verbal, but not in the verbal IQ test. Secondly, both samples scored within the normal range in metal rotation tests and memory capacity (Cavanagh-constante) although their visual digit-span differed.

Table 1. Intelligence quotient (IQ), mental rotation (scores), Cavanagh-constant (ms), memory span for digits, working memory capacity and response latencies (sec) for mathematically gifted and normal subjects (SD = standard deviation)

Test–score p-value			gifted	Normal	Z
IQ nonverbal		mean	124	107	-
2.79	0.004				
		SD	10	10	
IQ verbal		mean	112	105	-
0.997	0.34				
		SD	13	4	
Mental rotation		mean	17	14	-
1.317	0.193				
		SD	3	5	
Cavanagh-constant		mean	295	247	-
0.821	0.438				
		SD	116	77	
emory span for dig		mean	7.36	6.03	-
3.45	0.000				
		SD	0.957	0.489	
Response latencies		mean	57	101	-
3.95	0.001				
		SD	46	52	-

All were free from nervous diseases or injury and had no abnormality in their rest EEG. Informed written consent was obtained from each subject after the procedures had been fully explained. The local Ethics Committee gave approval for the experiment.

Stimuli. The subjects were presented with one of the following tasks: The length of the diagonal of a square is 5 cm. How long is the length of an edge of a square with double the area? The task requires computation based on a computational or an imagery resolution strategy (Fig. 1). Our subjects had to solve 10 different tasks with this condition.

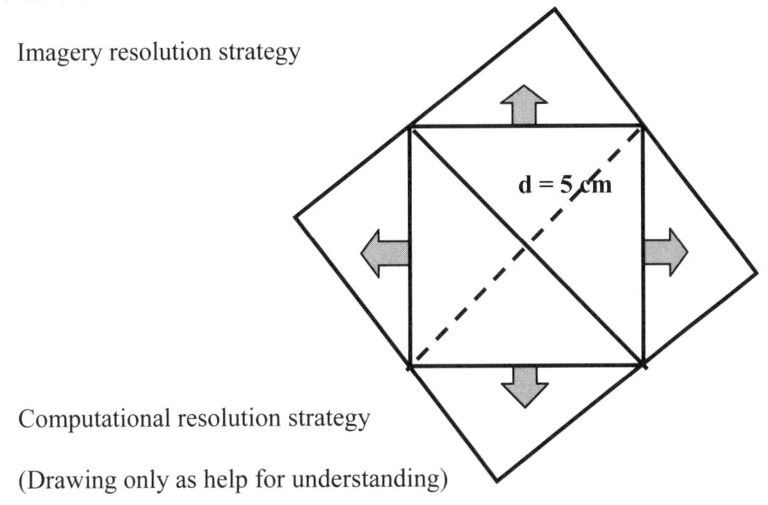

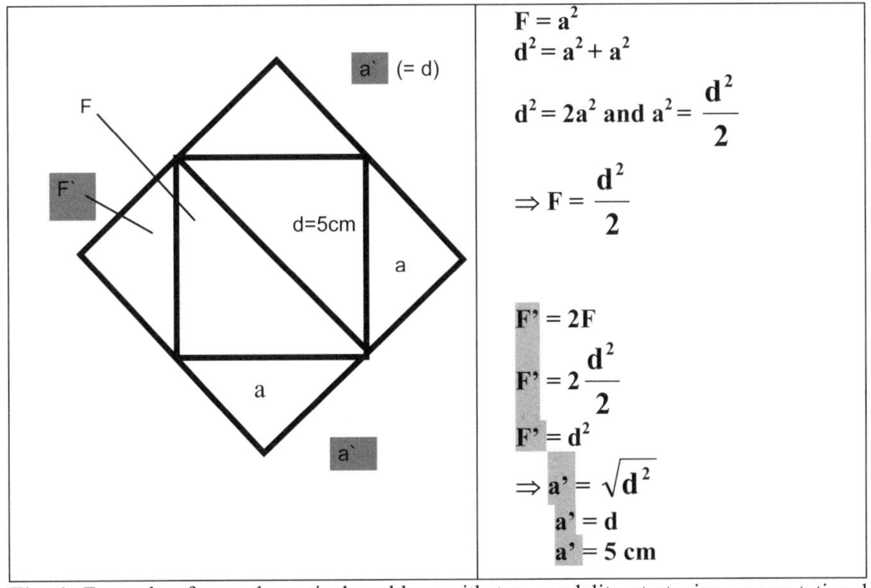

Fig. 1. Example of a mathematical problem with two modality strategies: computational strategy and imagery resolution strategy (Heinrich, 1997)

Procedure. When confronted with tasks with two modality strategies, subjects should activate one of the two possible strategies of computation/imagination . Subjects had to compute the results and give the answer aloud. At the end of each trial, a careful debriefing was carried out individually. Each subject was asked to explain the resolution strategy(ies) used during the solving process.

EEG recording. The EEG was recorded using 19 scalp electrodes. The standard Ag/AgCl electrodes were placed according to the international 10/20 system (Jasper 1958). All cortical electrodes were referenced to two linked ear lobe electrodes during the EEG recording. Data were collected using the Brainstar EEG software linked to a computer for stimulus presentation. The high and low frequency filters were set to 70 and 0.3 Hz, respectively. The data recorded in each channel were digitized at a sampling rate of 256 Hz. The EEG was analyzed in the time interval between the end of text reading and the answer, artefacts were excluded.

Data analysis. Instantaneous coherence analysis (Schack et al. 1995, 1999, 2001) was performed for 30 electrode pairs. The 30-dimensional vector of time courses of band coherences (13-18 Hz[1]) for each single trial was subdivided into segments with stable coherence values. Afterwards, the segments were clustered into six classes. The corresponding cluster centers were designated as microstates. Shannon's entropy with regard to the distribution of the different microstates gives evidence of the disorder of their occurrence. The hidden order may be described by transition probabilities and quantified by the conditional entropy of the occurrence of a microstate by observing the preceeding microstate. The difference between Shannon's entropy and the conditional entropy is denoted by entropy reduction H_{red} :

Shannon‘s entropy is defined by

$$H(i) = -\sum_{j=1}^{n_0} P(j/i) \cdot ld(P(j/i)) \tag{1}$$

where j, j = 1, ... , n_0, are the numbers of microstates.

The conditional entropy of occurrence of states under the assumption of i as the present state is defined as

$$H(i) = -\sum_{j=1}^{n_0} P(j/i) \cdot ld(P(j/i)), \quad i = 1, \dots, n_0, \tag{2}$$

where ld denotes the logarithm dualis. The difference

$$H_{red} = H - \sum_{i=1}^{n_0} P(i) \cdot H(i) \tag{3}$$

[1] This frequency band was sensitive in earlier experiments (Krause et al. 1998; Schack et al. 1999).

is called entropy reduction and reflects the sequential structure of the states. This indicates a strong sequential structure or concatenation of the states. The depth of the higher order sequential effect was calculated by means of a configuration frequency analysis (Lautsch and Weber 1995). Equation (3) was extended to higher order dependency up to five steps back and tested for inequality. To avoid alpha-inflation the Bonferroni-Holm correction was successfully performed on the level of 5 %.

Results

Behavioral data. Response latencies (RLs) for each sample are given in Table 1. Response latencies were significantly longer for the normal subjects as compared to the gifted subjects in problem solving. Moreover, the samples differ in their percentage of errors (gifted subjects: 15%, SD = 12%; normal subjects: 49%, SD = 23%; Z = -3.49, p = 0.001). Additionally we obtain results of a post-experimental debriefing.

At the first glance, the task-dependent response latency differences might be explained by the task-dependent different type of resolution. According to the double representation hypothesis, both types of resolution (the computational and the imagery resolution strategy) are accessed by gifted subjects. As a consequence, gifted subjects are able to select the most efficient strategy and to apply it. The double representation hypothesis also postulates that only one resolution strategy is preferably stored in the memory and used by normal subjects. This particular strategy, used by most normal subjects, is less efficient, leading to a longer response latency (Fig. 1, Table 1).

In contrast to the double representation hypothesis, we found in the post-experimental debriefing, that all subjects indicated familiarity with both modality strategies (e.g. computational and visual mental imagery resolution strategy). They reported they had learned application of both modality strategies at school; and had used them in the experiment. Against this background it is worth recalling that *behavioral data can not explain the difference between gifted and normal subjects* and does not answer the question of better cognitive process. Therefore we identify the mental representation and the mental process by means of instantaneous EEG-coherence.

EEG data

Sequential properties of microstates. Essentially, cognitive processes can be described by the microstate sequences derived from the segmentation and clustering of EEG-coherence (Schack et al. 1995; Krause 2000; Schack et al. 2001; Krause et al. 2001). Microstates are EEG-coherence maps which are stable over time. We examine dependencies of microstates in their temporal regime to investigate the difference in performance between gifted and normal subjects.

Entropy reduction H_{red} is a quantitative measure of the strength of concatenation of microstates according to equation (3). Accordingly, we looked for a difference in the entropy reduction between gifted and normal subjects. The gifted subjects exhibit a mean entropy reduction $H_{red} = 1.2$ (SD = 0.13) and thus achieve a significantly better result than their normal counterparts, whose entropy reduction H_{red} was 0.9 (SD = 0.19). The Z value for this comparison is $Z = -2.99$ ($p < 0.01$). The entropy reduction (Fig. 2) is calculated based on the concatenation of microstates with self-transition.

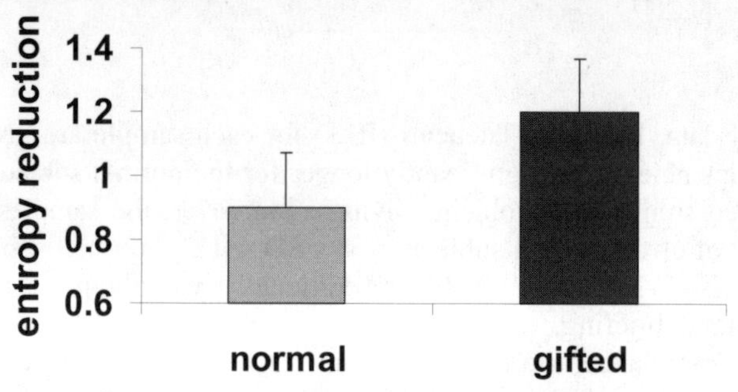

Fig. 2. Response latency and entropy reduction for gifted versus normal subjects (with self-transition of microstates)

Further support for our interpretation of the trend described above can be derived from the calculation of entropy reduction without the self-transition of the microstates. If we exclude the self-transition, the entropy reduction is reduced to concatenation of different microstates. In this case the gifted subjects ($H_{red} = 0.61$, SD = 0.09) exhibit a significantly higher entropy reduction than the normal ($H_{red} = 0.48$, SD = 0.06) subjects ($Z = -3.31$, $p < 0.01$). The sequence of microstates may be considered as a Markovian chain. The concatenation of microstates without self-transition of the microstates for a gifted and a normal subject is shown in Fig. 3.

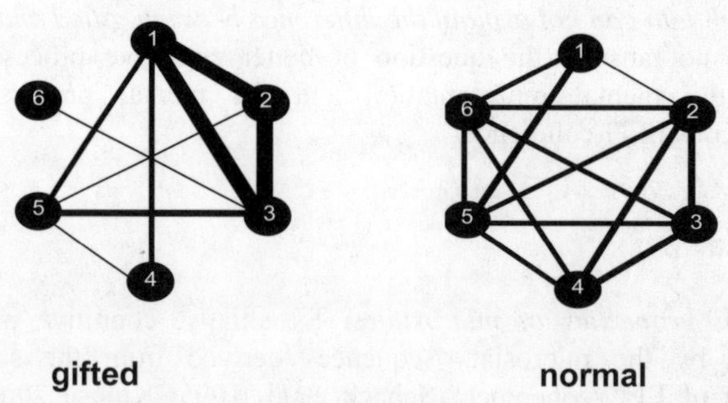

Fig. 3. Sequence of microstates (without self-transition) for one gifted versus one normal subject. The 6 microstates are marked by black circles with numbers. The strength of concatenation is coded by the thickness of the edges

As seen in Fig. 3, the gifted subject exhibits a stronger concatenation of subsequences of microstates.

Higher order sequential effects can be considered in microstate sequences as well. Based on a configuration frequency analysis (Lautsch & Weber 1995) we determine the depth of the sequential effects both for gifted and normal subjects in mathematical problem solving. The result is shown in Fig. 4. The difference of the mean maximum depth of the sequential effect between gifted and normal subjects is significant ($Z = 2.74$, $p = 0.006$). This corroborates the assumption that sequential properties of microstate sequences can differentiate the mathematically gifted from the normal subjects.

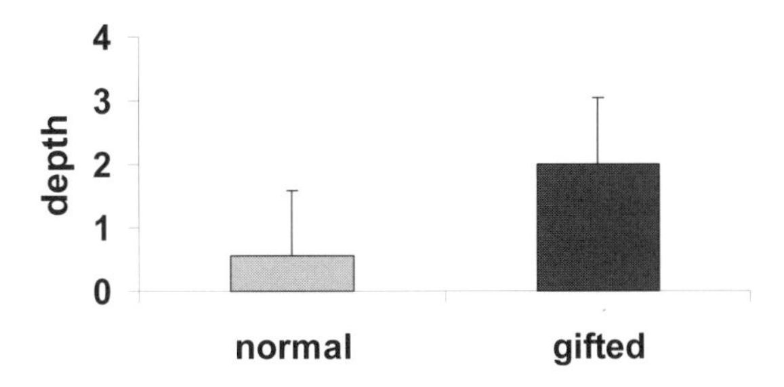

Fig. 4. Mean maximum depth of microstate-sequences (without self-transition) for gifted and normal subjects

Topographical properties of microstates. As mentioned above the two modality strategies in our experiment (computational component of numerical mental calculation and imagery component of figural mental calculation) represent two basic resolution strategies that could be subserved by neural networks belonging to different domains (verbal versus visuospatial). Zargo et al. (2001) give support the idea that numerical and spatial representations are intricately intertwined in the parietal and frontal lobes and that these numerical and visuospatial representations are involved at different levels of mental calculation in normal subjects. In our experiment we find that for gifted subjects the same cortical areas are active: The six microstates (Schack et al. 1999; Seidel 2000) were ranked by means of their self-transition probability for subject groups. For data reduction we restrict our analysis to those microstates with high self-transition. Based on this restriction, Fig. 5 shows the difference coherence maps (gifted minus normal subjects).

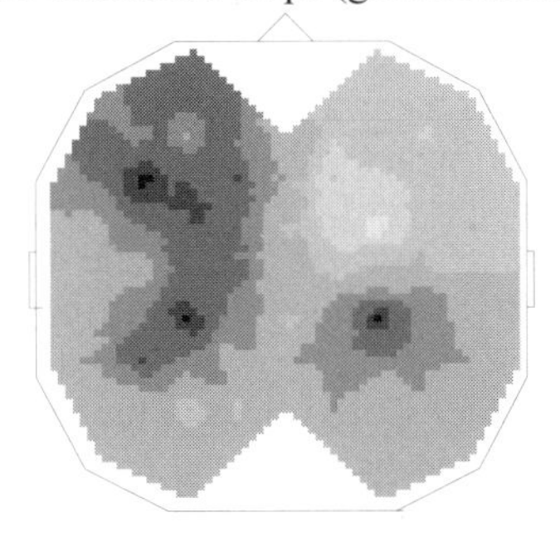

Fig. 5. Difference coherence maps (gifted minus normal subjects) of microstates with high self-transition. The coherence values are scaled between – 0.10 (white colour) and + 0.10 (black colour)

As seen in Fig. 5 gifted subjects exhibit a significantly higher coherence at the electrode pairs F7/F3 ($Z = -2.23$, $p < 0.05$), C3/P3 ($Z = -1.86$, $p < 0.1$) and C4/P4 ($Z = -1.93$, $p < 0.1$). In an unusual comparison of two different types of neuroscientific methods, namely PET – and EEG-studies, this result is in agreement with neural correlates of simple and complex mental calculation by Zago et al. (2001). Zargo et al. found that computing specifically involved two functional networks: a left parieto-frontal network responsible for the holding of the multidigit numbers in visuospatial working memory and a bilateral inferior temporal gyri related to the visual mental imagery resolution strategy. No additional network for mental calculation in gifted subjects is observed in our EEG-study.

Discussion

The main result is focused on the difference of the concatenation of microstates, represented by the sequential properties of the microstate sequence. By solving tasks with two modality strategies, our gifted subjects exhibit a higher entropy reduction and a higher order sequential effect than the normal subjects. The results indicate the gifted subjects' superiority over their normal counterpart in our mathematical problem solving experiments; this is clearly linked to a more deterministic concatenation of different microstates. This higher concatenation of microstates in the case of the gifted subjects might be interpreted as a better context-dependent application of the most efficient modality strategy. The result does not exclude, however, the possibility that in special cases other components may also contribute to sequential dependencies.

The difference between mathematically gifted and normal subjects appears in the topographical properties of the microstates as well. The activation of the computational network in gifted subjects is distinctly increased when tasks were processed with two modality strategies, suggesting that gifted subjects process tasks with two modality strategies better than normal subjects. Looking for the reason for this different activation, one might speculate that gifted subjects, in contrast to their normal counterparts, are able to better access both modality strategies, although their debriefing reports do not differ. This finding corroborates the double representation hypothesis and stresses the sensitivity of this method.

One could argue that the higher coherence at the electrode pairs C3/P3 and C4/P4 (Fig. 5) observed for gifted subjects might be linked to a better visualization and is not a feature of the double representation hypothesis in giftedness. Evidence for this assumption comes from the study by Jaencke et al. (1998) in which mental rotation tasks were presented and the fMRI activation was registered. Subjects with higher score values in the mental rotation test achieved a higher activation in the left and right parietal cortical areas and vice versa. The score values in the mental rotation test of our gifted subjects and their normal counterparts do not differ (Table 1). Therefore we reject the visualization hypothesis.

The present study has introduced a new method of investigating problem solving using microstate sequences based on EEG-coherence. As an alternative to the IQ testing, this method may be extended to measure intelligence more objectively.

Acknowledgements

We are grateful to Elizabeth Kley for her help in preparing the English version of our article. The development of the data analysis method was supported by the German Research Foundation (Scha 741/1-2).

References

Heinrich F (1997) Diskussionsmaterial zur Untersuchung der Doppelrepräsentationshypothese und ein Bemerkungen aus mathematikdidaktischer Sicht. (unveröffentlicht).

Jäncke L, Kleinschmidt A, Mirzazade S, Specht K, Freund HJ (1998) Mental rotation ability determines the activity in posterior cortical areas during the imagination, palpation and construction of 3D objects. NeuroImage **7**: 119

Jasper HH (1958) The ten-twenty electrode system of the International Federation. Electroencephalography and Clinical Neurophysiology 10: 371-375

Klix F (1992) Die Natur des Verstandes. Hogrefe, Göttingen

Krause W (2000) Denken und Gedächtnis aus naturwissenschaftlicher Sicht. Hogrefe, Göttingen

Krause W, Seidel G, Schack B (2001) Ordnungsbildung. Z. Psychol. 209: 376-401

Lautsch E, Weber S (1995) Methoden und Anwendungen der Konfigurationsfrequenzanalyse (KFA). Belz Psychologie Verlags Union, Weinheim

Schack B, Krause W (1995) Dynamic power and coherence analysis of ultra short-term cognitive processes--a methodical study. Brain Topogr 8: 127-136

Schack B, Grieszbach G, Krause W (1999) The sensitivity of instantaneous coherence for considering elementary comparison processing. Part I: the relationship between mental activities and instantaneous EEG coherence. Int. J. Psychophysiol. 31: 219-240

Schack B, Seidel G, Krause W, Heinrich F (2001) Coherence Analysis of the Ongoing EEG by Means of Microstates of Synchronous Oscillations. Proceedings-23rd Annual Conference-IEEE/EMBS Oct 25-28, Istanbul

Seidel G. (2000) Ordnungsbildung und Doppelrepräsentation im Denken mathematisch Hochbegabter. Sequentielle und topologische Eigenschaften von Mikrozustandssequenzen. Doctoral thesis, University of Jena.

Zargo L, Pesenti M, Mellet E, Crivello F, Mazoyer B, Tourio-Mazoyer N (2001) Neural Correlates of Simple and Complex Mental Calculation. NeuroImage 13: 314-327

Apperception, content-based psychology and design

Pertti Saariluoma, Cognitive Science, University of Jyväskylä, Finland

On contents of mental representations

A core area of scientific thinking is explaining. This means answering to the "why-questions and how questions" (Hempel 1965). Why does Sam have a fewer? Why did an organization fail abroad? Why a structure is able to support the weight of snow? How more effective valves for an engine can be designed? How to make computer games more attractive for female users? These are typical examples of design problems, all of which should be based on scientific explanation, i.e., what should be answered based on the laws of nature or as is becoming increasingly more evident, based on the laws of the human mind.

It is of the first important to designers to find an effective way of explaining. There may be numerous questions requiring adequate explanations. Naturally, the psychological concepts associated with the mental representations of designers form the basis of explaining. If designers make errors, for example, it is natural to assume that something was wrong in their mental representations and that answering such why-questions presuppose concepts of mental representation. However, it is far more difficult to find what kind of explanations one should use, i.e., what would be an effective conceptualization for investigating the problems of designers thinking (Simon 1969).

The key problem is finding an adequate *explanatory framework*. This means the set of concepts that can provide us with explanatory concepts and principles. A key problem is that psychology does not necessarily provide us with a unified framework, which could answer all types of questions relevant in designing. The reason for this is the nature of representations.

All representations must have at least two dimensions to be representations. They must have, as Peirce (1955) pointed out long ago, a physical sign or representation vehicle and information contents. A traffic sign or a book could serve as examples. The two properties are all too obvious but this is also the case with mental representation. Brain substance is the material basis on which mental contents of individuals is constructed biologically as well as by life long learning. There are no representations without having these two properties. Therefore, any considerations regarding mental representations should also take into account the two dimensions of mental representation.

Although it may immediately appear quite natural to reduce contents to brain stuff in looking for explanatory grounds, this will lead to difficulties. The human brain only offers a framework, which, in interaction with reality, may take unlimited numbers of individual forms. A young person can initially learn a language forget it and the use another as mother tongue. Our brain provides a platform for learning, but interaction with the environment determines what the outcome shall be. The brain cannot predict its environment and this is why it has to be open with respect to mental contents. Like a computer, the brain provides only the hardware for content but an explanation for mental contents. We obviously need a scientific language regarding mental content independent of brain-based languages.

References to neural phenomena are very rare, when reading organizational or design literature. One may think that this is due to the early stage of neural research, but it may as well be caused by the fact that neural arguments do not provide an efficient explanatory basis for investigating the content of thought (Saariluoma 1999). However, we do have a class of basically neural explanations, which are important in investigating thinking and design. This explanatory framework is human capacity.

Research into human information processing and thinking has shown that increased complexity of mental representations increases errors but, at the same time, that increasing expertise helps people to avoid these problems (e.g., Anderson, Farrell and Sauers 1985; Chase and Simon 1973; Johnson-Laird 1983). Nevertheless, this explanatory framework is not effective when we think about problems of mental contents, because human capacity is not a content-based notion. We can fill human attention or working memory with any kind of contents. This is why its explanatory power is limited in this sense (Saariluoma 1997). This means that it is necessary to discuss how the fact of mental contents can be theoretically and empirically captured.

It would naturally be rather strange to suggest that mental contents could not explain some aspects of human behavior. However, it is also very problematic to consider what kind of paradigms one could effectively use to analyze mental contents and its role in explaining human behavior in such domains as the human dimension of design. One can hardly say that human thoughts do not have contents. In fact, the opposite thesis, i.e., showing an idea without any information contents would be much more difficult. However, understanding that mental contents are important does not yet mean that we would really understand how to work with them.

An explanatory framework, constructed around the notion of mental contents, is evidently different from neural - or capacity-based – explaining and is intended to provide answers to very different types of problems. When capacity is about the complexity of representations, content-based thinking offers an explanation based on mental contents. Naturally, it is necessary to construct some theoretical concepts around it to be able to work with the psychological problems of mental contents.

From perception to apperception

The construction of mental representations is a crucial stage in thinking. From a content point of view, it can be argued that this process is not psychologically equivalent to perceiving. Firstly, we have in the contents of our mental representations during design, for example, non-perceivables, i.e., content elements, that cannot be reduced to any percept. This means that there is no current physical and receptoral (e.g., retinal) correlate for the active representational content. Typical examples could be: tomorrow, past, possible, infinity, n-dimensional space, construction laws, electrons etc. Secondly, it is important to make the well-know difference between "seeing" and "seeing something as something". A novice can see all the elements of a construction plan, but he or she does not necessarily understand their meaning or role in a plan. Construction plan elements may even be poorly conceived by experts. Finally, we have to consider the role of emotional information in mental representations. For these kinds of reasons, it is more consistent to use term apperception instead of perception, when the construction of representations are discussed (e.g., Kant 1787/1985; Saariluoma 1990, 1992, 1995, 2001; Stout 1890; Wundt 1913)

There are naturally many as possibilities to empirically investigate apperception as there are numerous possibilities to investigate attention, imagery or memory. Here, operationalization is based on investigating the content-elements of mental representations and using their properties in explaining relevant behavioral phenomena. One must collect data about the contents of apperceived representations and look their explicit as well as tacit content elements to find out which elements could explain behavioral features in question.

In investigating the economy of human thinking, for example, a set of tacit regularities with specified contents could explain, why human problem spaces are so small compared to computational spaces (Saariluoma 1990, 1995). These regularities proved to be an example of a larger class of regularities, important also in design that could be called functional content schemata.

This content structure apparently organizes elements in representations. Each element of mental representations makes sense or is senseful (Saariluoma 1990, 1995). This means that we can find a reason, explicit or tacit, for elements in representation explaining why they are included into this representation. When thinking about such problems as the economy of thinking and the organization of elements this schema seems to be particularly interesting. The most important explanatory thing in this content structure is its abstract contents, which presupposes that content elements are related to each other in terms of functional reasons explaining why the element is in the whole or which function it serves in the structure.

Apperception is thus a process that defines meanings of the elements and integrates them functionally to each other. In a wider sense, it is naturally important to look for other types of contents, which may increase our understanding of mental contents and how an explanatory framework can be constructed around it can be used to explain human behavior.

Errors in designing economic activities

An example of the content-based thinking can be found in research in economic thinking. As is well-known, design and planning are very important activities in economics. The question concerns not only about artifacts but also organizational and business concepts. To conduct business one must often have products material or immaterial, and they presuppose planning. However, it is equally important to design organizations to get production going and make plans for marketing the products. This means that thought processes typical in design form one very essential part of business thinking (Saariluoma, in press).

Business and work life are complicated task-environment for thinking. There are lots of uncertainties and possibilities for error. When asked, people suggested that from ten to twenty percent of operation's outcomes worse that expected (Saariluoma, Maarttola and Niemi 1998). In this research, we also interviewed 75 people about their most unforgettable thinking errors. In analyzing the text if was noteworthy that people tacitly or explicitly ascribed schematic thought procedures, which can be termed as thought models. As the topic of the interviews was faulty thinking, the thought models were naturally rather risky and negative (Saariluoma 2003). Typical examples are induction bias and overgeneralization.

Induction bias

INDUCTION BIAS is a classic thought model. It suggests that things continue to develop as they have done before.

"I find it difficult to understand that the guidelines, how to inspect such leaks were from the sixties. A little old ones.. Nothing had ever happened.".

Here, the interviewed person describes, how a risky way of acting was supported by inductive arguments. Nothing had happened before, so there was no reason to think about or update security procedures, even though the technology involved had changed. Consequently, one factory hall blew up and two workers were permanently injured.

Overgeneralization

OVERGENERALIZATION is another example of a risky thought model. This is very close to induction bias. It neglects making necessary distinctions.

"The strength of the MAIN COMPANY is that things have been carefully considered on the USA level. However, they naturally think that there is one marketing area, Scandinavia, and they are unable to see that the structures of businesses are very different here."

OVERGENERALIZATION means that people ignore essential distinctions (Beck 1976; Wills and Sanders 2000). People hide details and avoid thinking in the necessary conceptual detail. They do not look for the differences between conditions and pay attention to their similarities only, which may eventually be very superficial or insignificant.

These two examples illustrate that we have risky ways of thinking, when designing economic and business activities. In fact, such problems are known and one can find examples of fallacies in literature (Hamblin 1970). They are dangerous because they entail risky elements. Safety norms are constructed to eliminate risks. Logically, breaking these norms is risky. Nevertheless, people make such errors.

Working with contents

There are psychologically relevant problems that can best be resolved in the explanatory framework provided by mental contents. The problem of biased thought models is a typical example. However, apperception psychology is not the only approach to contents and this is why the last part of this paper shall concentrate on investigating, the differences between various content-oriented approaches.

Contents in the minds have naturally long been a problem in psychological thinking. Aristotle's work, "Sophistical refutations," for example, provides us with a collection of fallacies people may do in their thinking (Aristotle 1984; Hamblin 1970). Categories people use in their thinking have been of real interest in modern time epistemological work such that all perennial philosophers have discussed about human ability to become aware of and represent the world (e.g., Cassirer 1974). Since the relationships between theories of knowledge and psychology were not equally determined as today, these considerations entail many valuable ideas about the contents of the mind in the psychological sense.

In this paper, two types of mental contents have been presented, which play a role in design. They are the functional content structure, i.e., the property that design elements are connected by reasons. We know also that pseudo-functional reasons, i.e., reasons with correct form but misplaced contents, can explain design errors

(Saariluoma and Maarttola, in press). Another type of problem for correct design thinking is formed by risky thought models.

Nevertheless, the main goal of this paper has been theoretical. It has outlined a framework for working with design problems associated with mental contents. We know very little about the role of mental contents in design, but it seems that this explanatory framework provides an important platform for investigating the nature of thinking rearding design. Designers construct complex mental representations and these representation include very different types of information contents. Some are very domain specific but some are very general. The problem is to find ways of exploring this information in detail.

The procedure is very straightforward. A researcher interested in apperception and mental contents should firstly collect qualitative knowledge about the mental contents during the designers thought processes. This may be done by protocol analysis, surveys or interviews. Documentary analysis is also a natural way of approaching problems of mental contents. In this way, it is possible to find certain types of mental contents that may provide us with important overall explanatory information about mental contents.

This kind of psychological thinking may violate the principles of natural scientific thinking in psychology. This is not necessarily a serious problem, because much in natural science is qualitative. We, for example, classify plants and bacteria. Secondly, there is no necessity to think that the psychology of thinking and planning should be thought only in natural scientific terms.

References

Anderson, JR Farrell R and Sauers R (1984) Learning to program lisp. Cog Sci 8: 87-129.

Aristotle (1984) Sophistic refutations. In: Barnes J (ed.), The complete works of Aristotle. Princeton University Press, Princeton.

Beck A (1976) Cognitive therapy of emotional disorders. Penguin Books, Harmondsworth.

Cassirer E (1974) Das Erkentnisproblem in der Philosophie und Wissenschaft der neuern Zeit. Wissenschafliche Buchgesellschaft, Darmstad.

Dym, CL, Little P (2000) Engineering design: A project based introduction. Wiley, New York.

Hamblin CL (1970) Fallacies. Methuen, London.

Hempel C (1965) Aspects of scientific explanation and other essays in the philosophy of science. Free Press, New York.

Hiilamo H (1995) SKOP - a brief history (in Finnish). WSOY, Porvoo.

Iacocca L (1984) Iacocca – car autobiography (in Finnish). WSOY, Porvoo.

Johnson-Laird P (1983) Mental models: Towards a Cognitive Science of Language, Inference, and Consciousness. Harvard University Press, Cambridge: Mass.

Kant, I. (1787/1985) Kritik der Reinen Vernunft. Stuttgart: Philip Reclam.

Pahl G, Beitz W (1989) Konstruktionslehre (in Finnish) MET, Porvoo.

Peirce CS, (1955) The philosophical writings of Peirce. Dover, New York.
Rowe PG. (1987) Design thinking. MIT-Press, Cambridge, Mass.
Saariluoma P (1990) Apperception and restructuring in chess players' problem solving. In: Gilhooly K.J, Keane, MTG, Logie, RH, Erdos G (eds) Lines of thought: reflections on the psychology of thinking. Wiley: London, pp 41-57.
Saariluoma P (1992) Error in chess: Apperception restructuring view. Psychol Res 54:17-26.
Saariluoma P (1995) Chess players' thinking. Routledge, London.
Saariluoma P (1997) Foundational analysis: presuppositions in experimental psychology. Routledge, London.
Saariluoma P (1999) Neuroscientific psychology and mental contents. Lifelong learning in Europe 4:34-39.
Saariluoma, P (2001) Chess and content oriented psychology of thinking. Psihologica 22:143-164.
Saariluoma, P (2003) Thinking in work life: from errors to opportunities (in Finnish) WSOY: Porvoo.
Saariluoma P, Maarttola I (in press) Stumbling blocks in novice architectural design. Journal of Arch. and Plan. Res.
Saariluoma P, Maarttola I, Niemi P (1998) Ajatteluriskit ja kognitiiviset prosessit taloudellisessa toiminnassa. Helsinki: TEKES.
Wundt W (1913) Grundriss der Psychologie. Kröner, Stuttgart.
Simon HA (1969) The sciences of artificial. MIT-Press, Cambridge, Mass.
Stout, GF, (1890) Analytical psychology. MacMillan: London.
Wills F, Sanders D (2000) Cognitive therapy: transforming the image. Sage: London.

Sketches for Design and Design of Sketches

Barbara Tversky, Masaki Suwa, Maneesh Agrawala, Julie Heiser, Chris Stolte, Pat Hanrahan, Doantam Phan, Jeff Klingner, Marie-Paule Daniel, Paul Lee, and John Haymaker
Stanford University, Stanford, CA, USA
Chukyo University, Toyota, Aichi, Japan
Microsoft Research, Redmond, WA, USA
LIMSI, Orsay, France
NASA-Ames, Mountain View, CA, USA

Sketches for and by Design

It is said, though not without controversy, that what distinguishes design from art is function. Design is for a purpose, usually a human one. As such, design entails both generating ideas and adapting those ideas to intended uses. This occurs iteratively. Form and function. Studying how people go about both these tasks gives insights that can facilitate the design process. Two relevant projects will be described. The first investigates how designers and novices get ideas from sketches and applies those insights to suggestions for promoting generation of ideas. The second seeks to develop computer algorithms for designing individualized visualizations, algorithms that are informed by cognitive design principles.

Insights from Sketches

Why do designers sketch? The simple answer is that they are designing things that can be seen. But this simple answer underestimates the contributions of sketching to the cognition underlying design. After all, designers could construct things in their minds in three dimensions, and to varying extents, they do. But the mind rarely has sufficient capacity to contain an entire object of design; sketches can overcome this limitation. The mind may not notice inconsistencies or incompleteness; sketches demand some consistency and completeness. The mind may not have the capacity to construct, hold, and evaluate a design; sketches hold the constructions in view of the designer, freeing the mind to examine and evaluate. Thus, sketches, like other external representations, relieve short-term memory, demand consistency, and augment information processing. They are also public representations of thought, so they can be shown to others and reasoned on collectively. What the mind does in evaluating sketches to promote design has fascinated designers and cognitive scientists alike. Our own investigations have included experts and students of architecture and design as well as laypeople.

They have included analyses of the spontaneous, detailed, step-by-step reports of the thoughts of designers as they designed a building complex as well as experimental manipulations of interpretations of sketches. We review some of those studies and their results here.

Role of Sketches in Design Ideas. In contrast to other visualizations, such as diagrams and graphs, sketches, especially early ones, are replete with ambiguities. They are, after all, "sketchy;" that is, vague, committing only to minimal global arrangements and figures. Rather than inducing uncertainty or confusion, ambiguity in design sketches is a source of creativity, as it allows reperceiving and reinterpretating figures and groupings of figures. A designer may construct a sketch with one arrangement in mind, but on inspection, see another arrangement enabling a new, unintended interpretation (e. g., Goldschmidt 1994; Schon 1983; Suwa, Gero and Purcell 2000; Suwa and Tversky 1997). Both beginning and experienced designers are facile in making new inferences from their own design sketches. However, experienced designers are more adept at making functional inferences than novices, whose inferences are primarily perceptual (Suwa and Tversky 1997). A functional inference is seeing the flow of pedestrians in a sketch of a plan whereas a perceptual inference is seeing new spatial relations among structures. The facility of seeing function in structure is a hallmark of expertise in numerous domains from chess (Chase and Simon 1973; de Groot 1965) to mechanical devices (Heiser and Tversky, submitted).

What enables designers to see new implications in sketches, especially their own? The analysis of the protocol of one experienced architect as he designed a building complex is instructive. After perceiving new perceptual configurations in his sketch, he was more likely to get a new design idea than after interpreting the sketch in the same way. Getting a new design idea in turn led to perceiving new perceptual relations in the sketch, and so on, a productive cycle (Suwa, Gero and Purcell 2000).

Stimulatingng New Design Ideas.

Can the strategy used by the expert architect to enable new design ideas be explicitly adopted by others to same end? To see if searching for new perceptual relations could be used purposefully to enable new interpretations by a larger population, we gave undergraduates the ambiguous sketches shown in Figure 1and asked them to come up with as many interpretations as possible for each, a procedure adapted from one used by Howard-Jones (1998; Suwa, Tversky, Gero and Purcell 2001). Participants generated ideas for four minutes for each drawing. About two-thirds of the undergraduates, either spontaneously or by suggestion, adopted a strategy of attending to the parts of the sketch, either focusing on different parts or mentally rearranging the parts of the sketch, in order to see new interpretations. Participants attending to parts produced more interpretations, on average, 45 for the different parts group and 50 for the rearrange parts group, than the others, who did not adopt that way of interacting with the sketch and who generated on average 27 interpretations in the four minutes.

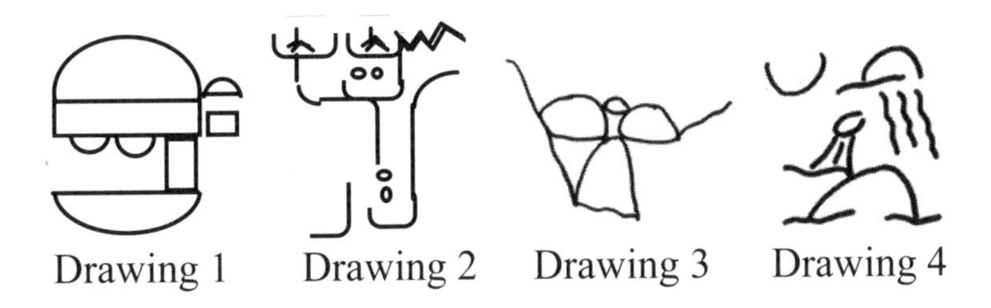

Fig. 1. Four Ambiguous Drawings

One factor that plagues designers and problem solvers in general is fixating on old ideas. Early in the design process, designers generate a flurry of ideas, but later in the design process, they find it harder and harder to see a design differently and generate new ideas. Adopting one of the parts strategies also provides protection against fixation. Undergraduates who adopted one of the parts focus strategies for generating new ideas produced relatively more ideas in the second half of the session than those who did not adopt a parts focus strategy.

Comparing Experts and Novices in Generating Interpretations of Sketches.

We replicated this experiment on groups of practicing designers, design students, and laypeople (Suwa and Tversky 2001). The practicing designers produced more interpretations and were more resistant to fixation than any of the other three groups, design students and laypeople, who did not differ from each other. Participants reported in detail the strategies they used to generate new interpretations. Primary among them were regrouping parts and changing reference frames. Participants also reported reversing figure-ground relations in the service of generating new ideas, but used this strategy less often. Both experienced designers and novices of all types used the same strategies, but the designers succeeded in generating more interpretations and resisting fixation. Perceptual reorganization is only half the process of generating ideas; the second, critical half is finding meaningful interpretations, a process which is conceptual in nature. This suggests that what separates experienced designersfrom novices and laypeople is the process of linking perceptual reorganizations to conceptual interpretations. This conclusion echoes the results of the protocol analyses of experienced and novice architects discussed earlier (Suwa and Tversky 1997). In that study, a major difference between practicing and novice architects was in facility of seeing functional implications of designs.

Constructive Perception

Designers appear to deliberately adopt perceptual strategies for reorganizing parts of ambiguous sketches in the service of generating ideas, a process we call *constructive perception* (Suwa and Tversky 2001). We believe that constructive perception can be fostered, and are experimenting with how to do it. Can constructive perception serve as a model for creativity in other domains? It seems likely. Even abstract domains that cannot be sketched have parts, which can be reorganized into new configurations and reconceived from new points of view.

Cognitive Design Principles

For design, beauty is not sufficient. The ideas must serve a user. Teapots should be easy to fill and easy to pour, and they should not burn the user. Instructions should be easy to apprehend and follow. To design a better teapot, we need to study how people use them. Similarly for instructions. Careful investigations into human cognition can provide guidelines for effective design. The domain we have chosen is visual instructions. Within those, we have selected two common and familiar cases: route instructions and assembly instructions. Visual instructions are a challenging domain because realism is not paramount. Effective visualizations omit irrelevant information and highlight, even distort, the relevant information. What is relevant depends on how people think of the task. To design effective visualizations for routes or assembly, we must know how people think about routes or assembly. Cognitive experimentation can elucidate people's mental models of routes or assembly or other domains. Cognitive experimentation can also elucidate how people perceive and interpret visualizations of these procedures and explanations. For these reasons, the contributions of cognitive experimentation go beyond traditional user testing. The cognitive experiments give insights into how people conceive of routes and assembly and how depictions and language can compatibly convey those conceptions. The conceptions as well as their diagrammatic and linguistic expression are principled. These principles serve as design principles. Let us illustrate how this happens in practice.

Route Maps

In order to design effective maps for guiding someone from one location to another, the first step is to know how people conceive of routes. From this understanding, design principles for the automatic construction of route maps can be inferred.

How do People Think About Routes?. A number of years ago, just before dinner, we approached students outside a dormitory and asked them if they knew how to get to a nearby fast food restaurant (Tversky and Lee 1998). If they did, we gave them a sheet of paper and asked them to either sketch a map or write instructions how to get there. We analyzed both sketch maps and route directions according to a scheme developed by Denis (1997) for the structure of route directions. He found that route directions consist of sequences of four kinds of segments: start point, reorientation, path progression, and end point. For example: you leave the station, you turn right, you go down Bahnhofstrasse, until you come to the cathedral. We found that this scheme, which had been developed to account for verbal directions, also characterized sketch maps. The similarities of syntax and semantics of route maps and directions suggest that both derive from the same underlying cognitive structure. Both route maps and route directions took a number of liberties with the Euclidean world. Degree of turn was approximate, around 90 degrees. Curves in roads were straightened. It turns out that these distortions of the Euclidean world also occur in memory for environments, maps, and spatial arrays; they are a consequence of normal perceptual organizing principles used in establishing and retrieving mental representations (e.g., Tversky, in press). In the sketch maps, long distances with no turns were shortened and short distances with many turns were enlarged, so scale was used to reflect the spatial information needed about the world rather than the spatial information of the world. Simply put, people think of routes as paths and nodes, where the nodes typically indicate change of direction.

Designing Route Maps. The sketch maps we obtained were typical of those that people draw for one another for navigation. Sketch maps such as these have undergone generations of informal user testing. They are quite different from highway maps or from the maps that can be downloaded from popular websites. Those maps suffer from clutter, too much extraneous information, and also from uniform scale, so that some important details may not be discernable. Agrawala and Stolte (2001) instantiated the design principles derived from cognitive research in the construction of computer algorithms that produce sketch-like route maps on demand. These maps, which users have praised, can be found at mapblast.com (Linedrive maps).

Assembly Instructions

Assembly instructions are the bane of do-it-yourselfers, who bring home boxes with enticing photos of barbeques or desks and contain dozens of parts. The instructions typically consist of a single detailed exploded diagram of the desired object. Consequently, the small parts are usually hard to distinguish, as is how to attach them. The order of assembly is typically not indicated. It is no wonder that not only is assembly frustrating, but at the end, do-it-yourselfers sometimes find themselves with extra parts.

How Do People Design Instructions? Users know better. Designing effective assembly instructions entails knowing how people think about the object to be assembled, how they think about the assembly process, and how visualizations can effectively convey both. To uncover cognitive design principles for assembly instructions, we ran a series of experiments, using a TV cart as a paradigm case (Heiser, Tversky and Daniel, in preparation; Heiser, Tversky, Agrawala and Hanrahan 2003). In the first experiment, participants assembled a TV cart using only the photograph on the carton as a guide. Afterwards, they generated instructions to assemble the TV cart. The visualizations produced varied widely across individuals. Those lower in spatial ability, as assessed by mental rotation performance, tended to produce 2-D menus of parts. Some also produced structural diagrams, showing the parts assembled, but without using perspective. The diagrams constructed by high ability participants were dramatically different. They typically produced step-by-step action diagrams in 3-D perspective. The steps corresponded to the major parts to be assembled. The action indicated the manner of assembly, typically using arrows and guidelines. The perspective chosen was that the showed the parts and how to assemble them. These diagrams went beyond structure to show construction, often by using extra-pictorial, diagrammatic devices such as arrows.

The characteristics of the diagrams constructed by high ability participants are good candidates for cognitive design principles. Their efficacy was tested in two further experiments. In the next, a new group of participants assembled the TV cart, and then rated the previous instructions for quality. Highly-rated instructions in fact had the qualities characterizing the diagrams of the high spatial participants: they presented one step at a time; they produced 3-D perspective views showing the assembly; they enlarged the small parts to be discernable; they used extra-pictorial features such as guidelines and arrows to indicate manner of assembly. A third group of participants used instructions varying in rated quality to assemble the TV cart. For high ability participants, instructions made no difference; in fact, those participants could and often did rely on the photograph on the carton to guide assembly. For low ability participants, however, the quality of instructions had the expected effects: good instructions enabled assembly that was faster and more accurate. Note that the participants were students at a highly-selective university, so that the low ability participants who benefited from effective visualizations are probably more representative of the general population.

Applying Cognitive Design Principles. These qualities of superior instructions were instantiated as design principles for automatically generating visualizations for assembly. The algorithms have produced elegant step-by-step visualizations for the assembly of furniture, including our paradigm TV cart, Lego, and other objects (Agrawala, Phan, Heiser, Klingner, Haymaker, Hanrahan and Tversky 2003). The algorithm decomposes a model of the object into assembly parts, and selects views that maximize the visibility of the parts to be assembled. Order of assembly is normally only partially constrained by the mechanics of assembly, that attaching some parts must be done before others. An innovation of the present approach is that the algorithm further constrains assembly order by selecting an order that maximizes the visibility of the assembly steps. Planning the procedures is thereby intertwined with presentation of the visualization. More generally, the ease of comprehending and implementing instructions should constrain order of assembly. This has general implications for design: to increase user ease, products should be designed in concert with instructions for their assembly or their use.

Summary and Implications

Cognitive research can inform and facilitate design. Studies of the kinds of design ideas that experts and novices generate from sketches have shown that new design ideas are frequently a consequence of reorganizing, then reinterpreting, the parts of a design. Using reorganization strategies in the service of generating new design ideas, or constructive perception, has two components, seeing a new configuration and connecting it to a new conceptualization. Experts seem to do this better than novices, suggesting that constructive perception may be cultivated. Constructive perception may have generality beyond design of visual objects.

For visualizations such as route maps and assembly instructions, even novices create visualizations that simplify and distort the visual information in ways that increase their usefulness by streamlining their message, for example, by enlarging small but critical elements and regularizing uninformative irregularities. The simplifications and distortions suggest cognitive design principles that can be implemented in algorithms to automatically generate visualizations. Such algorithms enable rapid inexpensive production of individualized visualizations.

Cognitive science has provided two messages for designers. Sketches benefit design. Design benefits sketches

Acknowledgements:

We are grateful for the encouragement and support of the following grants: Office of Naval Research, Grants Number NOOO14-PP-1-O649, N000140110717, and N000140210534 to Stanford University.

References

Agrawala, M., Phan, D., Heiser, J., Klingner, J. Haymaker, J., Hanrahan, P. and Tversky, B. (2003) Designing effective step-by-step assembly instructions. *Proceedings of SIGGRAPH '03.*

Agrawala, M. and Stolte, C. (2001). Rendering effective route maps: Improving usability through generalization. *Proceedings of SIGGRAPH '01,* 241-250.

Chase, W. G. and Simon, H. A. (1973). The mind's eye in chess. In W. G. Chase (Editor), *Visual information processing.* N. Y.: Academic Press.

de Groot, A. D. (1965). *Thought and choicein chess.* The Hague: Mouton.

Goldschmidt, G.: 1994, On visual design thinking: the vis kids of architecture, *Design Studies*, 15(2): 158-174.

Heiser, J. and Tversky, B. (submitted). Mental models of complex systems: Structure and function.

Heiser, J., Tversky, B., Agrawala, M., and Hanrahan, P. (submitted). Cognitive design principles for visualizations: Revealing and instantiating

Heiser, J., Tversky, B., and Daniel, M.-P. (in preparation). Constructing instructions.
Howard-Jones, P. A.: 1998, The variation of ideational productivity over short timescales and the influence of an instructional strategy to defocus attention, *Proceedings of Twentieth Annual Meeting of the Cognitive Science Society,* Hillsdale, New Jersey, Lawrence Erlbaum Associates, pp. 496-501.
Jansson, D. G. and Smith, S. M.:1991, Design fixation, *Design Studies*, 12(1), 3-11.
Schon, D. A. : 1983, *The Reflective Practitioner*, Harper Collins, USA.
Suwa, M., Gero, J. and Purcell, T.: 2000, Unexpected discoveries and S-invention of design requirements: Important vehicles for a design process, *Design Studies*, 21(6): 539-567.
Suwa, M. and Tversky, B.: 1997, What do architects and students perceive in their design sketches?: A protocol analysis, *Design Studies,* 18(4): 385-403.
Suwa, M., & Tversky, B. (2001). Constructive perception in design. In J. S. Gero & M. L. Maher (Eds.) *Computational and cognitive models of creative design V*, Sydney: University of Sydney , pp.227-239.
Suwa, M, Tversky, B, Gero J. & Purcell, T. (2001). Regrouping parts of an external representation as a source of insight. Proceedings of the 3rd International Conference on Cognitive Science (pp.692-696). Beijing, China: Press of University of Science and Technology of China..
Tversky, B. (In press). Functional significance of visuospatial representations. In P. Shah & A.Miyake (Eds.), *Handbook of higher-level visuospatial thinking*. Cambridge: Cambridge University Press.
Tversky, B., & Hemenway, K. (1984). Objects, parts, and categories. *Journal of Experimental Psychology: General, 113*, 169-193.
Tversky, B., & Lee, P. U. (1998). How space structures language. In C. Freksa, C. Habel, & K. F. Wender (Eds.), *Spatial cognition: An interdisciplinary approach to representation and processing of spatial knowledge* (pp. 157-175). Berlin: Springer-Verlag.

Dynamic aspects of individual design activities. A cognitive ergonomics viewpoint

Willemien Visser, Inria - National Research Institute for Computer Science and Control

1. Introduction

In a collection of papers presenting the *Developments in design methodology*, Cross (1984) presented "prescription of an ideal design process" as the first of the four stages that he distinguished. This stage, qualified as the period of "systematic design", was reflected by texts dating from 1962 to 1967. Stages two, three and four were "description of the intrinsic nature of design activity" (1966-1973), "observation of the reality of design activity" (late 1970s), and "reflection on the fundamental concepts of design" (1972-1982). Nearly twenty years after 1984, normative, prescriptive models of design (exemplified by Pahl and Beitz 1977, 1984) are still very powerful —even if the *Human Behaviour in Design* Symposium (2003, this book), organised in Pahl and Beitz' Germany with its particularly strong methodological tradition, testifies to a movement that might correspond to the second and third stages identified by Cross (1984): Are they stages of a new cycle?

Few models of actual design have been developed, however, and the proposals that have been made generally focus on particular aspects of the design process (Dorst 1997; Simon 1999; Visser 2002). In order to formulate such models, it is necessary to observe the actual activities carried out by designers. Studies conducted in cognitive psychology and cognitive ergonomics may collect such data. It is this approach to design that is central to this paper which presents a cognitive ergonomics viewpoint on individual design activities.

In the rest of this text, unless specified otherwise, the "design" referred to (through expressions such as "studies on design", or "design is opportunistic") is the actual design activity implemented by designers during their work on design projects —as opposed to normatively based descriptions or prescriptions of design methods or design processes.

Outset of the paper. We will describe our approach and illustrate it by data concerning two design strategies adopted by individual designers, i.e. the opportunistic organisation of design, and reuse. In our Conclusion, we will discuss the reluctance of industry to accept results that do not tally with design methods that are to be used in industrial design projects.

1.1 Relevance of data concerning individual design activities

Until some ten years ago, most design research was concerned with individual design. In more recent years, an important shift in research on the cognitive aspects of design has consisted in studying, perhaps even to a greater degree, collectively conducted design activities (Blessing, Brassac, Darses and Visser 2000; Cross, Christiaans and Dorst 1996). This change in focus, which has also been observed in other task domains, has been accompanied by the acceptance of design as a valid field of research in cognitive ergonomics for which analysis methods are being developed (Darses, Détienne, Falzon and Visser 2001). Yet, a majority of studies on design are being conducted in artificially restricted laboratory conditions, in which the design situation is rather different from that in professional design situations (see e.g. most studies in the 1997 Special issue of Design Studies on Descriptive models of design). Nevertheless, compared to other problem solving activities, design has started to be examined rather frequently in actual working situations. Recently, design is even being studied in large-scale industrial settings (Détienne and Falzon 2001; Visser 1993).

The present text is restricted to individual design. The continuing relevance of data concerning individual design can be justified on at least three grounds. First, even if many design projects are undertaken by large teams involving big numbers of designers, engineers and other people; even if discussion, negotiation, cognitive and operative synchronisation play a crucial role in the generation and evaluation of solutions; an important proportion of design activity remains the work of single individuals, especially during distributed-design stages. In addition, even during co-design stages, cognitive activities in collective design are those implemented in individual design to which are added activities that are specific to co-operative work. We see no evidence to suppose that co-operation modifies the nature of the elementary problem solving processes implemented in design, i.e. solution development and evaluation processes (Visser 1993). Finally, the development of appropriate work environments for designers, such as shared and private work spaces in computer-mediated design, requires the analysis of the links between the different forms of reasoning implemented in both individually and collectively conducted activities.

1.2 Dynamic aspects of individual design activities

Within the domain of individual design, this paper focuses on its dynamic aspects. Examples will primarily come from our own work. For some 20 years now, we have been collecting data on the use of knowledge, generally via design strategies. We did so through empirical studies, often conducted in the context of industrial design projects in different application domains, mainly software development, mechanical and industrial design.

The focus on dynamic aspects of design activities, i.e. on the *use* of knowledge, requires some comment. Many studies do indeed concern knowledge that is used in design, or knowledge that may be used in design. However, these studies seldom concern the way in which this knowledge is actually used, i.e. the modalities and conditions of their use in design activities.

In one of our first methodological studies on software design, we compared these two aspects of design knowledge use (Visser 1985, see also Visser and Morais 1991). We confronted two types of data that we had collected using different data collection methods. Four of the methods proposed (interviews and analyses of the result of the activity) may reveal knowledge possessed by designers, but it remains hypothetical *if* and *how* they use it. Only one of the methods (observing designers during their activity) may provide data on knowledge possessed by designers *and* on its actual use. The observational method is expensive and can usually only be applied to a few designers. It thus requires, in general, independent validation of its results.

2. Actually implemented design strategies

The "systematic design" movement in industrial, engineering and architectural design has its counterpart in software design, in the form of the "waterfall" model, and other "structured" and "stepwise refinement" methods. Early empirical studies conducted on design, especially in the domain of software design, generally characterise designers' activity as following such methods, i.e. as well structured and even as hierarchically organised, in other words, as following a pre-established plan. They assert that designers' global control strategy consists in decomposing their problem according to a combined top-down, breadth-first strategy. Both are general problem solving strategies that are not specific to design, but can be used in nearly any problem-solving task. These are the strategies identified, analysed and detailed in classical problem solving research, from Newell and Simon (1972) on.

In these early design studies, a trend seen was to "conflate prescriptive and descriptive remarks" on the activity (Carroll and Rosson 1985). Rather than to consider what the activity was really like, researchers focused on what it should be (Visser and Hoc 1990). One example is the study by Jeffries, Turner, Polson and Atwood (1981). According to these authors, "a reasonable model of performance... ought to be related to accepted standards of good practice" and "most expert designers are familiar with this literature and may incorporate facets of these methodologies into their designs" (p. 256).

More recent studies, however, observing designers in realistic situations or even in real work situations, show that the strategies implemented by these designers deviate from the top-down and breadth-first prescriptive model, and lead rather to an opportunistically organised design activity. In our own studies on software design we observed top-down and breadth-first decomposition strategies to be implemented, but only locally. Their combination did not seem to be the control strategy of the design activity at the global, organisational level. Other strategies were implemented at a local level. These could be strategies already identified in the problem solving research literature, e.g. simulation, or strategies that had not yet been presented in the literature, e.g. reuse.

The two examples of strategies actually implemented by professional designers when working on their design projects and presented in this paper, are the global strategy used by designers in order to organise their activity, i.e. the opportunistic-organisation strategy, and a more locally applied strategy, i.e. reuse.

Data collection. Data collection in our studies referred to below was based on observation (by the cognitive ergonomics researcher) and simultaneous verbalisation (by the designers). This approach is typical of cognitive ergonomics research, in that it provides data on the activities implemented by designers in their actual work situation (rather than on prescribed and/or idealised processes, or on actual activities observed in artificially restricted conditions).

2.1 Opportunistic organisation of design

Already in 1980, Green (1980) concludes a discussion of structured programming methods by advancing the idea that "good programmers.... leap intuitively ahead, from stepping stone to stepping stone, following a vision of the final program; and then they solidify, check, and construct a proper path. That proper path from first move to last move, in the correct order, is the program, their equivalent of the formal proof." (p. 306) Green notes that Wirth, who introduced the concept of "stepwise refinement", is himself "quite explicit; having described his stepwise refinement, [Wirth] says 'I should like to stress that we should not be led to infer that the actual program development proceeds in such a well organised, straightforward, top-down manner. Later refinement steps may show that earlier ones are inappropriate and must be reconsidered.'" (Ibid.)

The qualification that empirical studies on actual design activities use for the way in which designers organise their activity is "opportunistic" (Visser 1988a). Analysing as a design activity, the errand planning modelled by Hayes-Roth and Hayes-Roth (1979) as an opportunistically organised activity, we followed these authors' approach, and qualified the organisation of design activity as "opportunistic". We did so because designers' selection of consecutive design actions is determined by an evaluation function that is primarily based on the "cognitive economy" criterion, rather than by a pre-established plan, be it hierarchical or otherwise. Such plans may play a role, and often their role will be important, but they are only one of several possible resources that provide opportunities for action.

We observed this opportunistic design in different domains, e.g. in a series of studies conducted on three consecutive stages of an industrial design project: mechanical designers defining the functional specification for programmable-controller (PC) software design (Visser 1988a); a software engineer designing the PC software (Visser 1987); and a team composed of a software engineer and mechanical designers "testing" the PC software (but in fact also redesigning it) (Visser 1988b). In these studies, we identified six categories of data that could be "taken advantage of" as factors leading to the opportunistic organisation of design.

In this paper, only one category will be presented through an example from the observations made on `the functional-specifications design (for the other categories and examples, see Visser 1994). Designers can, for example, take advantage of mental representations of a design object related to the representations that they are using for their current design action. Analogy is one example of such a relationship. Considering second-phase tooling operations (in order to finish the rods) as analogous to the first-phase tooling operations (in order to shape the rods) on which he is currently working, the designer continually switches between their specifications. Often he takes advantage of the specification of a first-phase tooling operation O1 to specify, adapting this O1 specification, its corresponding second-phase tooling operation O2. Frequently, an O1 specification "makes him think" of an omission or error made on the corresponding O2 operation.

The observation that designers organise their activity in an opportunistic way is not restricted to inexperienced designers. On the contrary, it is something typical of expert designers. Nor is it the translation of a deteriorated behaviour that occurs only when designers are confronted with a "difficult" design task. Even when expert designers are involved in routine tasks, the retrieval of pre-existing plans does not appropriately characterise the organisation of their actual activity. An analysis of 15 empirical studies on design (Visser 1994) showed that

- even if designers possess a pre-existing solution plan for a design problem,
- and if they can and, in fact, do retrieve this plan to solve their problem (which is often possible for expert designers confronted with routine design),
- yet if other possibilities for action ("opportunities") are also perceived (which is often the case in real design)
- and if the designers evaluate the cost of all possible actions ("cognitive" and other costs), as they will do in real design,
- the action selected for execution will often be an action other than the one proposed by the plan: it will indeed be a selected opportunity.

Pre-existing plans that — if they are invariably followed — may lead to systematically organised activities, are only one of the various action-proposing knowledge structures used by designers. They may be interesting from a cognitive-economy viewpoint because executing an action for which such a schematic memory representation is available may cost relatively little if all schema variables relevant for execution have constant or default values. But if other knowledge structures propose relatively more economical actions, designers may deviate from such a plan. This is especially true for experts, who may be supposed to possess — or else to be able to construct without difficulty — a representation of their activity which allows them to resume their plan later on, when it once again becomes profitable to do so from the viewpoint of cognitive economy. Having to compare several action proposals and taking into account the cognitive cost of an action are two task characteristics which probably only appear in "real" design. This may explain why in laboratory experiments mostly systematically organised design activities were observed.

"Opportunities" must be "perceived": this perception is data-driven. It is on the basis of their knowledge that expert designers process the data that they perceive in their environment and that may take different forms: information (the state of design in progress, but also other information at the designers' disposal, information they receive, information they construct), permanent knowledge and temporary design representations (in particular, the designers' representations of the state of design). Taking advantage of these "opportunities" rather than following a pre-established plan will, indeed, lead to an opportunistically organised activity.

2.2 Reuse

All use of knowledge could be called "reuse" in that knowledge is based on the processing of previous experience and data encountered in the past. We reserve "reuse" (vs. other "use" of knowledge) for the use of specific knowledge (the "source" knowledge) that is at the same abstraction level as the "target" (the design problem to be solved) for whose processing the knowledge in question is retrieved. Thus, reuse of knowledge is opposed to the use of more general, abstract knowledge (such as knowledge structures like schemas and rules).

Reuse has been identified in various empirical design studies. The exploitation of specific experiences from the past is indeed particularly useful in design, especially in non-routine design (Visser 1993, 1995a).

Software-engineering researchers distinguish "design for reuse" from "reuse for design". The construction of reusable entities that are to be organised into a "components library" is often considered an independent design task, not necessarily executed by the designer who is going to reuse these entities. We are unaware, however, of any empirical studies conducted in such "design for reuse" situations. Existing empirical studies concern "reuse for design" and show that the two activities are not as separate as software-engineering researchers suppose.

A considerable proportion of the empirical, cognitive ergonomic research on reuse in design has been conducted in the domain of software, especially that of object-oriented (OO) software (Détienne 2002). Visser (1987) observed reuse in programmable controller-software design using a declarative type of language. She also studied reuse in other domains, i.e. mechanical (Visser 1991) and industrial design (Visser 1995b).

Several aspects of reuse have been examined in these studies. In this paper, we only discuss the question that is at the basis of reuse-based design, i.e.: When do designers decide to adopt reuse in order to solve a design problem, rather than base their problem solving on general knowledge, i.e. proceed to "design from scratch"?

Reuse takes place in, at least, five stages[1]: 0. construction of a representation of the target problem; 1. retrieval of one or more sources; 2. adaptation of the source into a target-solution proposal; 3. evaluation of the target-solution proposal; 4. integration into memory of the resulting modifications in problem and solution representations.

The construction of a target problem representation (stage 0) has seldom received attention in empirical studies. It is, however, during this stage that designers have decide whether they are going to (try to) adopt reuse in order to

[1] Our use of the term "stage" does not mean that such stages are purely consecutive, and that a previous stage cannot be returned to subsequently.

solve their design problem. We are not aware of any study informing us about designers taking into consideration the choice between design from scratch and reuse —rather than to "simply" design from scratch right from the start. Two individual studies seem to indicate that the "cost" of reuse is the main factor in the decision process whether to proceed to reuse (Burkhardt and Détienne 1995; Visser 1987).

In an experimental study, Burkhardt asked seven OO-software designers to describe elements they might want to (re)use. Half of his subjects mention that there are reusable elements whose actual reuse they would not envision because of the "cost" of their reuse. Data gathered by Visser (1987) present an example of a factor contributing to the cost of reuse — but only once candidate sources, i.e. reusable solutions, have been retrieved. This factor is the cost of required adaptation, itself a function of target-source similarity. Sources are indeed always to be adapted, in order to be usable as a possible solution to a target problem that is "similar" to the source problem that had been solved by the source. This conclusion coincides with a position adopted in several case-based reasoning (CBR) systems in which the selection of a "case", i.e. a reusable source, is guided by its adaptability (Smyth and Keane 1995).

The importance of the "cost" factor in the choice of a strategy such as reuse is completely in line with our identification of the primordial factor underlying the organisation of design, i.e. cognitive economy (Visser 1994). It is the relative cost of an action that determines its choice —if a designer is conscious of several possibilities.

With respect to the frequency of reuse, different authors in the domain of software engineering assert that 40 - 80 % of code is non-specific, thus reusable. Many authors advance the percentage of 80 %, mostly referring to Jones (1984), who summarises four studies conducted between 1977 and 1983. In 1989, however, Biggerstaff and Perlis assert that "over the broad span of systems, reuse is exploited today but to a very limited extent".

Empirical studies providing quantitative data all concern OO software, which, due to its mechanisms of inheritance, abstraction and encapsulation, and polymorphism, is considered to particularly "favour reuse", so the conclusions of these studies may be specific to this form of software design. As far as we know, the only empirical study observing "massive" reuse concerns OO software (Lange and Moher 1989). If empirical studies on other design tasks do not provide data on the frequency of reuse, experimental studies on analogical reasoning can inform us. Their general conclusion is that source retrieval seldom occurs "spontaneously", i.e. without having been suggested by the experimenter —which often occurs in these studies.

The data available on reuse-based design vs. design from scratch is thus still very meagre. Nevertheless, it is a central topic with respect to reuse, its role in design, and the possibilities to support designers in their reuse during design.

3. Conclusion

Our results concerning the organisation of design may inspire, at least, two approaches to design assistance: given the opportunistic organisation of the activity of designers working "in total freedom", one may either try to use tools to prevent such a way of proceeding, or offer designers tools that would assist them in their "natural" way of proceeding. As a researcher in cognitive ergonomics, we consider that assistance tools should be compatible with the actual activity, i.e. with the designers' mental structures and processes. If designers' activities are opportunistically organised because of reasons of cognitively interesting management, a support system which supposes — and therefore imposes — a hierarchically structured design process will at the very least constrain designers and may well even handicap them. Most tools continue, however, to be based on prescriptive or analytical, task-based models of design, i.e. not on data concerning the actual activity that is to be supported.

Results from cognitive ergonomics and other research that challenge the way in which people are supposed to work with existing systems are generally not received warmly. Abundant corroboration of such results is required before industry may consider taking them into account. The opportunistic organisation of design activity can be taken as one example of this reluctance. The results concerning this aspect of design have been verified repeatedly now, requirements for systems offering designers "real" support have been formulated on the basis of these results (see e.g. Visser and Hoc 1990), but only prototypes and experimental systems implementing some of these requirements are under development. An example of a software design environment based on the results concerning opportunism is GOOSE (Generalised Object Oriented Support Environment), the experimental CASE (Computer Assisted Software Engineering) tool for OO software design developed at the University of Keele by David Budgen and colleagues. One of the specifications for GOOSE was that it should enable its users to adopt an opportunistic strategy in developing their ideas. However, no commercial tools integrating these elements are available for use in industry, i.e. in "real" design, which is the focus of this paper.

Certain findings presented in this text may seem "obvious": one might think that any sensible person, especially a designer, might formulate them "simply" on the basis of their experience, or even of their common sense. One might think that no empirical studies are required to obtain the knowledge corresponding to these results. Studies in the domain of cognitive psychology and ergonomics teach us, however, that such "common-sense" judgements concerning "obvious" phenomena are fallacious, even if they correspond to the intuition of people who are proficient in the domain. Contrary to a widespread opinion, the fact that designers deviate from, or even do not follow at all, procedures that are prescribed by the majority of design methods, is not due to designers being nonchalant, making errors or displaying other deficiencies. These deviations or even abandoning of imposed structures have cognitive causes that are worth being examined and especially taken seriously in the development of design environments or other support modalities for designers.

Precisely because of their grounding in empirical studies conducted in real, complex situations, following the research methodology of cognitive psychology, these data are valuable as a basis for such system development. They constitute a basis that is more valuable than, on the one hand, psychological research conducted in restricted laboratory conditions, or, on the other hand, models based on introspection, norms, or other prescription-based models.

References

Biggerstaff, T. J., & Perlis, A. J. (1989). Introduction. In T. J. Biggerstaff & A. J. Perlis (Eds.), Software reusability (Vol. 1). Reading, MA: Addison-Wesley.

Blessing, L., Brassac, C., Darses, F., & Visser, W. (Eds.). (2000). Analysing and modelling collective design activities. Proceedings of COOP 2000, Fourth International Conference on the Design of Cooperative Systems. Rocquencourt: INRIA.

Burkhardt, J.-M., & Détienne, F. (1995, 27-29 June 1995). An empirical study of software reuse by experts in object-oriented design. Proceedings of Interact'95, Lillehammer (Norway).

Carroll, J. M., & Rosson, M. B. (1985). Usability specifications as a tool in iterative development. In H. R. Hartson (Ed.), Advances in human-computer interaction. Norwood, NJ: Ablex.

Cross, N. (1984). Developments in design methodology: Wiley.

Cross, N., Christiaans, H., & Dorst, K. (Eds.). (1996). Analysing design activity. Chichester: Wiley.

Darses, F., Détienne, F., Falzon, P., & Visser, W. (2001). COMET: A method for analysing collective design processes (Rapport de Recherche INRIA 4258). Rocquencourt: INRIA.

Design Studies. (1997). Special issue on Descriptive models of design. Design Studies, 18(4).

Détienne, F. (2002). Software design. Cognitive aspects. London: Springer.

Détienne, F., & Falzon, P. (2001). Cognition and Cooperation in Design: the Eiffel research group. In M. Hirose (Ed.), Human-Computer Interaction-Interact 2001 (pp. 879-880).

Dorst, K. (1997). Describing design. A comparison of paradigms. Delft: TU Delft

Green, T. R. G. (1980). Programming as a cognitive activity. In H. T. Smith & T. R. G. Green (Eds.), Human interaction with computers. London: Academic Press.

Hayes-Roth, B., & Hayes-Roth, F. (1979). A cognitive model of planning. Cognitive Science, 3, 275-310.

Jeffries, R., Turner, A. A., Polson, P. G., & Atwood, M. E. (1981). The processes involved in designing software. In J. R. Anderson (Ed.), Cognitive skills and their acquisition. Hillsdale, NJ.: Erlbaum.

Lange, B. M., & Moher, T. (1989). Some strategies for reuse in an object-oriented programming environment. Proceedings of CHI'89, Austin (USA).

Newell, A., & Simon, H. A. (1972). Human problem solving: Prentice Hall.

Pahl, G., & Beitz, W. (1977). Konstruktionslehre. Berlin: Springer.

Pahl, G., & Beitz, W. (1984). Engineering design. London: The Design Council.

Simon, H. A. (1999). The sciences of the artificial (3rd, rev. ed. 1996; Orig. ed. 1969). Cambridge, MA: The MIT Press.

Smyth, B., & Keane, M. T. (1995). Some experiments on adaptation-guided retrieval. In M. Veloso & A. Aamodt (Eds.), Case-based reasoning (pp. 313-324). Berlin: Springer.

Visser, W. (1985). Modélisation de l'activité de programmation de systèmes de commande [Modelling the activityof programming control systems] (In French). Actes du colloque COGNITIVA 85 (Tome 2), Paris.

Visser, W. (1987). Strategies in programming programmable controllers: a field study on a professional programmer. In G. Olson, S. Sheppard, & E. Soloway (Eds.), Empirical Studies of programmers: Second Workshop (pp. 217-230). Norwood, NJ: Ablex.

Visser, W. (1988a). Giving up a hierarchical plan in a design activity (Rapport de recherche 814). Rocquencourt, France: INRIA.

Visser, W. (1988b). L'activité de comparaison de représentations dans la mise au point de programmes. Le Travail Humain, Numéro Spécial "Psychologie ergonomique de la programmation informatique", 51(4), 351-362.

Visser, W. (1991). Evocation and elaboration of solutions: Different types of problem-solving actions. An empirical study on the design of an aerospace artifact. COGNITIVA 90, Paris.

Visser, W. (1993). Collective design: A cognitive analysis of cooperation in practice. ICED 93, the 9th International Conference on Engineering Design, The Hague (The Netherlands).

Visser, W. (1994). Organisation of design activities: opportunistic, with hierarchical episodes. Interacting With Computers, 6(3), 239-274 (Executive summary: 235-238).

Visser, W. (1995a). Reuse of knowledge: empirical studies. In M. Veloso & A. Aamodt (Eds.), Case-based reasoning. Berlin: Springer.

Visser, W. (1995b). Use of episodic knowledge and information in design problem solving. Design Studies, 16(2), 171-187.

Visser, W. (2002). A Tribute to Simon, and some —too late— questions, by a cognitive ergonomist. International Conference In Honour of Herbert Simon "The Sciences of Design. The Scientific Challenge for the 21st Century", Lyon (France), INSA.

Visser, W., & Hoc, J.M. (1990). Expert software design strategies. In J.M. Hoc, T. Green, R. Samurçay, & D. Gilmore (Eds.), Psychology of programming (pp. 235-250). London: Academic Press.

Visser, W., & Morais, A. (1991). Concurrent use of different expertise elicitation methods applied to the study of the programming activity. In M. J. Tauber & D. Ackermann (Eds.), Mental models and human-computer interaction. Amsterdam: Elsevier.

Individual Thinking and Acting: Summary of Discussion

Lucienne Blessing
University of Technology Berlin

Introduction

The participants in stream I consisted of a very good mixture of people from different disciplines, several of which had not met before. As a result, the short poster presentations, gave rise to numerous discussions, showing not only overlap in results and ideas, but also a strong discipline-dependent use of terminology. The poster presentations and the initial set of questions resulted in a large number of interesting issues, which were grouped and discussed, resulting in three main questions that formed the basis for the second part of the session.

Questions

The issues that arose from the poster presentations and initial discussions are listed below, the three main questions are highlighted.

Terminology:

- How to get to a common termonology of our units of analysis?
- Can images contain concepts?
- What is a concept?

Components of design activity:

- Can we distinguish between necessary vs. irrelevant operations and moves?
- Should we not define difficulty vs. complexity in relative terms?
- Is re-use a form of fixation?

Cognition

- Is cognitive economy a/the common denominator of our work?
- Is the subconsious process due to tacit knowledge?

→ *How does/can cognitive economy play a role in your findings/explanations*

Strategies

- Top-down vs bottom up?
- Is opportunism more successful or is it only more common?

- What is an opportunistic strategy (can a strategy be opportunistic)?
- Is an opportunistic procedure a knowledge driven procedur?
- Are systematic and opportunistic strategies complementary?
- How can we combine "positive" aspects of a methodic (systematic) approach with "positive" effects of "free design"?
- How do descriptive and prescriptive approaches differ? (experiments vs methodology)
- Contrast/compare real-life processes with theoretical models? (frame of reference)?

→ *What are the implications of the findings of strategy research on improving design?*

- Meta-cognition + design process vs. self-organising process?
- How can we predict "good" designs (results) from the activities/processes adopted during the design process? (prediction)
- What importance have for-links to the designer?
- Is there a correlation between for- and back links to the quality of design?

Research method

- What are the effects of data-collection methods on the findings?

→ *Where are we? What are we heading for?*

The following sections summarise the discussions that resulted from the attempt to answer the three identified questions. The statements as expressed by the participants are used in this summary whereeever possible, without explicitly citing and naming the particpants.

Question 1: How does/can cognitive economy play a role in your findings/explanations?

Discussion

Cognitive economy played a role in all presented research work, either explicitly or, for those not familiar with the term, implicitly. The feeling was that the concept of cognitive economy is an essential factor in explaining design behaviour. Cognitive economy is one of the reasons why we behave the way we do, and with an increasing age of the population, will become a factor of increasing importance.

But what is cognitive economy? From the discussion it appeared that the definition of cognitive economy is not clear. Is cognitive economy a purely mental process, revealing itself, for example in the difference between experts and novices, the former using experience to reduce trial and error? Or is cognitive economy the more general process of allocating and dealing with limited human resources, revealing itself, for example, also in the use of external representations? It was argued that the central reason for cognitive economy is too reduce mental load, which is necessary when the mental capacity of human beings turns out to be too limited to solve certain problems. This involves both mental processes. e.g. using experience, as well as external representations such as drawings or gestures. In particular in design, but also in other complex situations, we see that experience as well as the use of external represenations has a large effect on the quality of the design. A short discussion arosea around the aspect of divergent thinking, which at least at first sight is not necessarily effective, requiring more mental capacity, but is considered important for design. No conclusion was drawn.

Research has shown how, in more detail, experts deal with limited mental capacity. First, they form large chunks of knowledge, which they process as units. Second, they also use the contents: they know what the right chunks are. Third, they use, as all human beings, subconsious, settled rules to reduce the problem to a manageable size, e.g. by decomposition or by putting boundaries around the problem. It was emphasised that the use of subconsious rules, does not not imply a lack of purpose. Everything has a purpose, the human mind deals with things using functionality. It was argued that the finding that designers can say little about what they are doing, but can afterwards explain the design, shows that design is a purposeful activity yet starts with few explicit rules and thus must rely on other, subconsious rules. This was not accepted by all participants. The fifth was of dealing with limited mental capacity are the already mentioned external representations. These force one to be explicit, to be more complete. In a sketch, more is visible than what was originally drawn 'into the drawing'. Examples are relationships and connections.

The question then arose as to whether we can use and teach techniques to support cognitive economy, such as techniques to select suitable strategies and suitable external representations, and how this affects tools and methods. Can we not only use and teach something we do consciously, e.g. selecting and applying cognitive economy strategies consciously. Research suggests that we handle cognitive economy both consciously and unconsciously, i.e. that at least part of this could be supported.

This lead to the interesting question as to whether we should teach excellent designers or those that go for satisfactory solutions, i.e. whether we should focus on searching for the best or on efficient approaches. It was clear that the contents of 'excellence' has to be defined first. It seems that experts have many routines depending on the situation at hand. There is a difference when between designing something to meet complexity and something less complex. Much of what experts do is rooting.

The effect of these findings on the development and use of methods and tools. was considered to be unclear. Design methods and tools are developed to support designers, that is, to allow designers to solve design problems despite the limits of mental capacity. However, in particular computer tools may actually reduce the mental capacity available for designing because part of the capacity is used for operating the tool.

Proposals

1. The special features of the domain need to be determined: designers create artefacts, this can also include teaching methods, organisations, etc. In the discussion three keywords emerged for this domain: spatial, functional and goal-directed.
2. More needs to be known about ways of achieving cognitive economy as a way to reduce mental load.
3. Tools should be developed to help release mental capacity, and not increase the need for capacity, or disturb the mental processes. The tools should help with the operation, or come up with other operations that achieve the same release of capacity.
4. Sequenzing of strategies should be investigated for practice and for teaching purposes. It was suggested that a different framework is needed than currently available. Many functions are not deliberately "designed in". In preparing such a framework, it should be realised that some of the assumptions about the world are not correct, such as constancy. Designing changes the world *and* us. Physics knowledge does change us but not the world. Designs change our value systems and our behaviour and therefore the effects of our designs have to be considered. We do not capture the essence of science. We borrow ideas from lots of places, without understanding the situation in which these ideas have been generated. One of the essences of design is that it intends to change things. Interestingly, biological essence is not changing, but memory contents is changing all the time.

The following picture shows an attempt to link the issues dealt with in the discussion. The cause of design behaviour is the limited mental capacity of human beings. Cognitive economy is an answer to this cause because it will help reduce mental load and release capacity. The effects depend on the domain (an expert in one domain is not necessarily an expert in other domains). Cognitive economy is realised by several different, generic, strategies, such as chunking and external representation. In order to support these strategies, tools can be developed, as well as guidelines on sequences of strategies.

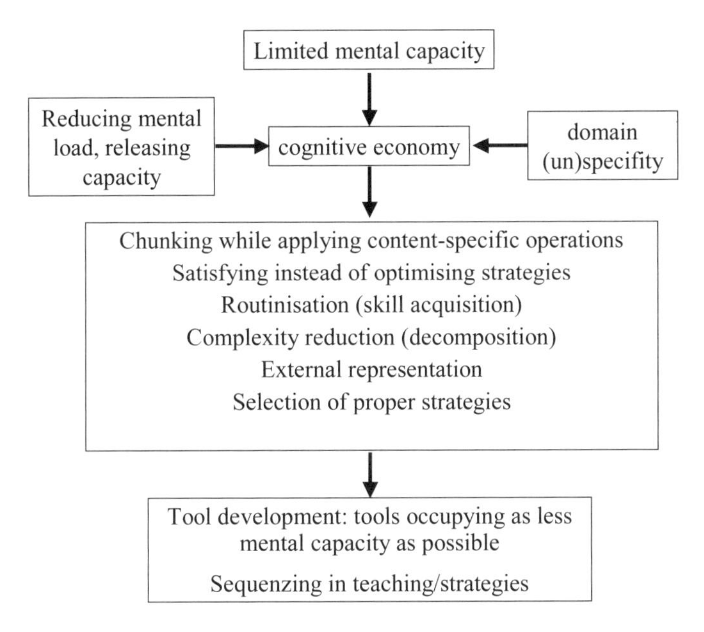

Fig. 1. Mental economy: cause, strategies and support

Question 2: What are the implications of the findings of strategy research on improving design?

Discussion

Several of the presentations referred to opportunism as the main design strategy, that is, as a plausible explanation for the observed behaviour. Designers typically use opportunities at hand to proceed, rather than follow a prescribed path. Humans were considered to be generally opportunistic in their behaviour, because they cannot do anything else but react to their environment But then, if all of us react to our environment, what is the use of methodologies: we have our view on things and organise our work accordingly.

The findings support the definition of strategy as knowledge-based regulation of action. If we do not have knowledge about the situation at hand, we cannot apply a strategy nor do we have a strategy. This would imply, that design strategies can only be planned shortly before they are to be applied and can only cover a period of time in which the situation does not change so much as to require a different strategy.

It was emphasised that although opportunism is the main strategy, designers are not just muddling through. Some studies indicate that a combination of opportunistic and hierarchical strategies is used. There is no one best strategy, but a changing between more hierarchical (often more global) and more opportunistic (often more local) strategies. Both types of strategies are based on cues, the results might be the same, but the cues are different.

It was argued that opportunism is nothing other than organising one's work and methods, but based on what is at hand, rather than on a predetermined plan. This implies the following conditions for opportunism to work: opportunities need to be recognised and ways of dealing with these opportunities need to be known. This distinguishes experts from novices and lay people.

What are the implications of these findings on the teaching of design? Should opportunistic strategies and procedures be taught, and if so, what are these? Can opportunism be taught at all? What are the objectives of teaching design, does one need an abstract concept, such as a process model, behind the cases?

Currently, teaching design involves teaching cases, general guidelines, principles and calculation methods which the students have to combine, or teaching design involves teaching methodologies as normative descriptions of the process of designing as well as the methods that can be used in the various phases. These methods and methodologies are considered to be too restricted. They should be more open to allow adjustment or deviation when opportunitisch arise, yet they should still support (it is always possible to break off a method, but then support is no longer available). The suggestion came up to consider less hierarchical but more networking methods.

Proposals

The two main strategies used in design can be typified using the following terms: :

- Opportunistic: bottom-up, situated, context dependent, experience-based, knowledge-driven, task-episode-accumulation (Ullman), local strategy, data-driven
- Hierarchisch: systematic, methodic, pre-planned, large mental load, structured, 'long'-term, global strategy, plan-driven

1. Both strategies should be used in combination to be able to have an overall direction, yet be flexible to adapt to the situation at hand. For teaching this implies that a methodology is taught to provide an overall framework, but that at the same time the students are pointed at task/situation specific modifications.
2. Research is necessary to obtain a better understanding of the link between suitable strategies and the characteristics of tasks/situations.
3. No proposal on how to teach opportunism could be formulated.

Question 3: Where are we? What are we heading for?

Discussion

What is design research? Our research is not to turn researchers into great designers, just as an English department does not produce Shakespeares. They lay the foundation for other people to achieve something near Shakespeare. In the same way, our research can assist us in understanding design and improve our educational system and the production of methods and tools.

Two types of tools need to be distinguished: analysis tools and 'design' tools. Currently, the latter are more documentation tools, not synthesis tools. The latter only exists in e.g. VLSI, but this even requires a very detailed specification.

Despite all our research we do not see much being taking up, not in research itself, not in industry and not in teaching.

A clear gap exists between researchers doing descriptive research (improving our understanding of design) and researchers/industries developing methods and tools, although the situation is slightly improving.

It could be argued that where this gap is closed, the gap between research and take up industry would reduce, as methods and tools based on the findings of descriptive studies should better take into account the actual situation and thus realise a better take up. However, even the research results of those that combine the two, are hardly exploited. The gap between research findings and industrial exploitation has been a topic of discussion for many years, but a satisfactory solution has not been found. A participant pointed out that some methods and tools involve so much user involvement that some tools people may not dare to use these, although they are quite happy for consultancies to use these. Other more automatic tools, people have no problem using.

The take up of research results in education has not been discussed, but a reference can be made to the paragrahs about opportunism. Considering the fact that researchers are those that teach, the impact of research results is rather limited.

Proposal

We should work on:

- more rigourous investigations, considering domain (un)specifity
- appropriate models/understanding of design(ing)
- link between experimental findings (human aspect) and development of methods/tools, link between descriptive and prescriptive
- implementation/ use of results in practice and teaching.

Conclusions

The picture of where we are as a community, was considered to be rather fuzzy. Much research is taking place, but an overview is lacking. It seemed that what is really needed, is an appropriate model of design that can be used to develop strategies, teaching, tools, etc. The aspects of cognitive economy and opportunistic behaviour should be taken into account. The question is whether a concept of flexibility and application of methods can be found, and whether there is a chance of having an appropriate model of design. To do so a framework is needed. The core idea behind the framework should be flexible, balanced design. Some of the existing models, in particular the process-based models may be useful as a starting point.

Topic II: Interaction between individuals

Herbert Birkhofer and Judith Jänsch, University of Technology, Darmstadt

1. Introduction

Engineering design in industry has changed rapidly in the last decades. Project work, simultaneous engineering, globalization and outsourcing are key issues which point to a substantial increase of partners involved in design work. The interaction between these partners varies substantially and covers local as well as global interaction. Interaction is a dominant factor for success of design processes as well as for the performance of the designed products.

This chapter summarizes the actual state of expertise in regard to teamwork and puts the contributions of section 2 "Interaction between individuals" into a framework. A main focus of this introduction is to obtain answers especially to the following questions:

- What is the meaning of interaction in design?
- How do individuals interact in design?
- How can we support interaction for designing?

The answers might bridge the gap between the view of individual designers (section I) and the methods and tools applied in design (section III).

2. Interaction in design

2.1 The role of interaction

Detailed investigations show that more than 30 percent of design work is carried out in teams. If teamwork in a global sense is understood as the interaction between individuals, it is evident that interaction is a main issue in design which highly affects time, costs and quality of processes and products.

Interaction occurs in a variety of milieus, ranging from discussions, meetings, reviews, and design conferences to global types of interaction referred to, for example, as "follow the sun" strategies. Therefore, interaction in design happens not only within the design department, but within the whole company on different hierarchic levels and also externally with suppliers and customers, etc.

2.2 The nature of interaction

2.2.1 Interaction as an action-perception process

Interaction between individuals happens in quite different styles: communication, co-operation, plays or contests, etc. It can be considered a mutual exchange of messages between people, which is based on actions and perceptions of the participants involved (Fig. 1).

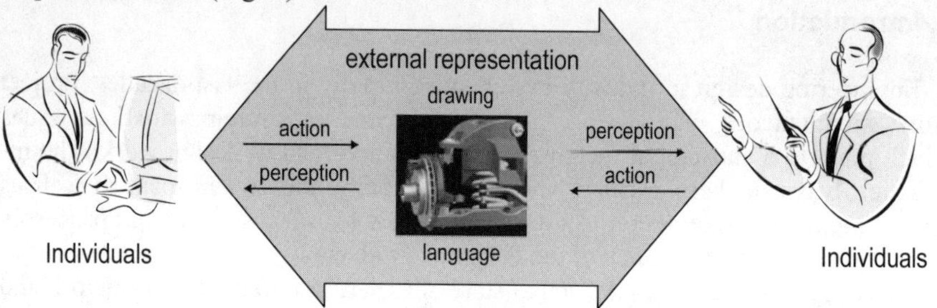

Fig. 1: Interaction as action-perception processes

Actions of an individual can be speech, gestures, facial plays, and sketches, and are intended to be a message to a counterpart, who perceives them. Interaction is often based on external representations, such as text or graphics.

2.2.2 Interaction, communication and information

It is important to point out that a main function of interaction in design work is the exchange of information. But there are also social, motivational and emotional effects (group dynamic effects) which have an important influence on interaction within teams. It is often observed that good interaction dramatically increases motivation within a team. Interaction influences personal relations between individuals and stimulates emotions in respect to the product or to the specifics of the design context.

Another basic function of interaction in design is the conscious or unconscious, spontaneous exchange of information between individuals. This transfer of information is often called communication. But communication should not be regarded only as an information transfer. Due to the change of information during the action-perception process, communication is information transformation, as well. Normally, communication in teamwork leads to an enlarged knowledge base summarizing the individual and different knowledge bases of all individuals involved. And last but not least, the abilities of individuals to communicate play a major role within communication.

2.2.3 The influence of individual on interaction

Individuals can also heavily influence interaction in teams. During the interaction process the individual perceives the design context and other individuals. They match the perceptions with their own mental models. A person's actions result from these processes (Fig. 2). The perception of other individuals' actions, the process of matching information with one's own mental models, and the expression of thoughts and concepts through actions is crucial for interaction, in general, and specifically within design.

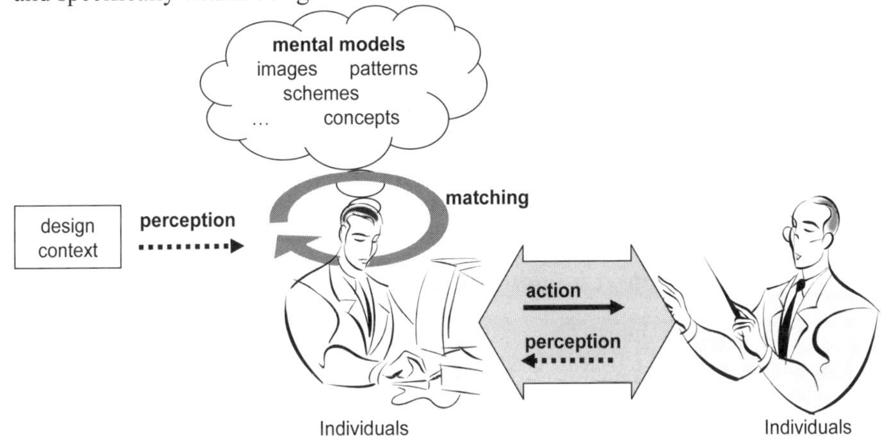

Fig. 2: Perception matching and action process while interacting

A main characteristic of an individual is his set of mental models. They are primarily responsible for how individuals perceive, think and act. A mental model might be understood as an actual picture of the real world captured by an individual. Mental models contain information about objects, processes, strategies or motivational aspects. If a set of these elements is put together and structured, we call them patterns, schemes, concepts or images. Mental models are created in a specific situation, where individuals perceive information, compare it to actual mental models and process it. Mental models develop steadily due to constantly changing situations and the individuals' ability to perceive them.

Perception implies a holistic, and for each individual a specific, impression of the design context and other individuals and their actions. Individuals might perceive the same situations quite differently due to their varying mental models.

Perceived information in an interaction process is compared with existing mental models and integrated, changed, extended and adapted. This kind of information processing can be understood as a matching process. There is also the possibility to create a completely new mental model if the information does not match an existing one. In the matching of information, the different views of the contents play a major role. It is those aspects in particular, which express contrary views on the same item that are most valuable for design work. These so-called "modalities" (list 1) are not just an element of human thinking, but rather an important tool for designing, in that the change of modalities causes creativity, flexibility, etc. which are indispensable for successful designing.

Part	⇔	Whole
Abstract	⇔	Concrete
Spatial	⇔	Temporal
Text	⇔	Graphic
Object	⇔	Process
Intuitive	⇔	Methodical
…	⇔	…

List 1: Modalities of human thinking

Information represented in mental models has to be "externalized" to initiate or at least to support communication. The process of externalization is, in general, an action process as part of a communication process. In communication individuals act by talking, sketching or drawing.

2.3 The design context influences interaction

As mentioned above, interaction takes place in a wide range of situations. For this reason it makes sense to define a context of interaction in design and a set of specific factors influencing interaction.

One powerful factor influencing interaction is the design situation, which is marked by the product, process and design task. Other factors are methods and tools which support designers on different levels of their work, e.g. creativity methods, design methods, meta plan techniques, tools for cooperative-work, tools for visualization, and tools for the obtaining of information, such as databases, CAD and VR-Tools.

The environment of design is provided by the company with its organization, suppliers, branch, culture and rules. It influences the interaction, for example, with hierarchical structures, languages and people's attitudes. Furthermore, there are constraints, e.g. the physical environment or limited resources in design, which can influence interaction. But doubtless the strongest factors in this framework are the individuals. They influence interaction with their personalities, experiences, knowledge, assessment, competence, motivation, etc.

3. A holistic view of interaction

Interaction between two or more individuals encompasses all of the actions, perceptions and matching processes in a specific design context (Fig. 3).

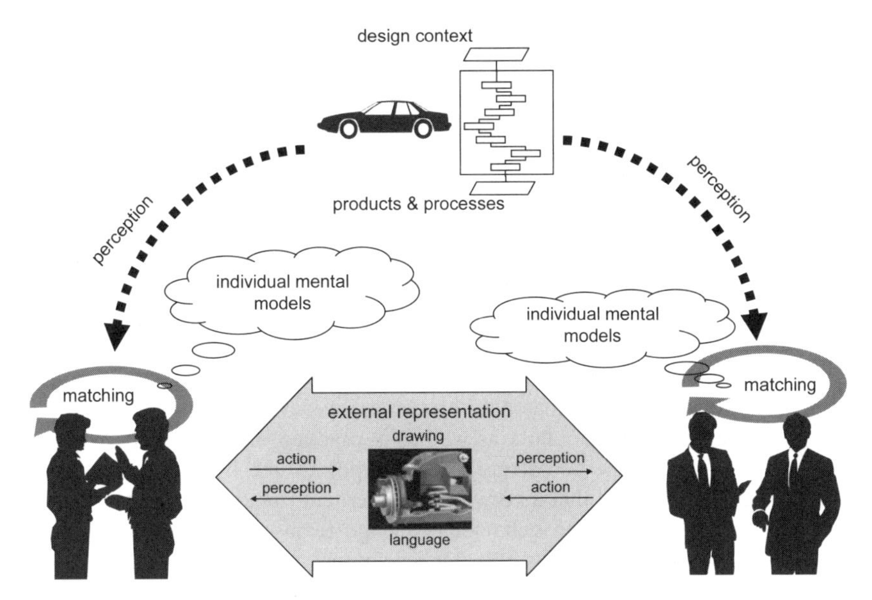

Fig. 3: The whole picture

Individual 1 (I1) acts within the design process by using external representations, like language, drawings or gestures. Individual 2 (I2) perceives the actions of I1 as well as the specific design context, and matches both within his own mental model. With the previous, adapted or new contents of his mental model, I2 acts towards I1. Obviously, I1 and I2 could have similar or completely different views of the same design context and the actions of the partner. It could be that I1 and I2 have a highly effective and efficient interaction because of their "shared mental models. On the other hand, interaction could be quite problematic or even useless, if there is not sufficient common ground. Furthermore, the effectiveness of interaction is based on the performance of producing proper external representations and the ability of the partner to recognize it.

4. List of contributions

The contributions within this chapter progress from the observation of asking questions to a collaborative product development. These short comments of the individual contributions should give a first impression if the research fields interaction.

- In the first contribution *Uday Athavankar and Arnab Mukherjee* explore the creative use of mental imagery in design by pairing students and blindfolding one of them while designing. This descriptive contribution creates a starting point to use mental imagery as a pedagogic tool. With this research they offer a possibility to support mental models and effectively use them (e.g. switching modalities).

- *Petra Badke-Schaub and Joachim Stempfle* investigate how design teams find solutions and what kind of difficulties occur during this process. Results of an empirical study are explained by a 'two-process-theory' relating to basic elements of thinking in design. This contribution mainly considers team work in respect to the design situation and environment, such as task planning and team organization.
- *Amaresh Chakrabarti* investigates the activities and mechanisms which are used by designers to identify requirements. He discovered which activities are most effective and in which stage of development they are applied. In his investigations he found sub-activities, such as problem perception and question asking, which comprised the most significant activities. In short, the paper analyzes activities and mechanisms of perception while identifying requirements.
- *Özgür Eris* identifies question-asking and decision-making as fundamental cognitive mechanisms (e.g. matching processes) in design activity. By observing design teams in the laboratory as well as in practice, and reviewing existing taxonomies of questions, he categorizes the types of questions designers ask. His findings demonstrate that a convergent-divergent thinking process is manifested in the question asking behavior of design teams, and that the effective utilization of this process results in better performance.
- *Ade Mabogunje* observes designers in industry and postulates that design is a question-driven activity. If this hypothesis is proven to be true, there should be strategies to support designers in question-asking. Therefore, he tries to build up a method which can modify the type of question being asked during the design process. For this purpose, he investigates the interaction process and classifies the utterances and expressions into categories and considers information retrieval. Mabogunje especially focuses on supporting the action process and question asking, in particular.
- How to create good conditions for collaborative product development is addressed by *Stig Ottosson.* He focuses particularly on personality profiles and knowledge of the people involved in the design process and how they communicate. He investigates how communication, and therefore, product quality are affected by the multitude of design context factors (physical environment, resources, organization, etc.)
- *Stephen AR Scrivener, Lai-Chung Le, and André Woodcock* deal with how to manage breakdowns in internationally distributed design projects. In this field, they focused on problems caused by trying to communicate across different disciplines, cultures and languages. They are working on ways to reduce constraints which originate in the design environment. By identifying problems in real project case studies, he devises intervention strategies and applies them in a second case study.
- *Ken Wallace & Saeema Ahmed* study how designers obtain information and interact with others to get that information. Within these studies they also consider the different strategies of novices and experts in obtaining information. This leads to conclusions about personal characteristics and their changes in designing. In a further step, they speculate how the currently best practice can be improved by the opportunities offered by modern IT.

Blindfolded Classroom: Getting Design Students to Use Mental Imagery

Uday Athavankar, Arnab Mukherjee
Industrial Design Centre, Indian Institute of Technology

1. Introduction

Due to its depictive nature mental imaging can handle spatial issues as good as sketching. Besides, imagery is a mental experience and is perhaps an ideal tool to synergistically work with the process of thinking – that too without the load of overt sensory-motor operations that the action of sketching demands. It is no wonder therefore that most designers talk about rich imagery experiences that they go through. Yet, mental imagery has not been used as a pedagogic tool in the design studio environment.

This two-part paper builds on the ideas developed in the earlier studies of creative use of mind's eye and explores their extension to the design studio (classroom) environment. Part I explores answers to the questions: Does mental imagery play an equally significant role in design problem solving in students (as it does for professionals)? Can it become an effective design tool for young students? Part II explores the extension of ideas developed in Part I to a classroom environment. We will return to these issues after a brief review of research on creativity and mental imagery as well as of the previous work on the use of mind's eye in design problem solving.

1.1 Studies of Mental Imagery and Creativity

Cognitive scientists have begun to meticulously explore the nature of mental images and its role in problem solving. Imagery, as well as the idea of "Mind's Eye," has now acquired legitimacy. Kosslyn showed that images are real psychological entities and identified the cognitive processes and operations that we can voluntarily perform on them (Kosslyn 1983). Roskos-Ewoldsen at el. have explored the relationship between imagery and creativity and have compiled the issues involved in imagery, creativity and discovery (Roskos-Ewoldsen et al. 1993). Working on mental synthesis, Finke showed how the manipulational, combinational and generational play that imagery affords is responsible for inventions (Finke 1990).

These studies do elaborate on the typical attributes of imagery like depictive qualities, spontaneity, its non-linear nature, and the possibilities of voluntary control that could potentially prompt creative pursuits, but most of them are concerned with inducing creativity within the framework of structured laboratory experiments using unrealistically simple tasks. None of them focus on the processes that support and dictate design problem solving.

1.2 References to Imagery in Design Literature

Imagery has two opposite uses in design problem solving. It affords simulation possibilities and thus permits the evaluation of ideas without liabilities to the real world. It also encourages fanciful play, which is an essential component of any creative act. Yet, the study of creative usage of mental imagery in design problem solving has not taken center stage in design research so far. Although some references to imagery have been made in design research literature (McKim 1972; Sommer 1978), these have been few and far between, indircct, and appear to be more often referred to in the context of sketching (Goldschmidt 1992 a–b, 1994; Fish and Scrivener 1990).

More direct studies on creative use of mental imagery in design problem solving come from a series of experiments where the designers/architects were asked to design when blindfolded. The first paper in the series addressed the question "Can designers design, if they are Blindfolded?" The answer was not only affirmative but also revealed the designers' amazing ability to manipulate their creations in their mind's eye (Athavankar 1997, 1999). The speed with which the creations in the imagery responded to the developing design thought showed its potential in supporting spatial arguments.

Other papers in the blindfold series include Singh, who attempted to "catch" the imagery experiences of the architect (Singh 1999). Later, Garde gave an identical design task to the three architects and compared the strategies generated by them to reason and arrive at visio-spatial decisions (Garde et al. 2001). These architects, completely immersed in their own visio-spatial mental simulations, stood "inside" and manipulated the mental scenarios they created. Another variation simulated the design process where two architects work at distant locations but are connected by a video-link. Both of them conjured up images in response to each other's words in the conversation (Gill et al. 2000). It should be noted that most of these studies referred to in this section used mid-career designers/architects with plenty of professional experience.

Part I

Keeping the educational perspective in mind, this part will start by studying how the younger architectural students, left to fully depend on their mind's eye, respond to its faculty of creative design. The results will later be extended to the classroom environment in Part II.

2. Pilot Experiments

2.1 Design Problem and Subjects

In the three pilot experiments involving three groups, six architecture students with over three years of classroom experience were randomly selected and put in pairs to form three groups (Groups 1, 2 and 3). Each group was required to design an independent three-bedroom house for a company chairman, but had no prior knowledge of the design problem. Besides the socio-cultural background of the client, a detailed list of requirements was provided that added to 248 sq. m. built up space to be accommodated in a 1035 sq. m. plot in a sub-urban area.

2.2 Structure of the Design Problem Solving Session

Students worked in pairs, but were initially separated and briefed on their roles. Later they were ushered in to the work area. They sat at right angle to each other sharing a tabletop.

- One of the students in the pair was blindfolded (referred to as BD: Blindfolded Designer in the discussions here). BD solved the design problem while thinking aloud and used gestures when necessary.
- BD's partner, (referred here as SD: Sighted Designer), closely followed the think-aloud session. Between the groups, the roles of SDs were varied. SD in Group 1 maintained a neutral stand and restricted himself to prodding BD for details and clarifications when required. He did not offer ideas or force BD into taking design decisions. BD had to totally depend on his internal resources for ideas and decisions. SDs in groups 2 and 3 made active interventions and contributed design ideas to the ongoing process. So the BDs, besides working on their own ideas, had to quickly evaluate new suggestions from SDs while completely depending on their mental imagery.
- At the end of the 45 to 60 minutes session, the two were separated and BD's eye mask taken off. Both were asked to sketch the final design ideas separately. The sketches were later matched.

2.3 Analysis

All the subjects had exposure to group work. While chemistry within a pair was important, it was somewhat controlled in the experimental procedure. In Group 1, the creative initiative was totally with BD. But in Groups 2 and 3, though both partners had to share the initiative, there were differences in the degree of active involvement. In Group 3, the BD dominated the nature and pace of design development, but did consult the SD when unsure. In Group 2, the initiative was more evenly distributed with a better exchange between the partners.

The next few sections present some of the key observations based on the analysis of the transcribed protocols of the three sessions. References to the time stamp in the later sections refer to the time counter of the video recording (Hr:Min:Sec). These observations are selected keeping the focus on strategies used in taking and sharing spatial decisions, abilities and difficulties in sharing ideas and concepts, different roles of SDs between groups, and problems in communication within the pairs.

3. Controlling the Contents of the Mind's Eye

3.1 Working Blindfolded is not a Constraint

Individual differences in adjusting to wearing an eye-mask and thinking aloud did exist. For example, BD in Group 3 was totally comfortable from the word "go", the BD in Group 2 took only a few minutes to reorient, while the BD in Group 1 took longer. Minor difficulties did persist. In Group 3, the BD spoke very fast and quickly moved from one decision to another leaving the SD little time to record and reflect on the ideas. This prompted the SD to request frequent repetitions. The BD in Group 1 on the other hand would lapse into silence and had to be prodded back into thinking aloud. Occasionally, the BDs would neglect to describe features they considered obvious. For instance, the BD in Group 3 neglected to discuss the orientation of the pitched roof. So, in their final sketches there was a mismatch.

Need to develop ideas in quick sequence made it difficult for both the partners to remember some of the previous decisions. The BDs had to either depend on memory or rely on their partners to fill in the details. In Group 3, the BD forgot about the decision to not have a spiral staircase. (See Fig. 1). The SD in Group 3 sketched very little of his, or his partner's ideas during session. As a result, he too had to unnecessarily rely a lot on his memory.

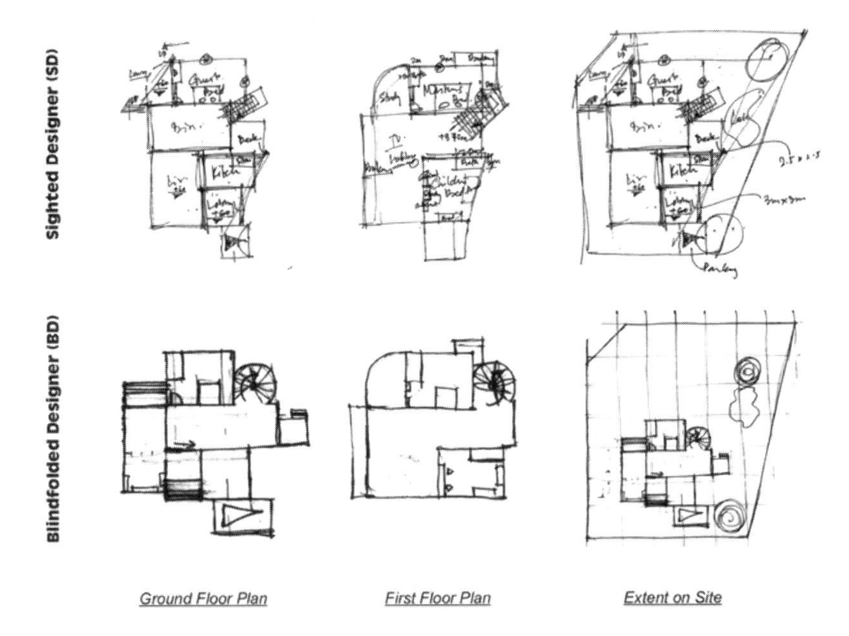

Fig. 1. Group 3 – Plans drawn at the end of the session. The BD continued with the earlier obsession for the spiral staircase, which the pair had agreed to drop midway. Although SD got the layout right, the extent of his house on the site is incorrect

Most were able to manage individual spaces and the interconnectivity at the same level, but the difficulty in coordinating spaces between two floors again showed individual variations. The BD in Group 1 had more difficulty coordinating than the other BDs. Surprisingly, despite the availability of sketching the SD in Group 3 had more trouble coordinating between floors than the BD. It should be noted that the BDs of Groups 2 and 3 had a trickier job as they had to heed to design suggestions from their sighted partners.

3.2 Mind's Eye Responding to the External Inputs

In Groups 2 and 3, the SDs were active participants in the design process. It was expected that active and creative contributions from SDs would make the BDs simulate these ideas in his mind's eye, comment on them, accept the good ideas and discard the rest. This was the only way available for BDs to share the SDs' ideas. How effective was the idea sharing? Protocols suggest that BDs did not find it too difficult to alter the contents of their mind's eye. In fact, they did so quickly and narrated their reactions to the SDs. To illustrate this critical point, we have selected two episodes from the early and later part of the protocol:

Group 3 – [0:08:44] Protocol clip from the early phase of designing:
BD – Huh? You want to put the living room back there? Won't we be putting it in the front? I don't think the size is that big that we can extend it back like that

(gestures) [SD - Make it double volume…double height...double height of the view from the bedroom] BD – Double height? You want the bedroom to be on the ground floor and have a double height? [SD – No, double height over the lake and the area around it] BD – Oh yeah!! (very excited) We can cantilever a bedroom over the lake, how's that?!

Group 3 – [0:58:04] Protocol clip from the later phase of designing, dealing with details:
BD – The bed is here, the balcony is here…alright? And…uh…where was the cupboard? [SD – To your left] BD – To my left? Along this? (gestures) [SD – Next…next to that bathroom] BD – But then that's the way to go to the study, no? [SD – Yes, so we'll have the cupboard wrap around the study door] BD – Oh! Smart guy! (laughs)

3.3 Fragile Imagery and the Conversation-mode

The images in the mind's eye are difficult to retain and focus on. The idea of pairing was conceived to overcome this problem. The SDs' presence and intervention was expected to help the BDs maintain focus. Did this happen? In Group 1, the SD was asked to remain neutral and only request BD for clarifications. Even this level of interaction with intermittent requests for details by the SD allowed the BD to maintain the image and work on it based on simulations in the mind's eye.

Group 1 – [1:24:35] Protocol clip of interaction. SD helps in maintaining focus: [SD – Which part - the clean end? Clean end along which part?] BD – *The East part...East part would be a clean end and inside would be...* [SD – How would it be? Clean end meaning?] BD – It would be a straight end [SD – Okay] BD – *The chamfer that happens, that would be along the building* [SD – And where would the chamfer be in terms of space?] BD – *It will be in the kids bedroom. That would become the area where they study* [SD – Okay]

3.4 Sharing a Common Architectural Space

Similarity of the final sketches indicates that ideas, concepts, and decisions were shared by the BDs and the SDs to a great extent in all the three groups. The degree of similarity in the final sketches does vary, but overall similarity is noticeable in sketches of all groups. (See Fig. 1, 2). Not that there were no differences in the final group outputs. The extent of the building on the plot was a major source of difference, primarily because of the difficulties in handling dimensions. Though in Group 3, the BD's final description showed an accurate understanding of the extent of the building on the plot, the SD got it wrong. (See Fig. 1). In Group 1, on the ground floor, the variations were in the location of study and the extent of the kitchen, store and deck. Some of these could be attributed to problems in communication between the pair.

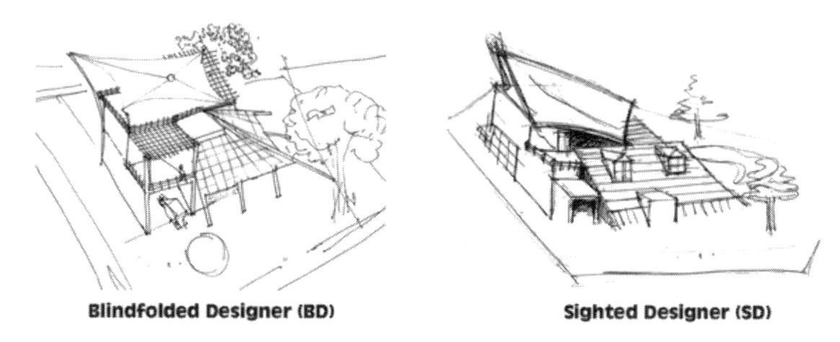

Fig. 2. Group 2 – Perspective drawings developed at the end. Despite the complexity of the roof, the similarity is apparent

3.5 Immersive Walkthrough as a Design tool

The BDs kept switching between the finger-drawing "plan" mode (where they used the gesture of drawing on paper) and the immersive "walkthrough" mode. The latter appeared to be the dominant mode of design, where the designers worked their way and organised spaces from inside. In Group 1, so detailed was the account of the inside spaces, that the BD had actually "seen" the water body reflect sunrays on the glass wall (Group 1 – 0:17:25).

Eye masks seem to have forced BDs to use walkthrough as a design tool. Having created a virtual site in their mind's eye, they walked through these spaces, continuously orienting themselves with respect to new design elements that they added along the way. Most popular site orientations were: [a] The way the site-plan was shown to them, and the north-south convention shown on the plan (this was understandable). [b] The direction from which each pair decided to enter the house. [c] Orientation with respect to the site landmarks. E.g.: Valley-side, Garden-side, etc. However, all this would suddenly change during the mental walkthrough, and that too without warning the partners. The orientation of the site often changed with reference to the BD's position at a given point in time during a walkthrough experience. As a result, the West, or valley side became "up," changed to "left," or "there" depending on where the BD imagined himself to be. In Group 3, even after setting up a directional convention, the partners invariably drifted from it (Group 3 – 0:07.26, 0:09:42).

The walkthrough mode of design had its price. In most cases, even when the partner consciously shifted the BD's attention to the outside of the building, the BD would soon return to the indoor spaces. (e.g. Group 3 - 0:51:20) The external details of the building often remained sketchy. Clearly they could stay "outside" just as easily, but they preferred to be indoors most of the times. For instance, while discussing issues related to the entry gate, the driveway and the entrance to the house, the BDs were "outside" the building. They delved into the "inside spaces" of the building more by a choice than due to the constraint imposed by blindfolding.

3.6 Words as Tool to Communicate the Nuances

To the BDs, words as a form of representation in team communication offered unintended advantages. They allow the essence of the central thought to be captured by using metaphors and analogies that evoke strong imagery. For instance, the BD in Group 1 conceived a roof that "snakes down" (like a snake hood) (0:15:30) and treated the space around the inner courtyard as the "heart of the house" (0:19:55). Words also often captured the early image driven obsessions that the BD experienced. An example of ambience as captured by the BD in Group 1: BD – *The feeling in it is very calm and...yet... there are those bright splashes of colour like the cerulean blue ramp and painted yellow wall* (Group 1 – 1:37:38) During the early stage of ideation, words accompanied by images allow complex visio-spatial ideas to be represented easily without careful detailing and design commitment. However, it is not easy to establish whether the words prompt the imagery or the other way round.

Part II

Part I discussed the results of the pilot experiments conducted to explore the technique of getting the architecture students to creatively use their mind's eye. The pilots offer reasonable evidence that the private experiences of the mind's eye can be shared and that the students can handle design tasks, at least of lower complexity, in their mind's eye. The next task was to test if a similar context could be created in a design studio. Part II explores the possibility of the idea being extended to the design studio environment (classroom) and addresses the question – "Can the mind's eye be used as a classroom tool by students of architecture and design?"

4. A blindfolded Classroom?

These explorations were conducted more as a design workshop with architectural students voluntarily signing up for it. The students were initially given puzzles that required mental manipulation (transformations like rotate, flip, move, etc.) of images in the mind's eye. The students were later explained how and why it is possible to use the mind's eye to perform number of spatial operations and the advantages of relying on it.

Students were divided into 17 pairs and were given a similar assignment to be solved. The problem solving session used the structure identical to the pilots. In nearly half the pairs, the SDs acted as active partners. Video camera recorded clips of the pairs working together, which was later used in a post-session discussion to recall selective experiences.

The situation changed dramatically in the classroom environment, something we had not expected at all. Some of the key observations are briefly stated below,

- Interactions within the pairs were more intense. Students as a group were less inhibited. While interacting, they used vigorous gestures and body movements. They were also not disturbed by the collective think-aloud din in the room, nor were they concerned about others. It was only after the session was over that they realised how exhausting it was to concentrate on the imagery for nearly 45 minutes.
- Most students found imagery to be a pliable medium to work with. One student mentioned how quickly he could shift walls and other elements and experience the change. In the virtual space that they created, students inserted and removed people at will. They saw the businessman (client) and children moving around in the spaces. Some students saw the details of the dresses in the imagery but the faces were indistinct. Some of this also bordered on involuntary fantasy. For instance, the BD in one pair reacting to the SD's decision of putting the guest bedroom on the ground floor, saw his guest stealing things and children sneaking out of the house.
- The normal convention in description of directions was in relation to landmarks of the site (e.g.: valley-side). Yet, BDs would fall back on (their) "left", "right" and SDs would often use north, south etc. One of the BDs turned around literally when explaining about the valley "behind" her. The SD followed suit.
- In most of the groups, the partners sat at right angles to each other or occasionally opposite each other, but BDs freely changed their seating orientations. Interestingly, one pair solved the problem of communicating the site orientation by sitting next to each other and facing the same side.

5. Conclusions

This paper does throw light on the ability of the student members of the pair to share a common architectural creation. Pairs seem to have reasonably similar visions, suggesting that the private experiences of the mind's eye can be shared. All the BDs could evolve strategies by creating a virtual site in their mind's eye, so that they could manipulate virtually built spaces around them. Immersive qualities of mind's eye experiences tended to be very addicting. While it ensured total and focussed involvement, it had also some limitations. Once inside the space, students did not easily come out and look at their creations from outside. Pairing also seems to help as partners force BDs to maintain a continuous focus and not drift away from current problem.

These explorations also suggest that the technique can potentially become part of the classroom interactions. At best these efforts can be treated only as a first step towards making mental imagery as a pedagogic tool. Post-session discussions gave an indication that students enjoyed this way of working and also found that it allowed them to fully concentrate on the problem. (Once the eyes are closed and the partner is helping you to focus, there are no distractions.) However, much will depend on whether the technique makes qualitative difference in the ideas and end results. Besides, how a teacher can intervene in the interaction and influence the direction of thinking or comment on is still ambiguous.

References

Athavankar UA (1997) Mental Imagery as a Design Tool. Cybernetics and Systems, vol 28 (1), pp 25-42

Athavankar UA (1999) Gestures, Imagery and Spatial Reasoning. In: Garo JS, Tversky B (eds) Visual and Spatial Reasoning. Preprints of the International Conference on visual and spatial Reasoning, (VR 99) MIT, Cambridge, June 15-17, 1999, pp 103-128.

Athavankar UA, Garde A, Kuthiala S (2001) Interventions in the Mental Imagery: Design Process in a Different Perspective. Proceeding of the 5th asian Design Conference, International Syposium on Design Science, Seoul National University, Korea.

Athavankar UA, Gill N, Deshmukh HS (2000) Imagery as a Private Experience and Architectural Team Work. In: Scrivener S, Ball LJ, Woodcock, Springer-Verlag (eds) Collaborative Design. London, pp 223-232

Finke R (1990) Creative imagery: Discoveries and Inventions in Visualisation. Lawrence Erlbaum, New Jersey

Fish J, Scrivener S (1990) Amplifying the Mind's Eye: Sketching and Visual Cognition. Leonardo, vol 23 (l), pp 117-126

Goldschmidt G (1992a) Serial Sketching: Visual Problem Solving in Designing. Cybernetics and Systems, p 23, pp191-219

Goldschmidt G (1992b) On Figural Conceptualisation in Architectural Design. Cybernetics and System research '92. World Scientifics Singapore, pp 599-606

Goldschmidt G (1994) On Visual Design Thinking: The Vis Kids of Architecture. Design Studies, vol 15 (2), pp 158-174

Kosslyn S (1983) Ghosts in the Mind's Machine: Creating and Using Images in the Brain. Norton, New York

McKim (1972) Experiences in Visual Thinking. Brook/Cole, California, p 972

Roskos-Ewoldesen, M J Intos-Peterson, and Anderson R (eds.) (1993) Imagery, Creativity and Discovery. North Holland, pp 123-150

Singh A (1999) The potential of mental imaging in architectural design process. In: Proceedings of International Conference on Design and Technology Educational Research and Curriculum Development, IDATER 99. University of Loughborough , England, pp 230–236

Sommer R (1978) The Mind's Eye: Imagery in Every Day Life. S. Delta book, New York

Analysis of solution finding processes in design teams

Petra Badke-Schaub, Joachim Stempfle
Institute of Theoretical Psychology, University Bamberg

1. Introduction

Technical development and globalisation impose problems on engineering design which require highly efficient processes due to time, quality and cost. Products, processes and knowledge change so quickly that there are hardly any standard procedures or routine solutions adequate for various situations. Overnight the best solutions may become inappropriate if parts of the problem context are changing. This situation is even more difficult because the designer is integrated in a social context, and thus, processes of division of labour, organisation of tasks, planning of interfaces, as well as continual linking individual work with group work are concurrent requirements in product development.

As a consequence, we can state that engineering design is a complex problem solving process with the necessity of coping with different types of Critical Situations related to the design context and to the social context (Badke-Schaub and Frankenberger 1999).

2. Theoretical Background

Commonly, a problem can be characterised by three components, a given situation, a goal and a barrier between both situations (Dörner 1996). Applying these global problem characteristics to *design teams*, we have to distinguish further between two focuses of action which we call *content* and *process*. With this differentiation we take into account that design teams are not only required to deal with the task itself, but also have to direct parts of their activity at structuring and organising the group process.

We propose that content and process-related activity of design teams can be described in similar terms, with the same steps of thinking operations underlying both processes (see also Fisch 1994). In table 1 the six steps related to the content and to the process are described.

Table 1. Steps in the problem-solving process, related to the content and the group process

Steps	Content	Process
goal clarification	dealing with the goal space: analysing goals and setting goals	goals related to the group process (e.g. distribution of tasks)
solution generation	proposals and solution ideas concerning the design task	proposals and solution ideas concerning the group process
analysis	questions and answers related to solution ideas	questions and answers related to the group process
evaluation	positive and negative evaluations related to the solution space	positive and negative evaluations related to the group process
decision	decisions for or against a solution or a proposal	decisions concerning the group process
control	control of the implementation of solutions	summary or control of group members' work.

3. Empirical investigation of design processes in groups

3.1 Data collection: field and laboratory studies

In a five-year-long field study engineers of the TU Darmstadt and psychologists of the University of Bamberg observed and analysed ten projects in four organisations (Frankenberger 1997; Badke-Schaub and Frankenberger 1999; Wallmeier 2001)[1]. In order to comprehend rules and deciding factors of cooperative design processes the observed design work was divided into routine work and types of Critical situations regarding their aim in the design problem context, (goal analysis and goal decision, solution search, solution analysis and solution decision) and in the social context (conflict-management and disturbance-management). Spread over all ten projects 895 Critical Situations were categorised and analysed. In relation to the frequencies, solution analysis turned out to be the most frequent type of Critical Situations. Thus, the question arises: What happens during solution analysis in groups? Although generalisations from student teams to design teams in industry must be drawn with caution, it is suitable to use laboratory studies for specific research questions. In the controlled setting of the laboratory we expected to gain insights into thinking processes which are not contaminated by additional unpredictable factors which occur in a field setting.

[1] This research was supported by the DFG (German Research Foundation): DO 200 12-1/2 and DaO 200 19-1.

Therefore the subsequent study was based on a the analysis of three laboratory (student) groups, solving a complex design problem extending over six hours. In cooperation with the Department of Product Development and Machine Elements of the TU Darmstadt[2] the teams, comprised of 4-6 students majoring in mechanical engineering, had to design a mechanical concept for a sun planetarium. This sun planetarium should be able to visualise the way of the sun across the sky for different positions on the hemisphere as well as for different seasons.

3.2 Data analysis and evaluation

Protocol analysis was used in order to capture the design thinking process in the groups (see also Goldschmidt 1996). Team communication has been completely recorded and analysed sentence-by-sentence in all three teams. Based on the assumption that communication provides a prime access to the thinking and problem-solving process of design teams, a multi-level coding system has been developed for the analysis of the recorded data. The coding system was designed for use not only in a design context but also for different problem-solving domains (economics, ecology, etc.). On the highest level, the coding system reflects the two main focuses of activity, *content* and *process*. Under each activity focus the steps in the problem-solving process, related to the content and the group process (Table 1) are being attached. Each step is then defined further by several actions. In this paper, only results from the top two levels of analysis, *activity focus and steps*, are reported.

4. Results

4.1 Analysis of frequencies

Analysing the frequencies in which the different communicative acts occur provides a basic understanding of the role of the different design steps in the design process. The important result about the frequencies of communicative acts under the two main activity focuses *content* and *process* is that all three groups show a similar distribution. This distribution can roughly be described by the "2/3-rule": In 2/3 of their communication design groups deal with the content whereas 1/3 of the group communication aims at structuring the group process. Similar results are reported from problem-solving groups in non-design contexts (Fisch 1994; Beck 2001).

On the level of steps in the design process (see Table 2, which displays the frequencies of steps as an average of the three groups in per cent) the distribution of communicative acts among the steps is quite similar in the three groups as well, with a medium correlation of .98 between the three groups. In the three observed

[2] We thank the participants of the study and our colleagues at the Technical University of Darmstadt, Dr.-Ing. Stefan Wallmeier and Dipl-Ing. Robert Lüdcke.

teams (groups averaged for these results), most of the team communication is concerned with the *analysis* of both the content (46%) and the process (17%,). The next frequent category is the evaluation of the content (13%) followed by goal clarification (7%) and by the evaluation of the process (5%). Overall, about 7% of team communication is related to the goal space, 63% is related to the solution space and in nearly 30% of their communication the observed teams deal with the group process. The two operators with the highest quantitative importance when dealing with the solution space are analysis and evaluation.

Table 2. Frequencies of design steps in the three groups as an average

Steps	Content	Process
goal clarification	7%	0,2%
solution generation	3%	0,8%
analysis	46%	17%
evaluation:	13%	5%
decision	1%	5%
control	0,2	1%

Regarding the frequencies of the categories related to the focus of activity and related to the steps in the design process the thinking processes of the observed design groups cannot be very much differentiated. The question that remains is how do groups go about in their design process? What are the similarities and the differences between the processes of the different groups?

4.2 Process analysis: A macro-perspective

The process analysis under a macro-perspectives is related to the occurrence of the activity focus and the design steps over the whole period of the design work. In all three groups a distinct period in which the group focuses mainly on the process can be discovered with the remaining time of group work dedicated to the content. However, the stage in which this period occurs differs between the three groups. In group 1, the process-centered period occurs at the beginning of group work; in group 2 this period occurs at halftime, whereas in group three this period does not occur before the very end of group work. This finding is rather remarkable. The three group's proceeding in the design task can be labelled from „chaotic“ (group 3) to „planned“ (group 1). Fig. 1 depicts the *sequence* of communicative acts according to the two main focuses of content and process.

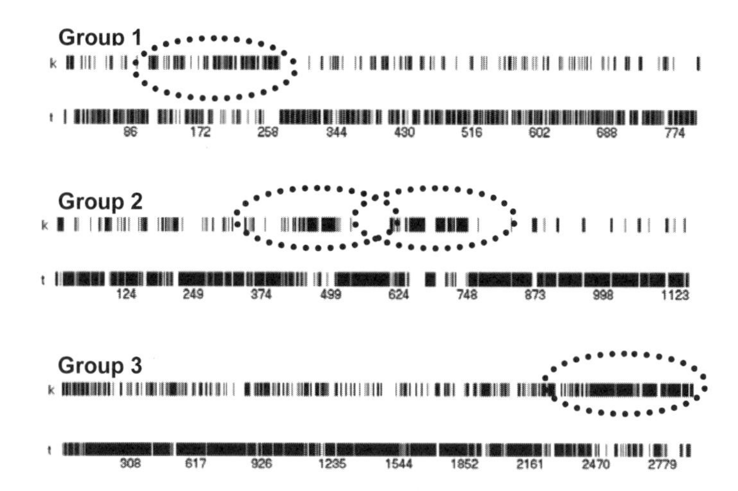

Fig. 1. Activity focus over the course of design work (k= process, t= content)

In empirical investigations Gersick (1995) found that project groups usually work with a muddling-through strategy until halftime of the task completion or until critical events occur which make a (re-)structuring process in the group necessary. These findings can not be replicated in our investigation. Obviously, the three groups, although they spend nearly the same amount of communicative acts on content- and process-related topics, evolve different strategies in order to solve the two main requirements content and process.

Stating the differences related to the activity focus of the groups the next question is about the distribution of design steps over time. In accordance with prescriptive accounts of design methodology and theories of problem-solving the data reveal that all three observed groups focus on the goal space in early stages of their work, whereas in later stages the focus shifts onto the solution space. Concerning goal clarification, however, a significant difference between the groups can be observed: whereas in group 1 and group 2 virtually no goal clarification takes place in the second half of the design process, in group 3 goal clarification decreases towards the middle of the design process but increases again in frequency towards the end. group 3 is the only group that, in the end, takes a second look at the requirements as stated in the goal requirements. Beyond the preference of goal clarification in the early phase of the work no systematic accumulation of design steps can be detected at any specific stage of the design process related to the solution space.

4.3 Process Analysis: A micro-perspective

In order to analyse the transitions in detail, two-step-sequences of design steps have been examined. The question to be answered is: Is the design process "chaotic" in the sense that any sequence of design steps is likely to occur or are there regularities with one step systematically following after another specific step? In order to answer this question, the transition probabilities between all steps have been calculated and then compared to the baselines of the steps. If, for example, content analysis occurs in 46% of all team communication, but after a content analysis in 55% of all cases another content analysis follows, this means that sequences of content analysis are highly likely to occur in team communication. A Chi²-test allows for calculating whether the observed transition probability is significantly higher compared to the baseline of the categories.

Fig. 2 displays *transition probabilities between the two communication focuses of content and process* in the three groups. In the figure a connection that ends with an arrow displays a transition that is significantly more likely to occur compared to the base rate (as calculated by a Chi²-test). A connection that ends with a straight line displays a transition that is significantly less likely to occur compared to the base rate. The first number behind the connection represents the transition probability, the second the base rate probability.

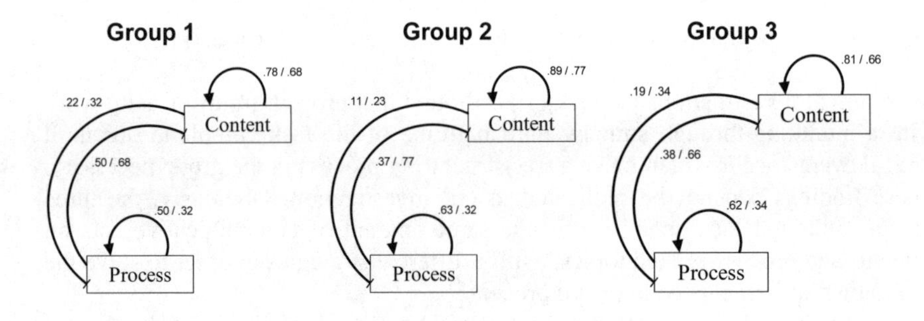

Fig. 2. Transitions between activity

As Fig. 2 reveals, in all three groups *a transition within the same focus of action* is highly likely whereas *a transition to the complementary focus of action* is highly unlikely. In other words: Once teams are dealing with either content or process, they tend to stick to the focus for several communicative acts before switching to the complementary focus. These findings reveal that the design process in teams is best described by *a constant interweaving of content-directed sequences with process-directed sequences*, both of some duration.

Furthermore, *transitions between design steps* have been analysed in the three teams. Concentrating on content-directed activities Fig. 3 displays transitions between design steps.

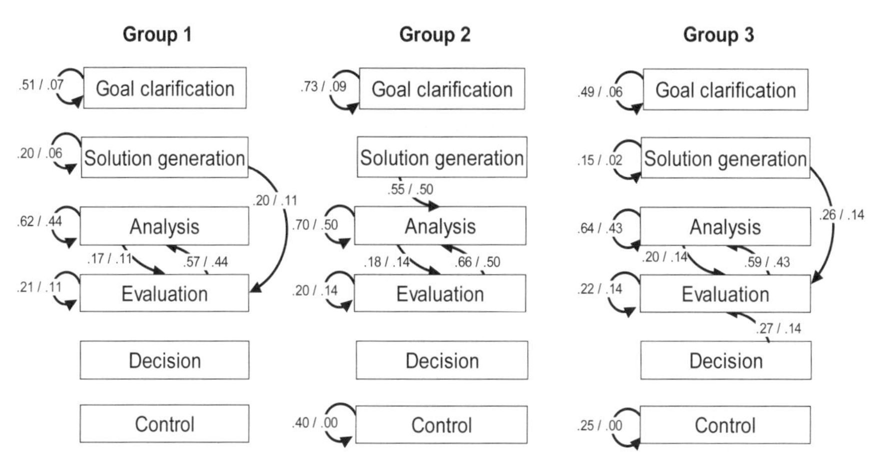

Fig. 3. Transitions between design steps related to content-directed activities

Three characteristics can be concluded from these data:

1. *Steps as significant elements in the design process:* For every design step – except the step 'decision' – a transition *within* the same step is highly likely to occur. The proposed design steps thus seem to be steps in the sense that groups tend to spend more than one communicative act on the same step before moving on to the next step.
2. *Feedback loops between analysis and evaluation:* There is a remarkable feedback loop between analysis and evaluation. The repeated loop of analysis and evaluation thus seems to represent the core of the thinking process in the observed teams. Analysis enables the teams to widen the solution space, evaluation serves to narrow it down again. The switch between analysis and evaluation might enable design teams to keep the size and thus the complexity of the solution space at an acceptable level by at the same time gaining information about the solution space.
3. *Fate of solution ideas:* In two out of the three observed groups, a new idea is highly likely to be followed by an immediate evaluation, not by analysis. Only group 1 frequently progresses from solution generation to analysis. From a methodological perspective, one would expect new ideas to be followed by a thorough analysis, not by an immediate evaluation. Our results show, however, that groups frequently do not follow such recommendations, but *do* evaluate solutions immediately without prior analysis. From a theoretical viewpoint, an immediate evaluation of solution ideas without prior analysis must be regarded as a severe drawback during the process of solution generation. Two errors are likely to occur. One is premature rejection of a solution because it does not seem to fit the constraints of the task structure. A second error likely to result from premature evaluation of solution ideas is the premature adoption of a solution that proves problematic later on. Both errors are grave. In the first case of premature rejection of a solution idea, an ingenious idea may be discarded and lost for the group. In the second case of premature adoption of a

problematic solution groups may spend considerable time working out a solution only to find out later on that the solution does not work. Given the proposed negative effects of premature evaluation of solution ideas, how can we explain the findings in groups 2 and 3?

5. Conclusions

5.1 A two-process-theory of thinking in design teams

The results described above disagree with theories on rational decision making which are based on the image of the decision process as a consciously deliberated course and thus sequence of steps. In contrast the data point to the fact that humans rarely strive for *optimum* but rather for *"satisficing"* solutions (Simon 1955). What humans mostly do is a sequence of several consecutive screening-processes consisting of frequency and availability heuristics in order to reduce complexity, thus enabling quick actions in complex environments (see Dörner 1996; Gigerenzer and Todd 2001). As a consequence, humans often generate only one, at best some few alternatives, which they know from similar situations in the past. An early rejection of solution ideas reduces complexity, time and cognitive effort and thus can save a lot of time and energy (see also Ehrlenspiel in this book). We call the sequence of solution ideas being followed by immediate evaluation *process 1*. The decisive factor of process 1 (see Fig. 4., left hand side) is that solution ideas are followed by an immediate evaluation. Only, if there are questions or misunderstandings in the group, an analysis might take place before evaluation. If the quick evaluation of solution ideas yields a positive result, the solution is accepted. If the solution is discarded, new solution ideas will be sought. If no more solution ideas can be found, however, then already discarded solution ideas may be reconsidered and analysed in more detail. This process provides at least two advantages: Firstly, a solution can be decided on very quickly. Secondly, process 1 will not threaten the collective self-efficacy of the group as only few questions are being raised and the problem seems to be solved easily. In contrast, thorough analysis brings up complexity and uncertainty and reduce the feeling of competence and self-efficacy in the group. But the more complex a problem, the more likely errors will result from applying process 1 because due to limited cognitive capacities of humans, it is unlikely that designers will be able to take into account different aspects in an immediate evaluation of the solution idea. This may lead to a premature adoption of a poor solution. For these reasons, we propose that process 1 is effective for well-defined problems, whereas with increasing complexity of the problem at hand it is more likely to result in failure. Under certain conditions however, teams employ a different thinking process which we call *process 2* (Fig. 4., right hand side). This process very much resembles a structured design process as stated in design methodology. In process 2, one or more ideas are generated, then analysed and only evaluated after analysis has taken place. If no workable solution is being found, either new solutions are generated or the already discarded solutions are further analysed in order to find possible transformations serving to turn them into workable solutions. As well as process 1, process 2 also has its specific advantages and shortcomings. One important shortcoming is that this process needs much more time and cognitive effort than process 1. Analysing solutions takes more time than a quick screening process. However, process 2 will minimise the risk of an erroneous solution. Thus, the more complex the design task, the more successful process 2 will be, compared to process 1.

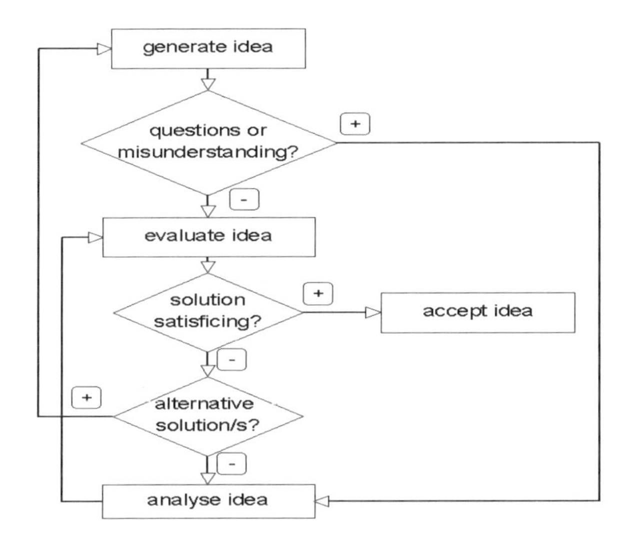

Fig. 4. Idealised thinking processes: process 1

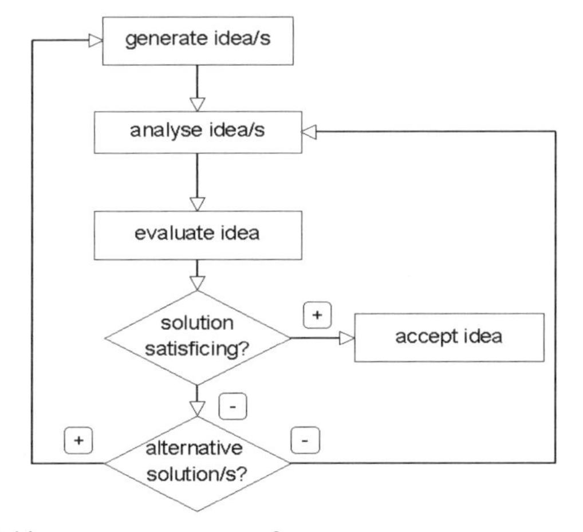

Fig. 5. Idealised thinking processes: process 2

If there are two distinct processes which teams can use in tackling a design task, what factors are responsible for the group's choice of which process to use?

This study does not provide enough data in order to be able to answer this question on an empirical basis. However, based on theoretical deliberations we want to propose five important conditions that will cause a group to shift from process 1 to process 2:

1. *Lack of common understanding:* In heterogeneous groups different levels of understanding provoke questions, thus causing the group to go into a thorough analysis prior to evaluation.
2. *Disagreement:* Challenging of ideas can lead to a more careful analysis of solutions.
3. *Failure of process 1:* If a team has problems coming up with a workable solution, the lack of new solution ideas often leads to a deeper analysis of the so far discussed solutions.
4. *Methods:* In order to avoid premature judgement methods such as brain storming and brain writing explicitly separate the processes of solution generation, analysis and evaluation.
5. *Self-reflection:* Reflecting one's strategies in tackling the problem may lead teams as well as individuals to consciously realise that they are stuck in a process-1-approach. From this point, teams or individuals can alter their proceeding and their strategies.

5.2 Implications for education and practice

Until now design methodology has not been as readily accepted in industry as design methodologists have expected. In our point of view, methodology should not start from a normative point of view, but from where practitioners are now, taking into account the constraints the practitioner faces in his everyday work, such as time constraints, financial constraints, cognitive overload through multiple projects, etc. Research conducted in cognitive psychology has provided many results that illustrate how humans deal with complex problems under varying conditions. These findings could provide the basis for a 'practitioner-centered' design methodology. What is necessary in education is to take into account aspects of group dynamics as well as the various conditions designers are confronted with in the industry. As a general recommendation, designers should not only be taught specific methods, techniques or tools that can help to structure the design process. What is much more important is to teach designers to reflect on their own strategies and heuristics in dealing with design problems. There are several studies which indicate that the analysis of one's own thinking process are a prerequisite for modifying inadequate thinking processes. Individual designers and design teams need to be able to assess the conditions of the given situation quickly and to flexibly adjust their own path of action depending on the requirements of the situation. This flexibility can not be taught, but must be learned through experience and self-reflection. However, education can emphasise the importance of continuous self-reflection, thus enabling the future designer to become a true reflective practitioner.

References

Badke-Schaub P, Frankenberger E. (1999) Analysis of design projects. Design Studies 20: 465-480

Beck D (2001) Sozialpsychologie kollektiver Entscheidungen. Westdeutscher Verlag, Wiesbaden

Dörner D (1996) The logic of failure. New York, Metropolitan Books

Ehrlenspiel K (2003) Integrierte Produktentwicklung. München, Hanser

Fisch R (1994) Eine Methode zur Analyse von Interaktionsprozessen beim Problemlösen in Gruppen. Gruppendynamik 25: 149-168

Frankenberger E (1997) Arbeitsteilige Produktentwicklung - Empirische Untersuchung und Empfehlungen zur Gruppenarbeit in der Konstruktion. VDI-Verlag, Düsseldorf

Gersick CJG (1988) Time and transition in work teams: Toward a new model of group development. Academy of Management Journal 32: 9-41

Gigerenzer G, Todd PM (2001) Simple heuristics that make us smart. New York, Oxford University Press

Goldschmidt G (1996) The designer as a team of one. In Cross N, Christiaans H and Dorst K (eds) Analysing design activity. Chichester, UK, John Wiley & Sons, pp 65-91

Pahl G (1999) Thinking and acting in the engineering design process – Results of an interdisciplinary research project. Konstruktion 6: 11-17.

Simon HA (1955) A behavioral model of rational choice. Quarterly J. Economics 69: 99-118

Stempfle J, Badke-Schaub P (2002) Thinking in design teams – an analysis of team communication. Design Studies 23: 473-496

Wallmeier S (2001) Potenziale in der Produktentwicklung. Möglichkeiten und Grenzen von Tätigkeitsanalyse und Reflexion. VDI-Verlag, Düsseldorf

Processes for Effective Satisfaction of Requirements by Individual Designers and Design Teams

Amaresh Chakrabarti
Centre for Product Design and Manufacturing, Indian Institute of Science

1. Background and Problem

The purpose of design research is to support development of better products by developing a better design process. A design process is initiated with the recognition of a need, leading to the establishment of requirements for the intended product. Therefore, establishing requirements is essential, and should be a central issue in design research (Chakrabarti 1994). A requirement is defined here as characteristics which a designer is expected to fulfil through the eventual design. Several categories of requirements have been proposed and their importance proclaimed (Birkhofer 1992). Several methods are suggested in design literature for aiding requirement establishment, from check-lists and QFD methods to combination of both as a software support ((Pahl and Beitz 1988; Pugh 1991; Kruger and Knoll 1994). Design process has been investigated in many studies and requirements remain part of these studies (e.g., Blessing 1994; McGinnis et al. 1989). However, detailed investigation as to how requirements get identified, analysed and used in the design process and how these influence the quality of its outcome - the emergent design - has not been undertaken before. The findings outlined in this article are based on four separate but related empirical studies of the design process. The first two studies (Morgenstern and Knaab 1996; Chakrabarti et al. 2002; Nidamarthi et al. 1997; Nidamarthi 1999; Nidamarthi et al. 2001) were aimed specifically at understanding how requirements are identified and satisfied during design, while the other two were focused on how shape evolves during design (Baumgartner 1995; Chakrabarti and Wolf 1995). In this article, results from these four studies have been brought together to identify the activities and mechanisms for requirements identification and application, and evaluate their effectiveness in developing products with a high degree of requirement satisfaction.

2. Research Method

The first, preliminary study (Morgenstern and Knaab 1996; Chakrabarti et al. 2002) used think-aloud video protocols of the product development processes of four individual designers, all of whom solved the same design problem under laboratory settings. Encouraged by the results of this study, a second, more elaborate study (Nidamarthi et al. 1997; Nidamarthi 1999; Nidamarthi et al. 2001) was undertaken. This involved use of protocol study and participant observation in the following settings:

- 2 individual designers, one common design problem, laboratory setting, protocols
- 5 design teams, another common design problem, laboratory setting, protocols
- 1 design team, existing design problem, industrial setting, participant observation.

Although the two studies differed in duration and detail, the objectives in both were to

- Identify activities and mechanisms with which requirements are identified during a design process
- Identify activities and mechanisms with which these requirements are (not) attempted to be satisfied in the evolving product
- evaluate the effectiveness of these activities and mechanisms as to how well they contribute to the development of a product in which the requirements are fulfilled.

The results of the second study not only corroborated the results of the first study, but also provided more detailed, generic and further understanding.

Of the other two pieces of research analysed in this article, the first (Chakrabarti and Wolf 1995) was a preliminary study aimed at understanding shape evolution during design. It was based on the analysis of the evolution of a sub-assembly of a device developed by two designers. The design process was recorded by formal means, e.g., design matrix (Blessing 1994) as well as informal means such as designers' notebooks and minutes of the meetings; research methods included document analysis and designer interviews. Encouraged by the results obtained, a more detailed study (Baumgartner 1995; Chakrabarti 1997) was undertaken that focused on the entire design process for this device. Although differed in detail, the central objectives in both these studies were to:

- Understand why designers change their designs in the first place
- Identify what the design changes are
- Understand why the designs are changed the way they are.

The results of this second study provided more detailed and further understanding than the first by contrasting, corroborating and extending results of the first study. Since requirements and solutions co-evolve and influence each other, as noted in several studies (McGinnis et al. 1989; Nidamarthi et al. 1997), these two pieces of research, although primarily focused on mechanisms for shape evolution, have generated insight into the requirement satisfaction process, and helped explain some of the central observations made by the two studies focused directly on requirement satisfaction.

3. Results

The activities and mechanisms for requirement identification and application and their evaluations are reported in this section.

Activities and Mechanisms of Requirements Identification

Design involves both problem understanding and problem solving (Chakrabarti and Wolf 1995; Nidamarthi et al. 2001). Problem understanding is the earliest stage of designing, characterized by generation of solutions that are not seriously carried on into more detailed stages (Nidamarthi et al. 1997); existence of these solutions is also noted by Baumgartner (1995). Problem solving is characterised by serious evaluation and detailing of the solutions generated. Requirements get identified throughout the design process (Chakrabarti 1994; Chakrabarti and Wolf 1995; Nidamarthi et al. 2001), but mainly at the problem understanding phase (Morgenstern and Knaab 1996; Nidamarthi 1999), see Figs. 1-2. The number of requirements identified gets increasingly less (Morgenstern and Knaab 1996) and solution-specific (Nidamarthi et al. 1997) at the later design stages.

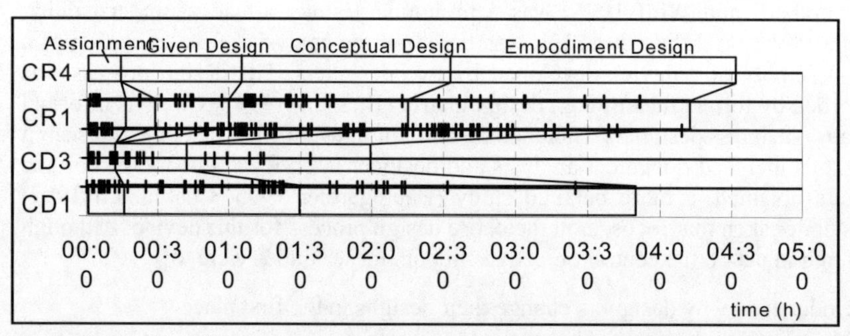

Fig. 1. Distribution of Requirements During Design (from Chakrabarti et al. 2002: CD1, CD3, CR1 and CR4 are identifiers for the four designers)

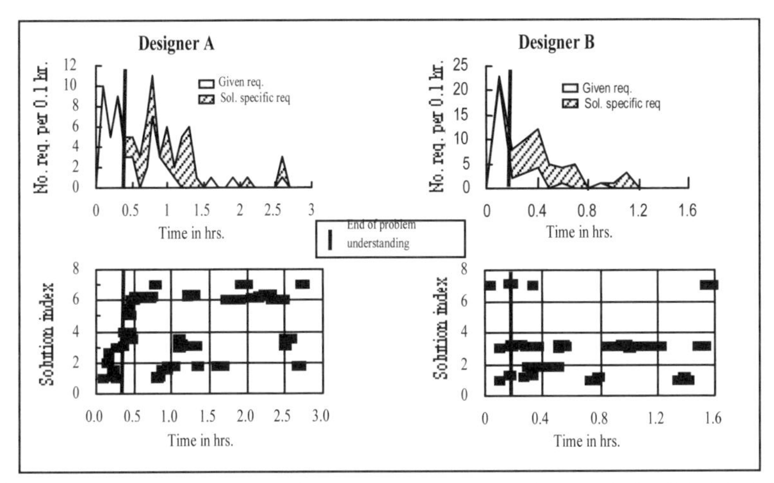

Fig. 2. Design history of Two Designers A and B: Comparative Analysis of Co-Evolution of Requirements and Solutions (from Nidamarthi et al. 1997)

The main source of requirements at the early design stages are the documents relating to requirements, e.g., design brief and available requirement specification (Chakrabarti and Wolf 1995; Morgenstern and Knaab 1996). Another source of requirements are design principles, e.g., simplicity, use of standard parts, reduction of material, etc., (Chakrabarti and Wolf 1995; Baumgartner 1995) that a designer believes any good design should fulfil. Requirements are also identified by considering various aspects of the solution that are important, and by considering the main, solution-neutral requirements, or design principles, in the context of the solution at hand.

Table 1. Useful and Harmful Activities and Mechanisms Observed in the Studied Cases (from Chakrabarti et al. 2002)

Generic activity	Sub-activities and methods	Quality
Identify	define requirements, study assignment and documents; inspect for completeness and consistency; identify related requirements; re-study	useful
Interpret	express in own words; transform units; summarise; recognise requirement's importance; imagine context, geometry, forces, motion, operation, production etc	useful

Evaluate	evaluate requirement, info, assignment, and document (qualitatively and quantitatively); study solution; realise misunderstanding, calculate, suspect gap in assignment, check requirements frequently	useful
Ask	ask questions, confirm suspicion, confirm understanding, clarify	useful
Remember	be reminded; note requirement/info; repeat requirement periodically	useful
Apply	detail requirement; recognise relation to other requirements, express requirement as solution; evaluate solution using requirement; search for faults in solution	useful
Others	use design principles, draw, sketch	useful
Not identify	not study document, not ask questions, vary importance of requirement without consulting or studying	harmful
Misinterpret	assume without consultation, evaluation or studying relevant materials; modify requirement in order to keep preferred solution	harmful
Forget	forget requirements, solution, assignment, info, documents or earlier decisions	harmful
Errors	copy errors, calculation errors	harmful

The main aspects considered that lead to identification of requirements are *functionality* (in particular geometry, force and kinematics), *manufacturability* (materials, production and assembly), *cost* and *use* (Baumgartner 1995; Chakrabarti and Wolf 1995; Morgenstern and Knaab 1996). The aspects were considered in a certain order: first feasibility (e.g., basic manufacturability) of an aspect is considered, followed by optimisation of the solution in that aspect (e.g., cheap manufacture) (Baumgartner 1995).

During the problem-understanding phase, requirements are identified through the broad activities of *identification*, *analysis* and *selection* of requirements, and sub-activities within each of these are as follows (Nidamarthi et al. 2001). Sub-activities in 'identification' are *perception*, *inference* (both these sub-activities are also noted in Morgenstern and Knaab 1996) and *modification*; those in 'analysis' are *visualisation*, *verification*, *weighing*, *questioning* (also noted respectively as *imagine*, *evaluate*, *find importance*, and *ask questions* in Morgenstern and Knaab 1996) and *relating*. The other activities noted are absence of, or mistakes in the activities, e.g., *not identify* a requirement, *misinterpret* a requirement etc. (Morgenstern and Knaab 1996), see Table 1.

As to how these activities change depending on whether designing is done individually or in teams is observed in Nidamarthi (1999). Most activities in individual design are limited to identification, with some analysis, and little in choice of requirements. Very little questioning is observed. The main activities in design teams are identification and analysis of requirements, with design effort in analysis slightly higher than in identification. Very little time is visibly spent in selection activities. Problem perception and questioning are the most significant activities.

Activities and Mechanisms for Requirements Application

Requirements are applied mainly at the problem solving (i.e., conceptual and embodiment) stages of development (Morgenstern and Knaab 1996; Nidamarthi 1999). The number of requirements applied is always less but proportional to the number of requirements identified (Morgenstern and Knaab 1996). Requirements at these stages are increasingly solution-specific (Nidamarthi et al. 1999), see Fig. 2.

Designers apply original, solution-neutral requirements by making them more solution-specific as they progress through the design (Nidamarthi et al. 2001). Requirements and solutions co-evolve (see Fig. 2.) and influence each other (Morgenstern and Knaab 1996; Nidamarthi et al. 1997; also observed by McGinnis et al. 1989). Problem solving happens in small episodes in which only one issue is considered at a time (Baumgartner 1995; Chakrabarti and Wolf 1995) until the solution reaches a satisfactory level (Baumgartner 1995). During these episodes, designers think in small steps, mainly at the component or sub-systems level, in terms of shapes, often drawing incomplete sketches focussing only on a feature relevant to the current issue (requirement); aspects of a solution are considered first in terms of its feasibility with respect to the aspect and then optimised, often for functionality first followed by manufacturability (Baumgartner 1995). Further requirements arise when a requirement is considered against the status of a solution in current, design context (Chakrabarti and Wolf 1995). Solution specific thinking is essential for application of requirements (Chakrabarti 1994; Nidamarthi et al. 1997).

During problem solving stages, requirements are applied through the broad activities of solution *generation*, *evaluation* and *selection*, and sub-activities found in each are as follows (Nidamarthi et al. 2001). 'Generation' involves *creation*, *modification* and *detailing*, 'evaluation' involves *characterization*, *relating*, *verification* and *questioning*, and 'selection' involves *comparison*, *decision*, or *identification* of further things to do. The other activities are absence of, or mistakes in activities, such as *not remember* or *not apply* requirements, etc., (Morgenstern and Knaab 1996) see Table 1.

As to how these activities change depending on whether designing is done individually or in teams is noted in Nidamarthi (1999). The main activity in problem solving for individual designing is generation of solutions, with the next being evaluation; visibly, the least amount of time is spent in selection of solutions. The main activities in problem solving for team designing are generation and evaluation of solutions, with evaluation effort significantly more than generation. Effort spent in selection activities is less, but it is still a significant proportion of the total time spent in designing.

Effectiveness of the Activities and Mechanisms

In general, the higher the effort in understanding and applying a requirement, the higher is its satisfaction in the product developed (Nidamarthi 1999). More specifically, requirement satisfaction success needs continuous attention to requirements. The failed requirements are those that have been neglected by designers for a considerable period of time; while the successful ones are considered continuously (Nidamarthi et al. 2001).

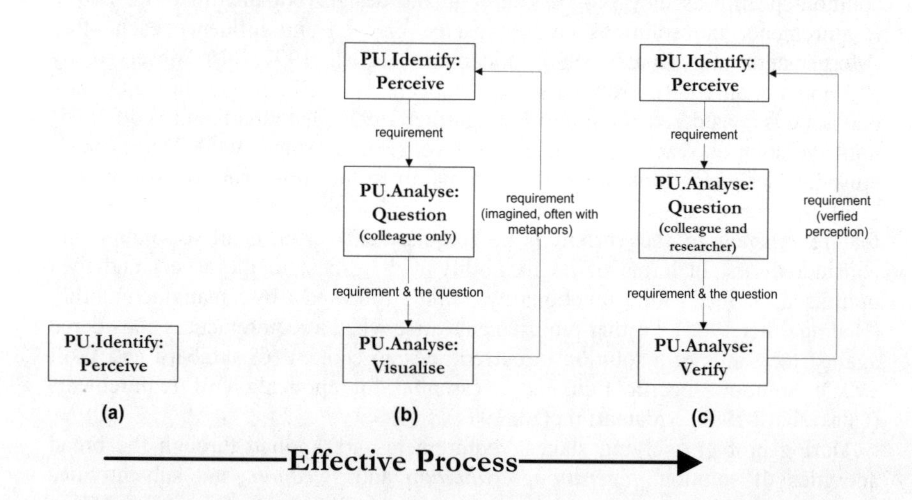

Fig. 3. Efficacy of Requirement Identification Mechanisms (from Nidamarthi et al. 2001)

This phenomenon can be explained from that fact that in a problem solving episode, only part of the design is changed using one or few of the requirements (Chakrabarti and Wolf 1995; Baumgartner 1995), and *therefore problem solving during an episode guarantees betterment of a design only with respect to these specific requirements. However, it does not guarantee that the design after the episode is better on the whole than the design before the episode, especially with respect to the requirements that were not used in the episode* (Chakrabarti and Wolf 1995). Therefore, as design progresses, the chances, of requirements that are not visited regularly, to continue to remain satisfied, as design gets more detailed, are increasingly less in the face of continuous, partial modifications.

Identification and analysis of requirements, especially using questioning, visualisation and verification are effective in problem understanding; Generation and evaluation, especially in terms of characterisation and verification before detailing are effective in problem solving (Nidamarthi et al. 2001), see Figs. 3-4. Absence of, or mistakes in these activities are generally harmful (Chakrabarti et al. 2002), see Table 1.

Morgenstern and Knaab (1996; also in Chakrabarti et al. 2002) identified chains of broad activities that represented sufficiency of identification and application of requirements. For instance, identification of requirements was seen as sufficient when *reading of the requirements from given document was followed by evaluation*, before the requirement was interpreted and internalised, while insufficient identification was marked by *reading of the requirement followed by no evaluation*. Nidamarthi (1999; also in Nidamarthi et al. 2001) (see Fig. 3.) identified more elaborate and precise, iterative clusters of sub-activities of various efficacy that were commonly applied during the design process.

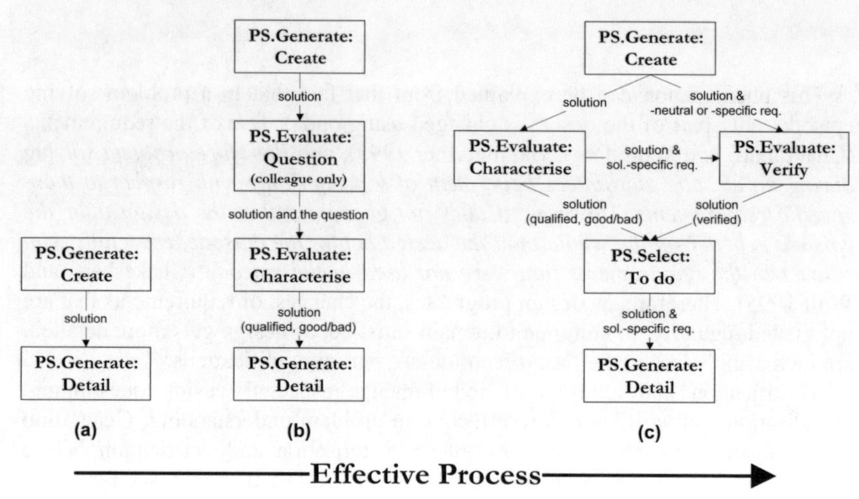

Fig. 4. Efficacy of Requirement Application Mechanisms (from Nidamarthi et al. 2001)

The most effective problem understanding mechanism was seen to be requirement *identification followed by its analysis by questioning and verification*, while the least effective one had only its *identification*, with *identification followed by analysis by visualization* being in-between. The most effective problem solving mechanism is characterized by *generation of a solution followed by its evaluation and selection* before detailing, while the least effective is *generation* only before detailing, with *generation followed by evaluation consisting of characterization* before detailing as the mechanism with effectiveness in-between (see Fig. 4.).

4. Potential Impact to Industry

In this article, a set of activities and mechanisms for requirement identification and satisfaction, that are common across design processes of individual designers and design teams, have been identified and evaluated for their influence on the eventual quality of the product in terms of its degree of fulfilment of the requirements. This knowledge is potentially useful for an organisation involved in product development must make use of, or take into account these so that the process of designing becomes more effective.

References

Design Research – Theories, Methodologies, and Product Modelling, Glasgow, 2001, pp237-244.

Pahl, Birkhofer, H. Höhere Konstruktionslehre, Vorlesungsskriptum, TU Darmstadt, 1992

Blessing, LTM A Process-Based Approach to Computer-Supported Engineering Design, Black Bear Press, Cambridge, UK, 1994

Chakrabarti, A Requirements Identification: a Central Issue in Design Research, Intl. East-West Conference on Information technology in Design, Moscow, Russia, 1994

Chakrabarti, A., Morgenstern, S., and Knaab, H. Identification and Application of Requirements and Their Impact on the Design Process: A Protocol Study, Research in Engineering Design, 2002 (in Press)

Chakrabarti, A. and Wolf, B. Reasoning with Shapes: Some Observations from a Case Study, Proc. ASME DETC (9th Intl. Conf. on Design Theory and Methodology, Boston, 1995), DE-Vol. 83, Vol.2, 1995, pp-315-322.

Chakrabarti, A. Reasoning with Shapes: A Case Study, Invited Speech, Annual Meeting of the CAD Group, Japan Soc. of Mech. Engineers, Tokyo, November 1997

Kruger, F., Knoll, R. QFD as a Base for feasibility Studies Within a Management Tool for Design, Studenarbeit, Cambridge University, UK and TU Darmstadt, Germany, 1994

McGinnis, B.D., Masme, B.S. and Ullman, D. G. The Evolution of Commitments in the Design of a Component", Proceedings ICED89, IME UK, 1989, pp 467-498.

Morgenstern, S, Knaab, H. A Descriptive Analysis of Requirements Identification Processes, Studenarbeit, Cambridge University, UK and TU Darmstadt, Germany, 1996

Nidamarthi, S. "Understanding and Supporting Requirement Satisfaction in The Design Process", PhD Thesis, Engineering Design Centre, University of Cambridge, UK, 1999.

Nidamarthi, S., Chakrabarti, A., and Bligh, T.P. The Significance of Co-evolving Requirements and Solutions in the Design Process, Proc. Intl. Conf. on Eng. Design, Tampere, Vol 1, 1997, pp-227-230.

Nidamarthi, S., Chakrabarti, A., and Bligh T.P. Improving Requirement Satisfaction Ability of the Designer, Proc. Intl. Conf. on Eng. Design ICED01, G, Beitz, W (1988) Engineering Design, Second Ed., Springer Verlag, London

Pugh, S (1991) Total Integrated Methods for Successful Product Engineering, Addison-Wesley, Wokingham

Manifestation of Divergent-Convergent Thinking in Question Asking and Decision Making Processes of Design Teams: A Performance Dimension

Ozgur Eris, Center for Design Research, Stanford University, USA

Introduction

Designing is question intensive. However, our knowledge of the role of question asking in design is limited. The research presented in this article is a summary of the significant findings of a doctoral dissertation that addresses this limitation (Eris 2002). When the findings are considered together, they constitute the conceptual framework of a question-decision centric design thinking model.

The subject of question asking processes of design teams first attracted the attention of the author during a video interaction analysis session aimed at hypothesis generation. In the session, data from a two week design project, where graduate engineering students designed, prototyped, and raced a paper bicycle, was considered. Close attention was paid to the questions raised in the interaction, and their influence on the design decisions that followed. Some questions seemed to strongly influence pivotal decisions and the performance of the resulting designs, whereas others dissipated and had no discernable impact. In either case, questions and decisions appeared to be tightly coupled at a conceptual level. One way of exploring that connection was to identify all questions and decisions, and construct a "question-decision" map. The intent was to test if such a representation might be useful in confirming the existence of a connection, and in discovering relationships between the nature and timing of the questions and the decisions they led to.

However, it was quickly apparent that constructing a detailed representation required a formal understanding of key aspects of question asking. Regardless, prior to focusing on questions to gain that understanding, it was useful to articulate the relationship between questions and decisions at a conceptual level.

The following steps were taken in carrying out this research:

1. Based on the observed interaction between questions and decisions, an argument in the form of a question-decision duality was presented and illustrated.
2. Published taxonomies of questions were reviewed, compared, and extended by the identification of a class of questions specific to design thinking.
3. A divergent-convergent thinking paradigm in the question asking processes of designers was identified, and discussed in light of the proposed question-decision duality.
4. Within that paradigm, a relationship between question asking and design performance was explored and demonstrated.

The Questioning Asking-Decision Making Duality

The validity of studying decision making as a rational process has been argued for in several academic fields. As Howard remarks, decision analysis is related to "the systematic reasoning about human action," and it "stands on a foundation of hundreds years of philosophical and practical thought" (Howard 1988). He defines decision analysis as "a systematic procedure for transforming opaque decision problems into transparent decision problems by a sequence of transparent steps."

Similar views on decision making have been proposed in the engineering design field, and the design process has been modeled predominantly as a decision making process. Hazelrigg argues (Hazelrigg 1999), "In order to ensure that engineering design is conducted as a rational process producing the best possible results given the context of the activity, a mathematics of design is needed. It is possible to develop such a mathematics based on the recognition that engineering design is a decision-intensive process and adapting theories from other fields such as economics and decision theory." An extension of this argument is the application of decision theory in evaluating alternative design concepts, where decision matrixes are constructed to determine the utility of concepts, and decision trees are constructed to map out concept selection processes (Dieter 1983, Pugh 1996).

While it is clear that the application of decision theory can improve design concept selection methods, there are limitations to treating decision making as *the* cognitive mechanism that drives design performance. Such approaches can be augmented by considering other cognitive mechanisms that influence design performance and their relationship to decision making.

When considering the utility of decision-centric approaches in design, especially decision trees that associate information with a decision/design process, it is beneficial to broaden the scope to the entire design cycle by asking: (1) How did the decision-maker reach a position from which he could map his knowledge onto a decision tree? (2) How is reaching that position related to the decision making process, and more importantly, to the design process as a whole?

These issues do not receive sufficient consideration from design researchers who take decision-centric approaches. That can lead to treating the decision making process *as* the design process—an unsound analogy. On the other hand, decision theorists acknowledge these issues by recognizing that decision analysis can only be practiced after that position is reached. Howard asks, "Is decision analysis too narrow for the richness of the human decision?" He then argues that "framing" and "creating alternatives" should be addressed before decision analysis techniques are applied to ensure that "we are working on the right problem." On framing, he states: "Framing is the most difficult part of the decision analysis process; it seems to require an understanding that is uniquely human."

The activities Howard identifies as being problematic, framing and creating alternatives, are inherent dimensions of designing. Researchers who study conceptual design have been attempting to formalize them for decades. Therefore, while design researchers have much to learn from decision theorists, decision theorists have much to learn from design researchers as well.

In light of this discussion, let us return to the first question that was posed, which can be answered by asking another question, and letting its answer point at a duality between questions and decisions. The question is: "How reversible is a decision making process?" In other words, "If one starts with a decision and works his way back through the cognitive events that led to that decision, what will he do when he reaches junctions in the decision tree that are associated with clusters of information and knowledge?" The answer proposed here is that one needs to consider the *questions* that made the acquisition or creation of that knowledge possible, and understand the question asking process of the decision-maker.

This view is illustrated with a data segment from a design experiment (Transcript 1), where teams of 3 graduate engineering students were asked to design and prototype a device that measures the length of body contours. In the transcript, the team is deciding on the number of the gear reduction stages between the sensor and the readout units of the device so that it provides a meaningful measurement.

All 14 questions are directly related to the decision the team is considering, and influence the 3.5 minute consensus building interaction by providing structure for the discussion, and generating and uncovering necessary information. (Several other questions, which lead to the concept of "gear reduction," precede this interaction.) The decision process is initiated by A, who brings up the need to make a decision on gear reduction in Q1. In Q4, B proposes to set the gear ratio so that a full rotation of the dial covers the whole measurement range. C makes the necessary calculations for that concept, and in Q8, asks others to consider the validity of his calculations, which leads B to think that they need 4 stages. In Q9, C considers the legibility of the dial, and asks others to interpret if the scale that would result from the gear ratio B suggested would be acceptable. A must have agreed with C's concern since she proposes a new dial concept—the dial rotating twice—in Q10. After the team considers that concept, C decides that 3 stages would be necessary if the dial rotates twice, and asks the others to judge. B agrees, and using 3 stages emerges as the decision. However, A challenges that decision in Q14 by questioning an assumption. C addresses her concern. The consensus is reached and the decision is made. Q2, Q3, Q5, Q6, Q7, Q11, Q12, and Q13 uncover information and knowledge relevant to the formulation of Q4, Q8, Q9, Q10, and Q14 and D1.

This example illustrates a strong relationship, a *duality*, between questions and decisions, which can be articulated with two axiomatic interdependencies:

1. Every question operates on decisions as premises since the questioner must make choices regarding the subject, object, concept, etc. of the question. Questions are formulated. There are no accidental questions even though questions can have unanticipated consequences.
2. Conversely, every decision operates on questions as premises since decision making entails creating, characterizing, rationalizing, comparing, and evaluating choices—decisions are devoid of meaning if a there is a single choice. Questioning is the enabling mechanism.

It follows that the quality of decisions a designer makes is coupled with the quality of questions he asks. This understanding forms the basis of a new unified question-decision centric design theory, where decision making takes place *during* question asking, and vice versa.

Table 1. Transcript 1. Designers A, B, and C are making a decision on the number of stages of gear reduction for their device. Under the right column, a decision and 14 questions are tagged

Time	S	Utterance	Tag
4:13	A	So, what kind of gear reduction did we decided we needed?	Q1
4:18	C	So, 0.25 inches…7…4…5…	
4:22	B	the circumference is… Do we wanna know the circumference then?	Q2
4:32	C	Right, not the area.	
4:33	B	The circumference is 2 Pi R?	Q3
4:36	A	Yep. [team calculates circumference together…]	
5:12	B	So we want something to only go around once?	Q4
5:17	C	Right, 50 revolutions.	
5:21	B	150?	Q5
5:24	C	Right. How many teeth are on these guys [gears]? This one has 5,6,7,8.	Q6
5:29	A	Or we can also do the belts. We can have rubber bands, yah.	
5:42	B	It seems like there are…Oh, it says on them actually. 24.	
5:47	C	That's 3. 3 to 1.	
5:52	B	And we need 50 to 1?	Q7
5:54	C	Yep.	
6:03	A	This is about a quarter of an inch, three quarters of an inch. [using ruler]	
6:08	C	So, we'd actually need 3 stages? Is that right?	Q8
6:16	B	3 times 3 to the 2 is 27…	
6:19	C	So that would still give us 2 revolutions.	
6:22	B	Yeah, we need at least 4 stages.	
6:30	C	That should be kind of hard to read, wouldn't it?	Q9
6:36	A	Well, maybe we can rotate around twice? I mean it's not hard to realize if it rotates around once, then we just need to aim for half of that.	Q10
6:47	C	So, which one of you has the smaller hands?	Q11
6:49	A	I have the smaller, probably smaller. I have long fingers.	
6:54	B	What was, what were yours?	Q12
6:57	C	40 inches…So, with the smaller hand if you go around, and if it's over 27 then it doesn't matter if it goes around more than once.	
7:09	A	After we could have it go…the indicator could rotate around twice and a little bit before it's hard to read. Do you know what I mean?	
7:21	C	Okay, 3 stages seems appropriate, right?	Q13
7:25	B	Yes.	D1
7:27	A	Is that assuming that we have a bunch of little gears though?	Q14
7:31	C	I'm kind of going under the assumption that we'll get about the same the gear ratio out of the rubber bands, too, since they're about the same size.	

Review of Taxonomies of Questions

Reviewing published taxonomies of questions was a natural starting point for exploring the nature of questions asked in design activity. Also, the insights gained would be useful in laying out the foundations of a framework that would serve as an analysis scheme for empirical research since the categories of a taxonomy can constitute natural units for a coding scheme. Therefore, relevant taxonomies from four fields were reviewed: philosophy (Aristotle), education (Dillon 1984, 1988), artificial intelligence (Lehnert 1978), and cognitive psychology (Graesser 1988).

Dillon's work was aimed at understanding more about the "kinds of questions that may be posed for research." His comprehensive review of twelve question categorization schemes yielded mixed results. He concluded that most of them did not operate on specific and consistent differentiating principles, and therefore, had limited utility. However, he perceived utility in Aristotle's approach. Aristotle's fundamental premise was to assume that our knowledge resides in the questions we can ask and the answers we can provide. Dillon interpreted the progression between Aristotle's four categories (Table 1, col. 1) as a "sequence of inquiry." He then constructed his own scheme based on "Aristotle's few, short, and encompassing propositions." His scheme distinguishes questions according to the level of knowledge entailed in the answers (Table 1, col. 2).

Lehnert's work was aimed at laying out the foundations of a computational model that could answer questions. She treated question answering as a process that can be broken down into two parts: understanding the question, and finding an answer. The first part has to do with interpreting the question, the second with searching memory for the best answer. The first part of her approach required the development of a taxonomy of questions. She argued that the most important dimension of a question that needs to be interpreted for it to be understood and answered appropriately is its conceptual meaning. She proposed 13 distinct semantic question categories (Table 1, col. 2).

Graesser focused on the relationship between question asking and learning in an education context. He adopted Lehnert's taxonomy, and extended it with five new categories (Table 1, col. 4). He then used the framework to analyze the frequency and type of questions students asked during tutoring sessions (Graesser 1994). The incidence of a specific class of questions correlated positively with student learning. He termed them "Deep Reasoning" questions (DRQs). Also, by comparing the extended version of Lehnert's taxonomy with Bloom's taxonomy of educational objectives (Bloom 1956), he argued that the DRQ categories mapped onto higher level of learning objectives.

For comparison, the four taxonomies were inserted into the columns of a table, and the semantically similar categories were aligned (Table 1, col. 1-4). Mutually populated rows point out synergy between the schemes. Since Dillon based his taxonomy on Aristotle's, and Graesser based his on Lehnert's, columns 1-2 and 3-4 reflect two independent evolutionary threads. The comparison resulted in seven distinct classes. The first four classes are the categories of Aristotle's classification scheme. The fifth class consists of the Generative Design Question categories, which will be discussed in the next section. The sixth class consists of the Judgmental category. The last class addresses questions that do not truly seek answers.

Since Aristotle's scheme abides by a differentiating principle and a meaningful hierarchy, his categories were treated as four baseline classes, and the other schemes were mapped onto them. Lehnert's scheme abides by a differentiating principle as well, but lacks an order for relating the categories. Dillon's categories constituted an extended baseline since his scheme was an expansion of Aristotle's.

Table 1 indicates a high degree of mapping between the taxonomies apart from a few missing categories (not considering the Generative Design Question categories). Conceptually, the extended version of Lehnert's scheme mapped onto Dillon's, and thus, onto Aristotle's schemes. That was a positive finding as it indicated strong agreement in the thinking of the authors, and suggested that Dillon's or Lehnert's taxonomy would constitute a sound basis for an analysis scheme.

Table 2. A comparison of taxonomies of questions. The types of questions termed "Deep Reasoning Questions" by Graesser are italicized. The types of questions termed "Generative Design Questions" by Eris are underlined

ARISTOTLE	DILLON	LEHNERT	GRAESSER	ERIS
Existence (Affirmation)	Existence Instance	Verification	Verification	Verification
Nature (Essence)	Definition		Definition	Definition
			Example	Example
Fact (Attribute/ Description)	Description	Feature Spec.	Feature Spec.	Feature Spec.
		Concept Complete	Concept Complete	Concept Complete
		Quantification	Quantification	Quantification
	Function Rationale	Goal Orientation	Goal Orientation	Rationale/Function
	Concomitance	Disjunctive	Disjunctive	Disjunctive
	Equivalence Difference		Comparison	Comparison
Reason (Cause/ Explanation)	Relation Correlation		Interpretation	Interpretation
	Conditionality & Causality	Causal Antecedent	Causal Antecedent	Causal Antecedent
		Causal Consequent	Causal Consequent	Causal Consequent
		Expectational	Expectational	Expectational
		Procedural	Procedural	Procedural
		Enablement	Enablement	Enablement
				Proposal/Negotiation
				Enablement
				Method Generation
				Scenario Creation
				Ideation
		Judgmental	Judgmental	Judgmental
	Rhetorical		Assertion	
		Request	Request/Directive	Request
	Deliberation			

A Convergent-Divergent Thinking Paradigm in the Question Asking Processes of Designers

Comprehensiveness of the reviewed taxonomies was considered by: testing the appropriateness of treating their categories as analysis units for coding questions designers ask; identifying questions that cannot be accounted for by those categories; introducing new differentiating principle(s) and categories to address unaccounted questions.

Questions were extracted from data collected in two design settings: the product development center of a US automobile manufacturer (Eris 2003), and a series of laboratory experiments where graduate engineering students designed and prototyped a device, a "bodiometer," that measures the length of body contours (Eris 2002). In extracting questions, a question was taken to be: "*A verbal utterance related to the design tasks at hand that demands an explicit verbal and/or nonverbal response*." When the published taxonomies were used to categorize the questions, 15.4% of them could not be accounted for. Analysis of the nature of the unaccounted questions resulted in the identification of an overlooked domain.

The common premise of the published taxonomies is that a specific answer, or a specific set of answers, exist for a given question. Lehnert and Greaser also seem to assume that the answer is known. Such questions are characteristic of *convergent* thinking, where the questioner is attempting to converge on "the facts." The answers to converging questions are expected to hold truth-value since the questioner expects the answerer to be truthful. DRQs are such questions. An example is: "Why does the moon rise at night?" where the questioner is seeking a rational and truthful explanation for the rise of the moon.

However, questions that are asked in design situations tend to operate under the diametrically opposite premise: for any given question, there exist, regardless of being true or false, multiple alternative known answers as well as multiple unknown possible answers. The questioner's intention is to disclose the alternative known answers, and to generate the unknown possible ones. Such questions are characteristic of *divergent* thinking, where the questioner is attempting to diverge away from the facts to the possibilities that can be generated from them. It is useful and appropriate to term these types questions "Generative Design Questions," or GDQs. An example is: "How can one reach the moon?" where the questioner wants to generate possible ways of reaching the moon, and, at the time of posing the question, is not too concerned with the truthfulness of potential answers.

A GDQ generally yields multiple answers, which satisfy the question to various degrees. Upon asking a diverging question, the designer's role is precisely to tackle that quality of it by investigating how each answer satisfies the question, and establishing criteria for favoring one answer over the others. That process of investigation, comparison, and evaluation constitutes decision making in design. And, as argued for earlier, it does not necessarily take place after the question is posed; it also occurs *while* the question is being formulated.

Extending the Taxonomies: Generative Design Questions

A comprehensive scheme for coding the questions asked while designing needs to account for the types of questions that fall under the GDQ concept as well. A good starting point for developing such a scheme was to adopt one of the taxonomies and extend it with GDQ categories. Two of the taxonomies reviewed in the previous section, Dillon's and Lehnert's, are articulate. Even though Dillon's taxonomy is more structured, it was more appropriate to adopt Lehnert's since it has been effective in coding questions in discourse, and its utility has been enhanced by Graesser's discussion on DRQs. Thus, Lehnert's taxonomy was adopted as the basis for the coding scheme used in this study. The questions that could not be accounted for by the taxonomies, GDQs, were analyzed further. Five GDQ categories were proposed as extensions (Table 1, col. 5): Proposal/Negotiation, Scenario Creation, Ideation, Method Generation, and Enablement.

Proposal/Negotiation: The questioner wants to suggest a concept, or to negotiate an existing concept. An example is: "How about attaching a wheel to the long steel beam?" The questioner is suggesting a concept, and expecting the answerer to supply her/his corresponding opinion(s), which would not be definitive. The questioner intends to establish a negotiation process by exchanging suggestions, and build toward new concepts. Suggesting a new concept usually requires a consideration of the hypothetical possibilities the new concept can lead to. These questions are significant because proposing an idea in the form of a question promotes consideration and feedback, and negotiation promotes synthesis.

Scenario Creation: The questioner constructs a scenario involving the question concept and wants to investigate the possible outcomes. An example is: "What if the device was used on a child?" The questioner wants to generate and account for as many outcomes as possible from the scenario(s) that can be constructed. These questions are significant because accounting for possible outcomes generates and refines design requirements.

Ideation: The questioner wants to generate as many concepts as possible from an instrument without trying to achieve a specific goal. An example is: "Are magnets useful in anyway?" The questioner does not intend to achieve a specific goal by using the magnets. He does not have a purpose other than to generate as many ways of utilizing magnets as possible. Ideation questions are significant because operating without a specific goal frees associations and drives concept generation.

Method Generation: The questioner wants to generate as many ways of achieving a specific goal as possible. An example is: "How can we keep the wheel from slipping?" The questioner wants to generate secondary conceptualizations, which, if realized, will cause the initial conceptualization—keep the wheel from slipping. These questions are significant because operating with a specific goal generates a set of methods for implementing concepts.

Enablement: The questioner wants to generate acts, states, or resources that can enable the question concept. This category is the GDQ version of the original Enablement category Graesser labeled as a DRQ. What deems it a GDQ is the assumption of multiple possible initial conceptualizations. For instance, "What allows you to measure distance?" is a GDQ if the questioner intends to *generate* resources for measuring distance. However, it is a DRQ if the questioner assumes the existence of a known set of resources and intends to identify them. That differentiation can only be made by considering the context of the question. Enablement questions are significant because identification of multiple resources promotes surveying and learning from existing designs.

Convergent-Divergent Question Pairs: a Design Performance Metric

In light of the arguments presented regarding the manifestation of convergent-divergent thinking in the question asking and decision making processes of designers, and the association of two specific classes of questions (DRQs and GDQs) with convergent-divergent thinking, it was natural to consider a relationship between the incidence of DRQs and GDQs and design performance. It was assumed that the quality of design decisions are directly coupled with the quality of questions that precede them, and that high quality decisions yield high quality designs. Then, it was hypothesized that question asking processes of designers correlate with their design performance.

The hypothesis was refined through the following observation: design teams seemed to discover more design concepts when they asked DRQs and GDQs, and subsequently, conceptualized more articulate and a higher number of designs. This was in agreement with Graesser's finding, the correlation between the incidence of DRQs and learning performance. However, the interaction he studied could not have promoted the type of learning that occurs in design activity since divergent thinking is less of a factor in tutoring. Therefore, it was postulated that GDQs might also be correlated with performance, in a design context, where divergent thinking is a strong factor. This led to a refinement in the hypothesis, and the incidence of DRQ-GDQ *pairs* were hypothesized to correlate with performance.

The refined hypothesis was tested through twelve 90 minute long laboratory experiments where 12 teams of 3 graduate engineering students participated in the bodiometer design exercise. Performance of each design team was measured by two methods: an objective scoring scheme that was a function of how well the prototypes met given design requirements, and a subjective scheme where three engineering professors rank-ordered the prototypes. Measurements obtained by these methods correlated strongly, $r^2 = .55$, $p < .01$.

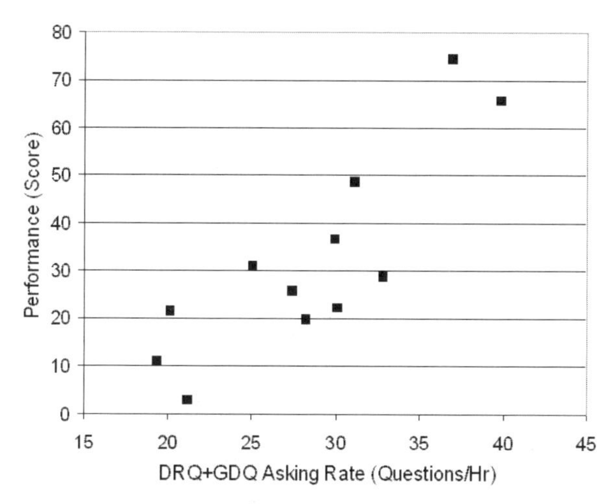

Fig. 1. DRQ+GDQ asking rates of the teams plotted against their prototype scores

When the combined DRQ+GDQ asking rates of the teams were plotted against their prototype scores, a linear relationship was visible (Fig. 1.). Statistical analysis revealed strong correlation with high significance, $r^2 = .68$, $p < .05$. There was no significant correlation between performance and the incidence of all questions, DRQs, GDQs, or any other single question category. These findings suggest that DRQs and GDQs need to be treated as *complementary pairs* when it comes to establishing them as a design performance metric. (For a detailed discussion on the experiments and the findings, see Eris 2002.)

Conclusion

The findings on the relationship between question asking, decision making, and performance demonstrate DRQ+GDQ utilization to be a mechanism designers rely on for managing divergent and convergent modes of thinking. When they are considered in the context of the design process as a whole, they can be mapped onto design phases (conceptualization, implementation, and assessment), and utilized to explain the structure of design thinking:

During <u>Conceptualization</u>, GDQs are used to preserve ambiguity by:

- Reframing previously recognized needs and understandings
- Generating alternatives
- Creatively negotiating proposed design concepts

During <u>Implementation</u> and <u>Assessment</u>, DRQs are used to reduce ambiguity by:

- Reiterating goals
- Focusing on deliverables

- Seeking and establishing causality
- Synthesizing/Reducing the number of proposed design concepts

Based on this understanding, a design thinking model, illustrating the transformation of design requirements into design concepts through asking Generative Design Questions, and then into design specifications and decisions through asking Deep Reasoning Questions, can be constructed (Fig. 2.).

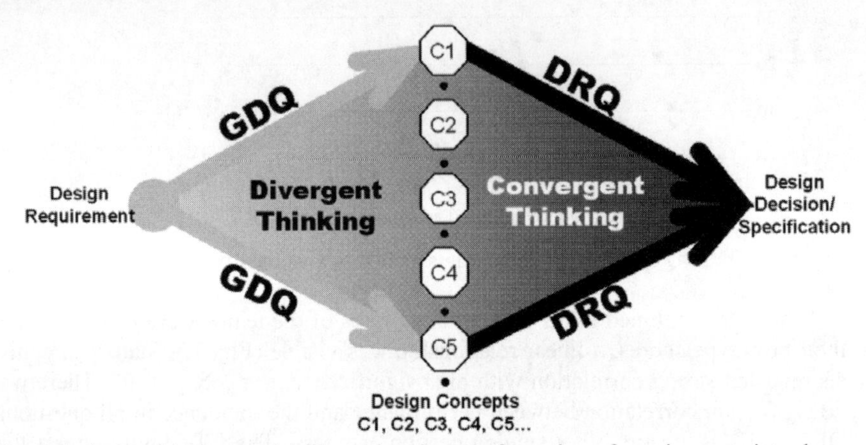

Fig. 2. A design thinking model illustrating the transformation of requirements into design concepts through asking GDQs, and then into decisions through asking DRQs

In conclusion, the manifestation of convergent-divergent thinking in the question asking and decision making processes of designers in the form of Deep Reasoning and Generative Design Questions constitutes a performance dimension in design.

References

Aristotle. Posterior Analytics.

Bloom SB, Editor (1956) Taxonomy of Educational Objectives, Handbook I: The Cognitive Domain. David McKay Company, New York, USA.

Dieter EG (1983) Engineering Design: A Materials and Process Approach. McGraw Hill, New York, USA.

Dillon TJ (1984) The Classification of Research Questions. Review of Educational Research, vol. 54, 327-361.

Dillon TJ (1988) Questioning in Science. In: Michel M (ed) Questions and Questioning. New York: De Gruyter, Chapter 4, p. 68-79.

Eris O (2002) Perceiving, Comprehending and Measuring Design Activity through the Questions Asked while Designing. Ph.D. Dissertation, Stanford University.

Eris O, Leifer L (2003) Facilitating Product Development Knowledge Acquisition: Interaction Between The Expert and The Team. International Journal of Engineering Education, vol. 19, no. 1, p. 142-152.

Hazelrigg GA (1999) An Axiomatic Framework for Engineering Design. Journal of Mechanical Design, vol. 121, p. 342-347.

Howard RA (1988) Decision Analysis: Practice and Promise. Management Science, vol. 34, no. 6, p. 679-695.

Lehnert GW (1978) The Process of Question Answering. Lawrence Erlbaum Associates, Hillsdale, New Jersey.

GraesserA, Lang K, Horgan D (1988) A Taxonomy for Question Generation. Questioning Exchange, vol. 2, No. 1, 3-15.

Graesser A, Person N (1994) Question Asking During Tutoring. American Educational Research Journal, vol. 31, no. 1, 104-137.

Pugh S (1996) Concept Selection-A Method that Works. Creating Innovative Products using Total Design, Addison-Wesley, Chapter 14, p. 167-176.

Towards a Conceptual Framework for Predicting Engineering Design Team Performance Based on Question Asking Activity Simulation

Ade Mabogunje
Center for Design Research, School of Engineering Stanford University, USA

Problem

The product development process has undergone significant changes in the last two decades. Consumers have become more sophisticated in their choices, products have become more complex, and the barriers to entry to competitors have been significantly lowered. All this has resulted in an increased emphasis on creative teamwork and shorter product development times. In turn this has led to an increased number of descriptive studies of the engineering design process in the research community as well as the development of various tools and methods aimed at improving the process. Success in this endeavor has been a difficult battle for the research community and several authors have made attempts to analyze the obstacles responsible for the difficulty. Blessing for example notes:

"… identifying whether this [a method or tool] indeed contributes to success is far more difficult and the results are not easy to generalize. Success is difficult to measure other than in a real, industrial situation and action research in an industrial situation is notoriously difficult, let alone comparative action research. Furthermore, the success of a method or tool depends on the context in which it is being used. This context is different for every design process, because every design project is unique" (Blessing et al. 1998).

I have quoted Blessing et al here because not only do they articulate in a succinct manner a common observation in the research community, but they do so in a way that suggests the types of solutions that will be needed (See also Dowlen 1997).

On close inspection, it appears that whatever solution set we adopt must meet five important conditions, stated here in no particular order. First, it must address the issue of product complexity. Second, it must address the issue of project uniqueness. Third it must be capable of accounting for multiple factors that have been known to influence design. Fourth, it must be repeatable and verifiable under similar conditions, and last but not the least it must address the problem of transfer to industry. The following question then arises: Will it be possible to develop a general research methodology along these lines using a scientific paradigm?

Blessing et al. [ibid.] proposed a method which they hope allows for comparison, classification, and assessment of design research, their claim is that such a method would lead to a more rigorous approach to design research. To complement their approach I wish to present here a second approach, one along the lines of a design paradigm. Rather than being rigorous, it is I hope playful. The reason is that it appears to me that the research problem we have is very similar to the design problems faced by our subjects, namely an ill structured problem with a high degree of complexity. Hence I am curious to see if borrowing some of their methods will give us valuable insight into our own problems. More specifically, I am using a paradigm from computer game design that specifies a 3-step process: Design-Build-Play (Private correspondence with Will Wright). Computer game design is at its heart a simulation, and offers a number of advantages to advancing design research if used strategically.

There are at least five advantages to be gained by representing our findings in computer simulation models. First it will allow us to represent many more variables and relations because of higher capacity of computers to store such information. Second it will allow us to easily handle dynamic variables and perform what-if scenarios based on our models. Third it addresses the cost issue associated conducting field studies, since we can in general run hundreds of simulations to explore alternative scenarios that could have happened in practice. Fourth a simulation model could serve as a platform for testing alternative models of the design process using a common base case. This means we can gain multiple perspectives on the process and also build on each others work and begin to generalize our findings. While I make this case for simulation models, I wish to acknowledge some potential disadvantages. In particular, I see a danger that these models may lead us to ignore those observations that are difficult to model. Hence I advocate a combination of descriptive studies and simulation studies.

Previous Work

There is a long established tradition of organizational simulations in the field of organizational behavior (Burton and Obel 1993). These simulations tend to model whole organizations at a level of granularity that is much larger than we see in design research. In the early nineties, a simulation program named Virtual Design Team (VDT) was developed to describe the activities of a project team within an organization (Cohen 1992). The conceptual model in VDT seeks to explain how the duration and quality of an engineering project are affected by actor variables, task variables and organizational variables. Actor variables include elements such as the skill of the actor with respect to a particular task, her preferences for using certain communication devices during task execution and her position within the organizational hierarchy. Task variables include a description of the level of complexity of a task and the degree of uncertainty associated with the task activity. Organizational variables include such variables as the structure and communication policy.

The relationship among these variables is based on a combination of Galbraith's theory of information processing in firms (Galbraith 1974) and heuristics for estimating the duration of tasks and quality of decision making during the execution of a project. Paraphrased, Galbraith describes knowledge work as routine work, punctuated by exceptions. These are situations where the information needed to complete a task exceeds the information available to the worker. He goes further to describe how the organization-nominally a hierarchy, but evolving to a matrix as needed-serves as an exception handling system with exceptions flowing along communication channels to other workers who have the information (and authorization) needed to complete the task. He explains how hierarchies can get overloaded with exceptions when uncertainty is high relative to the skill levels of the first line workers and describes alternative strategies for dealing with this information overload. The heuristics for estimating the duration of tasks and quality of decision making focuses on the process by which the first line worker handles exceptions. The worker can do this by doing nothing (default delegation), by reporting to a colleague (lateral communication) or by reporting to a supervisor (vertical communication). In communicating with the colleague or supervisor, a further choice is made with respect to the communication medium to use, a memo, a fax, a telephone, or a face-to-face meeting. These choices are constrained by the organizational structure and the communication policy within the organization and could have an effect on other members of the organization and ultimately the total time to execute a given task. Based on this model, the design duration could be estimated as the sum of the standard time to complete a task and the communication time.

$$t_{total} = t_{task} + t_{communication} \quad (1)$$

Using the concepts of VDT it has been possible to model design team activities in other situations yielding very insightful outcomes (Mabogunje et al. 1995; Hansen et al. 1997). As a result, the framework being developed here will build on this basic formula.

Method

Simulating design in order to explore how different factors influence the process and consequently the design outcome is like a jigsaw puzzle. There is a need to bring together in a coherent manner several pieces of work – which would include research findings and models that have been developed through descriptive studies. Hence there is need for an overarching theme or a finite set of overarching themes to tie the pieces together. At this early pre-theoretical stage of design research, the choice of the theme is quite arbitrary. Given our concern about the relatedness of research results to industry practice it would be helpful if the theme chosen is one that could have face validity in industry. The themes I have chosen are question asking activity and development time.

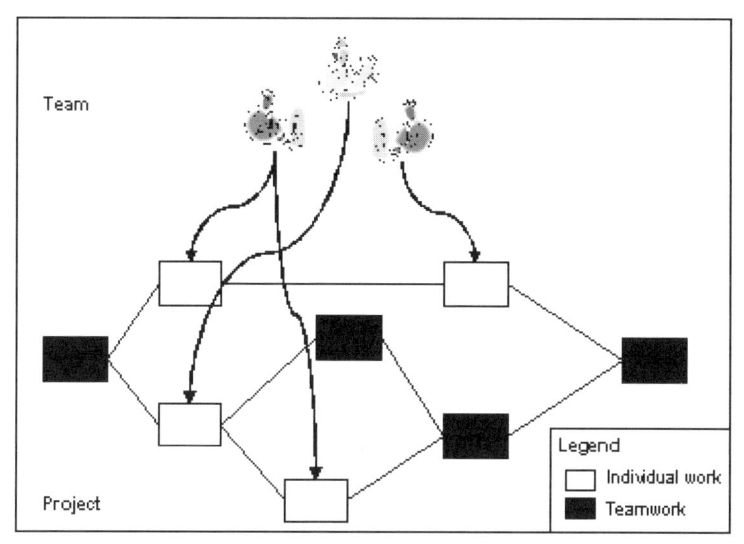

Fig. 1. Teamwork in design is a combination of working on individual assignments and working in groups. A proper study design must account for these two modalities of working

Question asking activity in design

My interest in the question asking behavior of designers grew out of the observation that several of the design for manufacturability tools that were adopted by American companies for example quality function deployment (QFD) and design for assembly (DFA) could be viewed as tools for modifying the types of questions being asked by engineers during new product development. This suggested a correlation between question asking behavior and product quality. An additional advantage to using a question-asking framework is that it provide an important behavioral link to information retrieval tools and communication tools in design – for examples queries for searching a data source, and motive for calling up a colleague or an expert during design. Thus the ubiquity of questions in design methods, tools, and design communications makes question asking activity a useful overarching theme.

Development time

The other overarching theme is development time. As mentioned earlier this has become an important basis for competition in industry. At the same time it is a dominant paradigm in the engineering field where there is a concern for the efficient use of resources. At a deeper level, time provides us a powerful mechanism to keep track of changes that occur during the design process and a common standard to make comparisons across different projects. From the VDT simulation, we know that:

$$t_{total} = t_{task} + t_{communication} \quad (1)$$

To extend this model of VDT and thereby demonstrate a generalized approach, I will use the results of two other descriptive studies of question asking activity (Bales 1950; Mabogunje et al. 1993) to further breakdown the task time. To integrate and accommodate the three models, it will be useful to think of the design process as consisting of a varied pattern of team work and individual work. As shown below:

Case Study 1

This first case study builds on the work of the sociologist Robert Bales (Bales 1950). He studied the interaction process in small groups and developed a coding scheme which classified the utterances and expressions of the participants into six reciprocal pairs (1. Asks for information, 2. Ask for opinion, 3. Ask for suggestion, 4. Give information, 5. Give opinion, 6. Give suggestion, 7. Disagrees, 8. Shows tension, 9. Shows antagonism, 10. Agrees, 11. Shows tension release, 12. Shows solidarity). He later grouped these into four main categories of problems an interaction system needed to resolve - task problems and socio-emotional problems. The first two categories in the problem solving sequence, namely questions and attempted answers, were used to handle task problems and the other two categories were used to resolve socio-emotional problems.

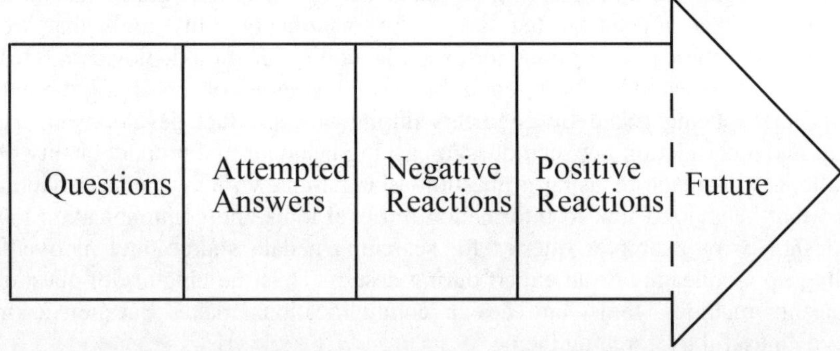

Fig. 2. (Source: Bales 1950) The interaction process of small face-to-face groups can be idealized as one consisting of questions, attempted answers, and either positive or negative reactions. The figure highlights the central role of questions in conversations about future possibilities and draws attention to the fact that answers are not the only responses to questions

Bales saw the interaction process as an alternating emphasis on these two areas and in his words:

"When attention is given to tasks, strains are created in the social and emotional relations of the members of the group and attention then turns to the solution of these problems. So long as the group devotes its attention simply to

socio-emotional activity, however, the task is not getting done and attention would be expected to turn again to the task area."

This view of questions for example suggests that the time required to complete a question-answer session must include the time to deal with the socio-emotional reactions (Fig. 1.). This implies that the total time can be subdivided into the task time and the socio-emotional time $t_{\text{emotion processing}}$. Furthermore the task time can be divided into time to pose questions and time to answer questions.

$$t_{\text{total}} = t_{\text{task}} + t_{\text{emotion processing}} \tag{2}$$

$$t_{\text{task}} = t_{\text{question asking}} + t_{\text{answer generation}} \tag{3}$$

The idea of these socio-emotional breakdowns is corroborated by the excellent study of factors that influence teamwork in design practice which resulted in the elaboration of the term "critical situations" (Badke-Schaub et al. 1997). In this study they used the categories of "disturbance management" and "conflict management" to account for these additional events that were outside the general problem solving process.

Case Study 2

The second case study involves the development of an information retrieval tool named Dedal (Baudin et al. 1992). This tool was developed within the larger framework of design rationale capture and design information reuse. Grubber and Russell provide a useful overview of empirical studies focused on design rationale capture and question asking behavior of designers spanning several domains: architecture, electromechanical device design, heating and air condition systems, user interfaces for software and hardware, and the design of instructional courses (Grubber and Russell 1996). In the information reuse study leading to the development of Dedal, the same design task was given to four designers, two of them used paper based documents and the other two used Dedal as a tool to retrieve information from the same set of documents now in electronic form. The tests showed an average reduction in the task solution time from about four hours to two and a half hours, indicating the potential impact of information retrieval tools on the design process time.

Table 1. The Impact of an Information Retrieval Tool on Task Completion Time

	Document Type	Task Completion Time
Designer 1	Paper	5 hr 30 min
Designer 2	Paper	3 hr 30 min
Designer 3	Electronic (Dedal)	2 hr 30 min
Designer 4	Electronic (Dedal)	2 hr

Of course this does not take into consideration the time required to build the tool or the time to index the documents. The impact of information retrieval tools in general represents a reduction in the total time it takes to search for solutions t_{search}, and from this point of view the total time could be expressed as follows:

$$t_{total} = t_{task} + t_{search} \quad (4)$$

The salient point in this study is that a classification of questions in the pre-study (when the paper based documents were used) revealed potential areas of time savings which corresponded with opportunities to introduce new tools and methods to support the design process (See Fig. 6. in Mabogunje et al. 1993).

Model

Building on the VDT model, we arrive at the following formula for total time:

$$t_{total} = t_{task} + t_{communication} + t_{search} + t_{emotion\ processing} \quad (5)$$

The reader is reminded that this total time pertains to each of the activities that could be represented on a pert chart of the design project. For some activities one or more of the time elements would be zero showing that the underlying factors are not at play. To complete the model, there is a need for a more specific description of the design task. In keeping with the theme of question asking, MacLean et al's design space analysis method appears robust enough to model the design task. The principal elements of this method are questions and options, criteria and assessments (MacLean et al. 1996). To paraphrase,

"… the method views each design feature as only one option available among a set of options. It uses questions for structuring the options and enumerates the criteria that determines the choice of a particular option."

This implies the following specific activities: Question asking, Option creation, Option assessment. Hence the task time is given by:

$$t_{task} = t_{question\ asking} + t_{option\ creation} + t_{option\ assessment} \quad (6)$$

Combining this with the model proposed by Bales equation #(3) we get:

$$t_{task} = t_{question\ asking} + t_{answer\ generation} + t_{option\ creation} + t_{option\ assessment} \quad (7)$$

and the total time is given by:

$$t_{total} = t_{question\ asking} + t_{answer\ generation} + t_{option\ creation} + t_{option\ assessment} + t_{communication} + t_{search} + t_{emotion\ processing} \quad (8)$$

Equation #(8) represents a model of development time based on a framework of question asking activity. The next step which is beyond the scope of this paper is to integrate this model with the VDT model. The VDT model uses a discrete event simulation, and it will be possible for example to change the simulation probabilities of the actors to reflect individual differences in cognitive style, temperament and experience and see how this affects the development time.

Discussion and Summary

Through modeling, a formula for development time was developed. What emerged was a model of time with seven elements – equation (8). Upon seeing this formula, the immediate question that comes to mind is the degree to which it is reasonable. Leifer has suggested that the trend toward distributed design has given us the opportunity to break up the design process and reveal a number of problems that were not previously apparent (Larry Leifer, private conversation 2002). From this view point it is possible to see several opportunities for time latencies. Along a similar line, Storath et al in their exploration of tele-engineering provide a pictorial representation of the product development process that helps to contextualize the variables in the equation (see Fig. 3.).

They identify four main components involved in tele-engineering: information, tools, cooperation, and design tasks. From this it is possible to make the following rough assignment:

1. $t_{emotion}$ and $t_{communication}$ belong primarily to the cooperation domain.
2. t_{search} belongs to the information domain
3. $t_{question}$, t_{answer}, $t_{option\ creation}$ and $t_{option\ assessment}$ belong to the design tasks domain.
4. All seven elements belong to the tool and method domain because of the potential impact they could have on the efficiency and effectiveness of these processes.

Thus the seven elements appear to be reasonable if not too few.

Finally, in deriving the formula for t total, models developed in several descriptive studies have been used. It is hoped that this illustrates the potential of simulation to serve as a method for the systemization of design research knowledge. In particular it can be customized for different projects to replicate their unique features.

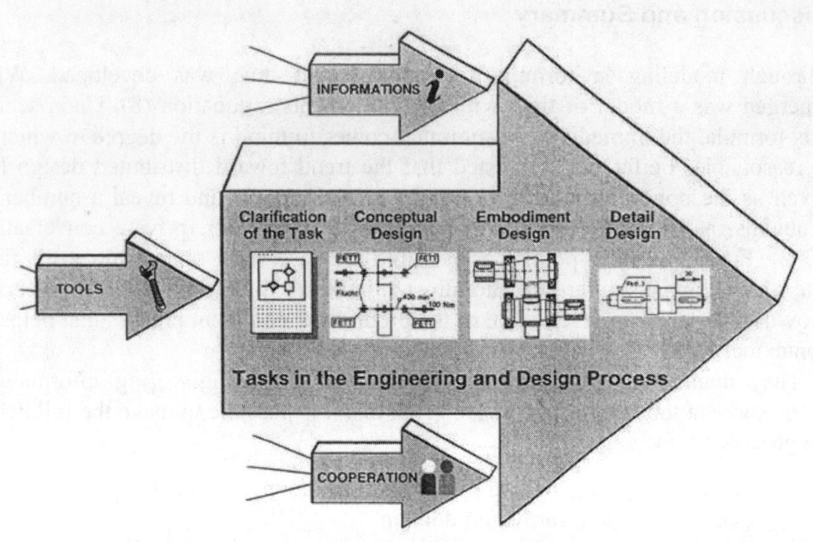

Fig. 3. (Source: Storath et al. 1997) describe the team interactions in a real sense. The model developed in this study has eight components which puts it within the same range. Distributed design provides the opportunity to observe process breakdowns and hence make the underlying process knowledge more explicit rather than tacit. If each of the four components shown in the diagram (information, tools, cooperation, and tasks) had a time latency associated with it, the model of development time will easily have four components of time to

Several trials can be run to address issues of repeatability. Several models can be integrated together to account for the multitude of factors that could influence the design process, and the simulation itself is a useful tool around which to "play" and hold conversations with designers in Industry about their process. There is still a lot of studies to be done, and models to be developed, operationalized, verified and validated but I believe a design paradigm as I have illustrated will be a useful one to merit further consideration.

References

Badke-Schaub P, Frankenberger E, Dorner D (1997) Analysing Design Work by Critical Situations: Identifying Factors Influencing Teamwork in Design Practice. In: Riitahuhta A (ed) Proceedings of the International Conference on Engineering Design, Tampere, Finland

Bales RF (1950) Interaction Process Analysis: A Method for the Study of Small Groups. Addison-Wesley, Cambridge

Baudin C, Gevins J, Baya V, and Mabogunje A (1992) Dedal: Using domain concepts to index engineering design information. In: Proceedings of the 14th Conference of the Cognitive Science Society, Bloomington, Indiana

Blessing LTM, Chakrabarti A, Wallace KM (1998) An Overview of Descriptive Studies in Relation to a General Design Research Methodology. In: Frankenberger E, Badke-Schaub P, Birkhofer H (eds) Designers - The Key to Successful Product Development. Springer-Verlag, London.

Burton RM, Obel B (1993) On the Validity of Simulation Models in Organization Science. In: Proceedings, The Next Quarter Century: Economics, Education, Management, A conference in Honor of Richard M. Cyert, Carnegie-Mellon University, Pittsburg, Pennsylvania

Cohen GP (1992) The Virtual Design Team: An Information Processing Model of the Design Team Management. Ph.D. thesis, Stanford University

Dowlen C (1997) Development of a Cognitive Framework for Design Science. In: Riitahuhta A (ed) Proceedings of the International Conference on Engineering Design, Tampere, Finland

Galbraith JR (1974) Organization Design: An Information Processing View. Interfaces 4:28-36

Gruber TR, Russell DM (1996) Generative Design Rationale: Beyond the Record and Replay Paradigm. In: Moran TP, Carroll JM (eds) Design Rationale: Concepts, Techniques, and Use. Lawrence Erlbaum Associates, Mahwah, New Jersey

Hansen PHK, Larsen JH, Mabogunje A, Leifer L (1997) Simulating Design Coordination. The Stanford-Aalborg Empirical Studies Part (1) In: Riitahuhta A (ed) Proceedings of the International Conference on Engineering Design, Tampere, Finland

Leifer LJ (2002) Professor of Mechanical Engineering, Stanford University, California, USA

Mabogunje A, Baya V, Brereton MF, and Leifer LJ (1993) 210-X: A Method and a Methodology for Improving Product development Performance. In: Hubka (ed) Proceedings of the International Conference on Engineering Design, The Hague, Netherlands

Mabogunje A, Leifer LJ, Levitt RE, and Baudin C (1995) ME210-VDT: A Managerial Framework for Improving Design Process Performance. In: Frontiers in Education Conference, Atlanta, Georgia

MacLean A, Young RE, Bellotti VME, Moran TP (1996) Questions Options and Criteria: Elements of Design Space Analysis. In: Moran TP, Carroll JM (eds) Design Rationale: Concepts Techniques and Use. Lawrence Erlbaum Associates, Mahwah, New Jersey

Storath E, Mogge C, Meerkamm H (1997) Teleengineering: Product Development in Virtual Design Offices Using Distributed Engineering Network Services. In: Riitahuhta A (ed) Proceedings of the International Conference on Engineering Design, Tampere, Finland

Wright W (2003) Chief Designer, Maxis, Inc., and Designer of the Computer Game SimCity

Collaborative Product Development Considerations

Stig Ottosson, Linköping University and Trollhättan-Uddevalla University, Sweden

Introduction

Product and process development is time dependent and *complex* e.g. as there are many "agents" taking part in the processes who occasionally can cause new situations to suddenly appear. Such agents are managers, internal and external specialists, team members, test persons, customers, users, inventors, controllers, etc. Product and process development is also *complicated* as there are many knowledge areas involved, such as engineering, business administration, computer technology, sociology, psychology, management, etc. Also to take in account when forming good conditions for successful development is that chaotic situations can develop rapidly as e.g. new players can occur by chance, can new technologies suddenly appear, can the financial support completely change the grounds for development, etc.

Research

The research behind this paper is mainly based on Insider Action Research (Ottosson and Björk 2003) from 1990 to 2002 in Swedish industry (e.g. Handiquip AB, Access Industries, Alert Invest AB, and Careva Systems AB) and on Action Research from 1993 to 2002 in student projects at Halmstad University, Linköping University and at Trollhättan/Uddevalla University in Sweden.

Overall considerations

When an individual on her/his own is going to develop a product or process in principle the most important factor that will influence the work is tools & methods. Depending on the product or process to be developed the need of resources will vary. Also the network of resource people – people to communicate with - is important for the outcome.

Already when two or more people together will develop a product or process, a difficult to plan chaotic system has been created meaning that more factors will influence the work than when a developer on her/his own makes the job. Important internal influencing factors in the organisation are for team work *management paradigms*, *the physical environment*, *resources*, and *tools & methods*. Also in this case the availability of *resource people* is of extreme importance as the main resource of information collection comes from dialogues with senior people (see the contribution of Wallace et al in this book). Fig. 1 shows schematically factors that heavily influence the outcome of a team work due to our experiences. External factors - as e.g. project directives given by a customer - will not be treated here.

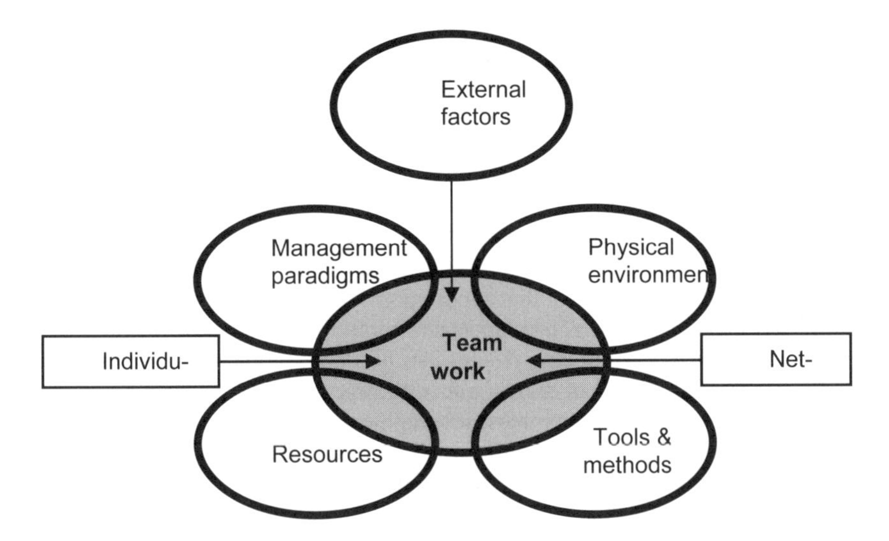

Fig. 1. Important factors influencing the outcome of a product or process development project

Thus when practical development is to be done in an organisation the outcome of that work is dependent of many factors that largely are outside the individual developer's ability to influence in any substantial way. It is in industry well known that e.g. the project directive can impose such limitations for the work that it will be very difficult to carry through it in a successful way without breaking rules. Such directives can be to work in consideration with ISO 9000, which often makes it hard to develop new products in an efficient way. This as ISO 9000 has a static mind set accompanied by a bureaucratic system that unfortunately focuses on form more than result.

In a team work especially *the management paradigm* used - the view of life - is very important for the speed of the development as well as of how innovative the solutions will be. Two basic principles exist that can be called classic and dynamic. The classic principle – also called Newtonian – builds on the view that almost everything can be planned carefully in advance and that detail planning should be done before any work starts. Simulations can, due to this view, exchange real tests if good enough computer programs are developed. In discrete points – at gates – the outcome is controlled. Traditional Integrated Product Development (IPD) – and Concurrent Engineering (CE)/Simultaneous Engineering

(SE) – are based on the classical view. Dynamic Product Development (DPD) on the other hand build on the view that it is rather wasted time to plan in detail more than for one week or less (Ottosson 2002). Instead a clear vision, rough long term plans, and detailed short term plans are used. As simulations give answers on questions raised before the simulations, real tests are also needed to find out unasked questions if good products are to be done.

Physical environment in Fig. 1 means the localities where the product development is performed and what is in them of artefacts, sound level, heat, light, etc. Scientifically not very much has been done to investigate what circumstances are better than the other for successful development activities. However we found (Branzell et al. 1997) in studies of students performing innovative product development at Halmstad University, that open areas were better than closed areas for innovative product development. Thus early in the development of new products/processes – during the innovative design phase – it is due to our experiences also in industrial development processes very important that the designers are localized together ideally without walls or stairways separating them.

Sitting together also means that a common vocabulary and a common understanding is created fast, which is important to avoid problems in the later development work. When parametric design can be done the need of sitting together should in principle not be as critical although investigations have shown that collaborative work with team members geographically distributed mainly communicating over internet so far often fail to reach wanted results.

Resources means all types of resources useful for the development work. Examples are money, people, machines, energy, computers, communication tools, etc. As a rule of thumb too little resources and too much resources are negative for the outcome. To bring forward a model of a mechanical solution and even to bring forward a functional prototype, we have seen that small resources in fact can be more helpful than hampering. Recently e.g. 55 students separated in 11 teams with five team members in each team at Trollhättan-Uddevalla University managed to bring forward 10 unique and functional models of lifting equipment for ambulance personal, rollators for elderly, etc. with a total budget of 100 EUR. The student work was done during 7 weeks.

Tools & methods are instrumental and can be divided in many sub-titles, which to some extent is covered elsewhere in the book. The use of the different tools & methods are e.g. dependent of if new solutions are to be developed - which is called *innovative design* - or if existing solutions are to be improved, which is called *parametric (or math) design*. These two types of design need circumstances that are very different from each other what regards competence, paradigms, localities, work principles, organisation, etc. When innovative design is to be done, one needs to bounce between an abstract and a concrete view as well as between a totality and a detail level. When innovative design is done the parametric fortunes of computers can not be used very much until solutions exist to optimize. Therefore creative abstract thinking, sketching and the making of simple models is important to make, followed by benchmarking before the advantages with parametric design can be utilised. Especially during the innovative development stage it is due to DPD important to in an iterative way - using the Pareto principle - make many tests and adjustments. First when a functional solution exist can parametric/math design be used in a fruitful way due to this view.

In Fig. 2 the shaded areas represent when parametric design is advantageous to use. Also when problems occur while performing parametric design the curtains in the "Design Window" have to be opened to help parametric design effort to make a leap.

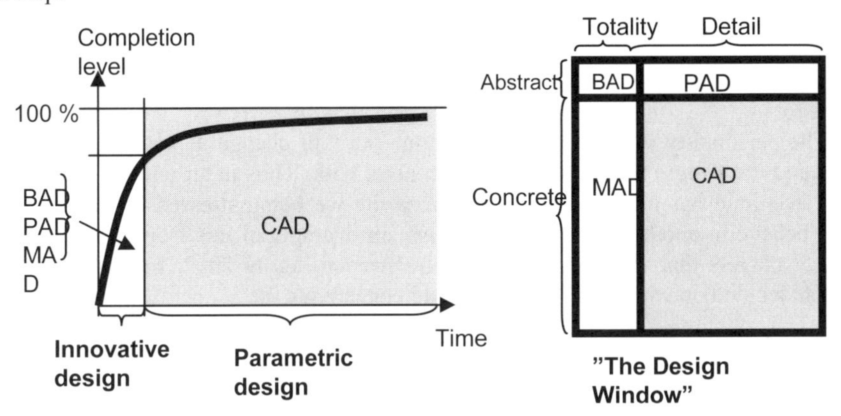

Fig. 2. Without innovative design – which mainly is non-parametric - precedes parametric design. The two stages need different tools (BAD=Brain Aided Design, PAD=Pencil Aided Design, MAD=Model Aided Design, CAD=Computer Aided Design)

Personality profiles

It is a common knowledge that when a development team is to be formed it is first of all important to find a manager that has a suitable personality profile for the mission both when things are running well and when unforeseen problems will show up – which they tend to do in development projects. The view of life, and the competence (knowledge and experience) are other important factors to take in consideration. The manager in turn has to find team members with suitable personality and knowledge.

When a team is to be set up, the different personalities that each team member has will be of great importance for the outcome of the work. Also, depending on the unique personality everybody has, she or he will be more or less suitable to be a designer. Examples of wanted personal profiles for designers are that they shall be active, social, flexible, creative, enterprising, positive to changes, caretaking and warm-hearted. However every organisation and every manager will have different opinions on which profiles that are important to focus on when new designers are to be engaged in a development team.

As everyone has sometimes experienced, the own personality can play some tricks now and then. This as our personality is inherent in our genes and formed by our environment and our experiences. Our own picture of our personality is often not the same personality that other people have experienced.

How well two or more people will interact is dependent of the task, the personality profiles that each of them have, and how well they will go together - how well the personal "chemistry" will work. Thus setting up a team with the best experts that exist in the world, or forming a soccer team with the best players, doesn't mean that a success will be the result – as classical Newtonian principles prescribe. Also it is important to match people with different profiles instead of matching people with similar personalities and knowledge, which we will come back to.

The personality is not constant over time but will change depending on situation and the people the individual is in contact with. Thus in an un-pressed situation everyone has one type of behaviour while we being stressed or threatened, will behave in another way. Table 1 shows an example of this from two tests on soccer players that we have performed in Sweden during 2002. In that case the coach decided on which profiles we should concentrate on.

Table 1: Important personality dimensions for soccer players. The results of one player is shown. The solid line shows his behaviour in relaxed situations. The dotted line shows his behaviour when he was physically and mentally exhausted

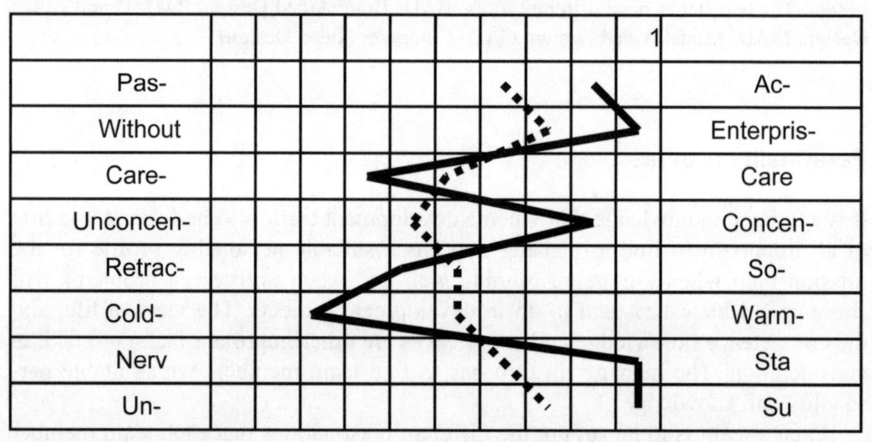

Knowledge& information generation

As everyone has her/his personality, everyone also has a different knowledge profile and different ability to increase the own knowledge. Discussions about the term *knowledge* have been a philosophical topic for the last 25 centuries. Already Greek philosophers (e.g. Aristotle/Socrates and Platon) discussed knowledge as:

- Techne (Practical - Productive Knowledge)
- Episteme (Theoretical - Scientific Knowledge)
- Fronesis (Knowledge as Practical Wisdom)

The meaning of Techne and Episteme is rather obvious to us as e.g. that is how we have separated the university system from the practical world. Thus making good research due to the classical/Newtonian view means that the researcher must have an outsider perspective not interacting with what she/he is studying. When people speak about *Competence* they usually mean a combination of Techne and Episteme.

Fronesis is a more difficult term as it grows when both Techne and Episteme grows. Dynamic and life experienced people with broad knowledge and intuition are wiser than people who lacks one of these characteristics. Therefore is an expert not also automatically wise.

There exist many opinions of what knowledge is, if knowledge and information is the same thing, etc. However it seems logical to regard *knowledge* as something personal that becomes *information* when we express it to someone else either as spoken words or in written form. A big difference between the two ways of giving information is that spoken information evaporates fast if one does not make notes or recordings from a speech or a two way communication while written information lasts. Estimations tell that less than 20 % of the information that we get is used to build up new pictures of the world while the remaining part comes from our earlier in our brains stored pictures (Wheatley and Kellner-Rogers 1996).

The personal ability to increase the individual knowledge is dependent of many factors as e.g. the own intelligence, creativity, ability and feelings in general. The ability to take in information and signals and to combine them with the own so far accumulated knowledge and know-how to new knowledge and know-how, is dependent e.g. of the mood, interest and motivation in each moment. This can be called a filtering effect. Being in good mood, having a large interest and being motivated means that the own filters are open to take in most information and a maximum of all incoming signals. Not being in good mood, being uninterested and being unmotivated means that no or very little information and few signals will be able to add to the own existing knowledge and know-how. Fig. 3 shows the complex reality in a simplified way.

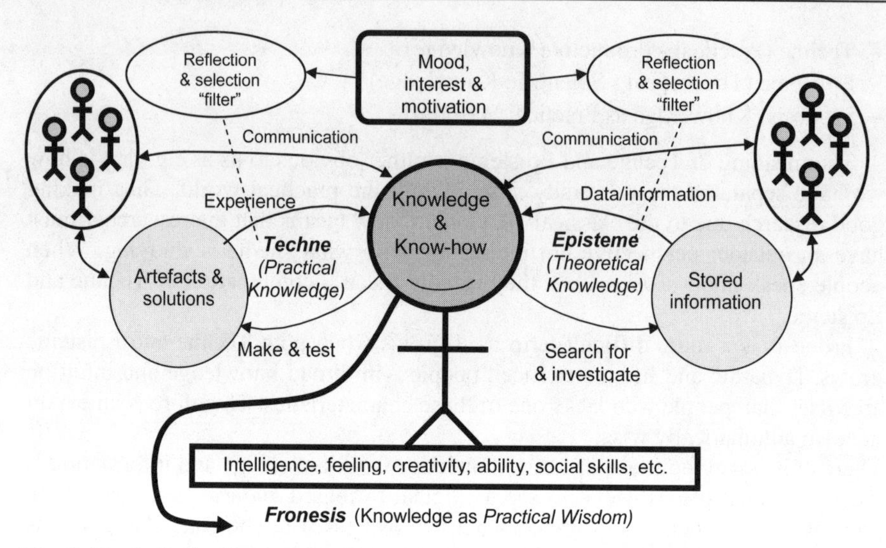

Fig. 3. The individual knowledge and know-how is generated due to many factors of which some important ones are shown in the figure. Stored information is here recorded information. Artefacts are e.g. models, prototypes, manufactured products

The nature of *artefacts & solutions* is often difficult to understand as it is complex. The knowledge we get and the simplified theories we can make we normally store as in different ways in books, on EDM/PDM files, etc. Depending on the own knowledge some *stored information* can be difficult to understand and evaluate. Other information is difficult or impossible to get as it can be qualified as secret material.

Every development process is both an iterative knowledge generation process and an iterative never ending information generation process. By going through stored information over and over again and by communicating with other people we can iteratively improve our knowledge.

As Fig. 3 shows, one important way to extend the own knowledge is to collect and process information from other people through the communication with them. If the individuals are geographically more than approximately 5 m away from each other however the total message transmitted and sensed with our five senses will be rapidly decreasing (Branzell 1995). This is an important reason why it is vital that the project team members during the innovative design phase are localized together in localities ideally without separating walls.

Common for all information is that it can – and is – manipulated to be what the author - of good or bad reasons - wants it to be. Also reality is constructed from our thoughts of reality and that there is no reality until that reality is perceived. Further no clear dividing line exist between ourselves and the reality we observe to exist outside of ourselves (Wolf 1989). Instead, reality depends upon our choices of what and how we choose to observe. These choices, in turn, depend upon our minds or, more specifically, the content of our thoughts and our mood, intentions, interest, our expectations, our desire for continuity, etc. Not to forget also is that the more we determine one side of reality, the less the other side is shown to us!

To deepen our understanding and knowledge of a "signalling" object or phenomena we also can - and must often - manipulate it or the environment around the object to see what kind of reactions we get from our different manipulation actions. To help us notice and evaluate e.g. weak signals we can also take help from "machines" that have been programmed to give us wanted information. We prefer often to call such information for "artificial" implying that machines can think in a similar way as human beings do, which is not possible. This as a thought is not only a phenomena in our brains but also a result of how our brains actively interpret our experiences in the confrontations with the surrounding world. This is probably why all trials to design machines that think as humans so far have failed (Rose 2002).

When two individuals interact (communicate) with each other to solve a problem (see Fig. 4.) their collective knowledge is larger than when they do not interact. The more equal personality and knowledge the individuals have in principle the lower the collective knowledge will be implying that one – to get maximal result - should strive to compose teams with different personalities, knowledge, experiences, sex, religions, etc. The reason why the collective knowledge will be larger when people with different characteristics are brought together is that our "sleeping" knowledge from the deeps of their memories will be pulled forward with the help of association tracks emerging in the communication and from our body language. This is e.g. why Brain storming can help to find new creative solutions.

Organisational considerations

In traditional line organisations the formal information flow is in one dimension (down – up). The manager decides who will get certain information, how much information she/he will get and when. Only the manager has the full overall information, which means that a risky situation is at hand if something should happen with the manager. Also the tempo will be slow when someone who does not have enough information for her/his job must seek information from the manager, who may not be at hand when the information is needed.

The operative knowledge of a team that is organised as a line organisation in principle is limited to what the most knowledgeable individual has. This as no holistic effects will occur between individuals as they have only fragmented information of the whole project. If e.g. a product development team is set up a project manager and four experts with different knowledge profiles they will have some overlapping knowledge, which normally is their bridge to mutual understanding independent of the information the manager chooses to give. They will act independently of each other within their specialities to solve the problems they have been asked to solve and not to engage themselves in problems other team members may have if these problems do not have a direct effect on the own job. That is why we often experience sub-optimizations in line- and matrix organisations implying the need of detail planning and gates to secure that no one gets behind or ahead of the other team members due to the planning.

If a project manager does not fragment information, if she/he decentralises decisions and encourages direct contacts between the experts, the line organisation will be transformed to a planetary organisation (see Fig. 4.). This type of organisation improves due to our experiences human relations within the team and will problems be solved faster and cheaper mostly without the involvement of the manager. As time in that way will be released for the manager, that time can be used by her/him to be one step before the team. However it is important to underline that the manager must not abdicate from being manager by distributing all resources to the sub-managers. If that is done the manager can not implement the advantage of being one step ahead. Also such a behaviour opens up for one or more sub-managers to be informal leaders, which can be problematic.

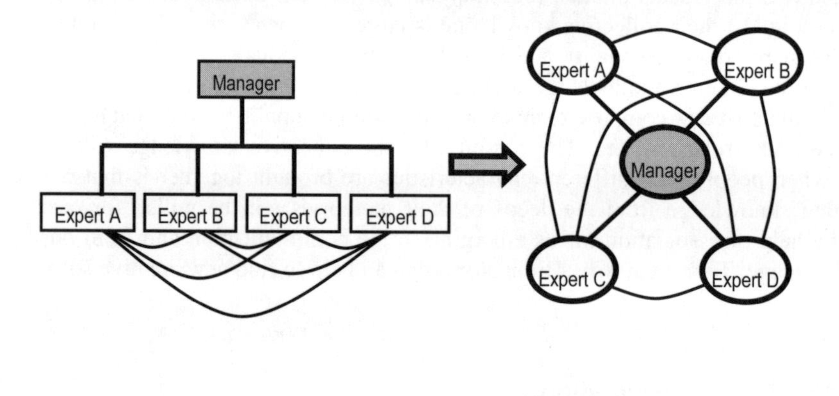

Fig. 4. If each team member is encouraged to inform each other and communicate directly the traditional line organisation is transformed to a planetary organisation. Note that the outsider position of the manager is transformed to a natural insider position if feasible localities are available, which is especially important during the innovative design phase (c.f. Fig. 2.)

Being an active, present and powerful manager who uses the planetary principles means due to our experiences that the operative knowledge will increase compared to when the line organisation thinking is used. This effect can be called holistic. However the more people that are being involved in a project the smaller the operative knowledge increase will be the result with every new member.

Conclusions

Early in the development of new products - during the innovative design phase – it is due to our experiences very important that the designers are localized together ideally without walls separating them. Choosing the planetary organisation model means that costs and time is reduced compared to when line organisations are used. Sitting together also means that a common vocabulary and a common understanding can be created fast, which is important to avoid problems in the later development work. When math/parametric design can be done after the innovative design has been completed, that work can be done on distance as the designers now can and will work individually for which the personality profiles are not that important.

References

Branzell, A., Ottosson, S. and Shapiro, G. (1996): *Localities for Creative Meetings,* Third International Symposium in Product Development in Engineering Education, Halmstad, pp 159-168

Ottosson, S. (2002): *Dynamic Product Development - DPD,* Accepted article to appear in Technovation - the International Journal of Technological Innovation, Entrepreneurship and Technology Management

Ottosson, S. & Björk, E. (2003): *Research Considerations,* Technovation - the International Journal of Technological Innovation, Entrepreneurship and Technology Management, in press

Rose, J. (2003): *I think – but not only with my brain (on Swedish),* Forskning och Framsteg, No 8 December, pp 14 – 20

Wheatly, M.J. & Kellner-Rogers, K.B. (1996): *Self-organization: The Irresistible Future of Organizing*, Strategy & Leadership, April, pp 18 - 24

Wolf, F.A. (1989): *Taking the Quantum Leap,* Harper & Row Publishers, New York

Managing breakdowns in international distributed design projects

Stephen AR Scrivener, Andrée Woodcock, The Design Institute, Coventry University, UK
Lai-Chung Lee, National Taipei University of Technology, Taiwan

Introduction

Globalisation has resulted in increased multinational cooperation. To compete in international markets, small and medium-sized manufacturers need to exploit their particular specialisms (e.g., in terms of personnel, knowledge and resources) by making strategic alliances with companies having complementary areas of expertise. By making alliances with designers located in primary international markets, indigenous manufacturers hope to achieve an in-depth understanding of the target market and their requirements. Cross-national alliances offer the potential for mutual benefits, such as sharing information and resources, reducing costs, enlarging markets, etc. In addition, the geographically dispersed team members may be supported by a number of communication infrastructures (e.g., networked, multimedia environments), which can not only reduce lead-time and the cost of product development, but also facilitate collaboration with partners (Siemieniuch and Sinclair 1999).

Our focus of attention is not large, collaborative ventures, such as NATO and EU alliances, but the smaller, short-term agreements, which are made, for example, between Far Eastern clients and European designers. We term these International Cooperative Design Projects (ICDPs), which typically include a design consultant, clients and mediators or facilitators.

Although little research has been reported, especially concerning ICDPs for product development, what evidence there is suggests that international design collaboration, combining external and in-house design expertise can benefit design. Examples include Samsung's design alliances with the IDEO design consultancy in California (Aldersey-Williams 1996), and IBM's Laptop PC case study of working with an industrial design consultant in Milan (Sakakibara 1998).

Unfortunately, such collaborations and alliances, whilst offering the potential for mutual benefits, can become fraught with managerial difficulties. Problems may arise when trying to communicate across different disciplines and languages, and in managing projects under the limitations of time and distance. ICDP members have to deal with such difficulties in communication, collaboration and management to ensure that the design matches the brief, to meet the project deadlines, to ensure smooth progress through the design process, and so on. Having little time or experience to solve these problems, these difficulties may seriously impact on the overall success of the project and the participants' satisfaction with it.

The problem

From the literature the major problems in international cooperative projects (including international cooperative design projects) were found as follows:

- Time constraints and geographical separation causing communication issues, such as "poor communication" (Enshassi 1994; Egginton 1996) and "waiting for a response" from partners (Aldersey-Williams 1996).
- Project management issues included "schedule delays" (Hetland 1994; Levcik and Stankovsky 1979; Dodgson 11993) and "underestimation of project size and complexity" (Hetland 1994).
- Cultural and language difference (Schneider 1995; Cleland 1994; Pandia 1994).

However, for those interested in understanding the nature of such problems and how best to avoid or resolve them, the literature provides an incomplete picture. For example, although Egginton's (1996) study confirms the importance of facilitators for large, multinational, consortium-based projects, their exact management role (e.g., coordinator or mediator) in ICDPs, especially small design projects, is not explained. No source data could be found on actual projects, perhaps due to confidentiality agreements. The literature does not reveal the stages in which problems arose, how they occurred and what communication media were involved. For example, from Sakakibara (1998) it is not possible to determine what tools were used, what problems occurred during the design project, or how these were overcome by communication technology. Few studies articulate or evaluate strategies for remedying problems.

Hence, although the literature clearly indicates that international distributed working adds layers of complexity, which require additional skills and commitment on the part of all team members to achieve project goals, it provides little direct guidance on how to recognise problems and how to avoid or deal with them. With this background we sought to understand the type of problems that arise in such cooperative endeavours as a prelude to developing and testing what might prove to be general purpose, low cost remedies. Here we describe the research programme but focus attention on the processes of uncovering problems and devising and applying intervention strategies.

Understanding and identifying problems in ICDPs

Our aim was to achieve understanding, identification and alleviation of problems of ICDPs. Understanding was acquired in a number of related ways, i.e., by reviewing the literature, examining ICDP project records and by undertaking a real ICDP.

The China External Trade Development Council (CETRA), a government organisation supporting ICDPs for Taiwanese manufacturers, has sponsored projects for a number of years. We were offered unlimited access to the project documents associated with 21 ICDP projects sponsored by them during the period 1993-1997.

The ICDP project (Case Study 1) involved three participants in the project: a product design consultant in the UK, with a mediator and a client in Taiwan. The client was an office product manufacturer having an in-house design team. A product design project was executed from preparation to detail development using a number of synchronous and asynchronous communication technologies. The client believed a mediator was required, as a translator, because he thought linguistic difficulties might effect the conduct of the work. The project might have been cancelled if the mediator had not been willing to act in this capacity. However, the role of the mediator went beyond that of a translator, as he also supported the installation of the communication technology, as the client had no familiarity or experience in this area. The client's company was a medium size enterprise in Northern Taiwan, producing thermal binding machines, laminators and pouches, hot melt glues and guns, and super glue. The case study design project was to create a two in one machine for thermal binding and lamination, which was seen as an innovative approach for office or Small Office/Home Office (SoHo) users as a space-saving product. Market research had not found any products offering this type of dual functionality. The project life cycle included the four stages: design briefing, design analysis, concept development and concept refinement. Fig. 1 shows both the distributed organisation and the technology used to support the project, and Figure 2 the resultant product.

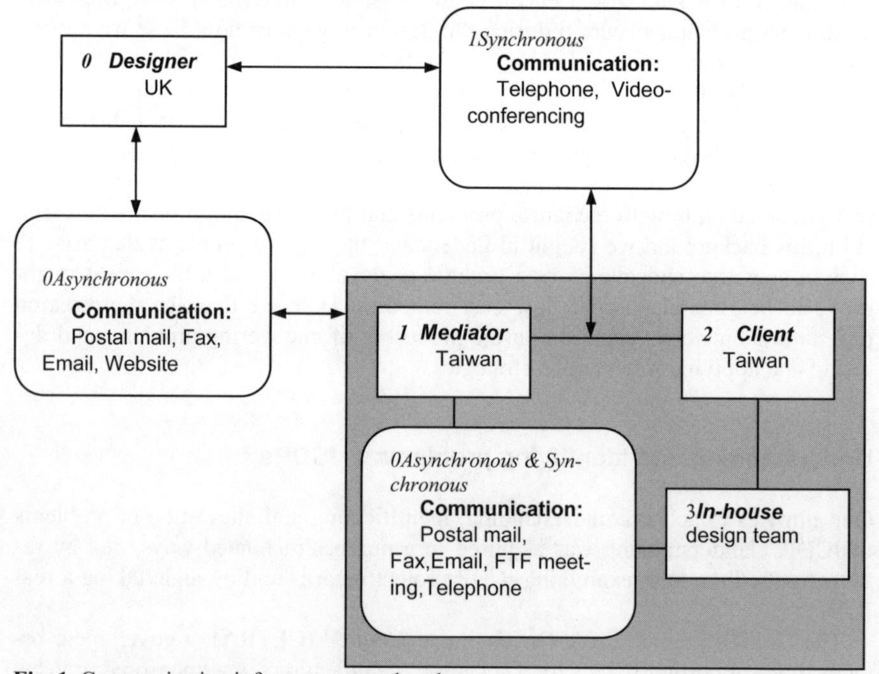

Fig. 1. Communication infrastructure and tools

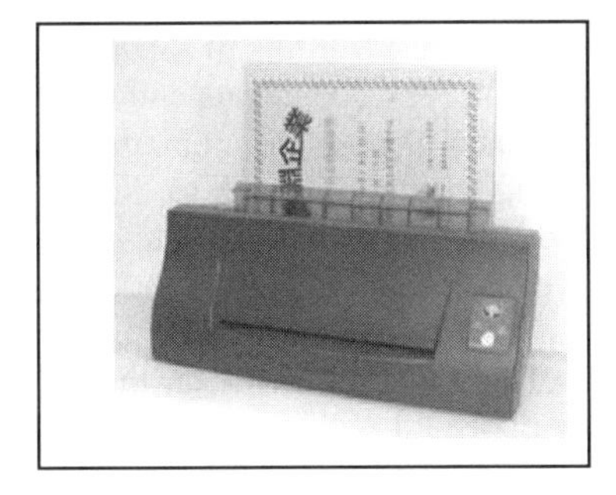

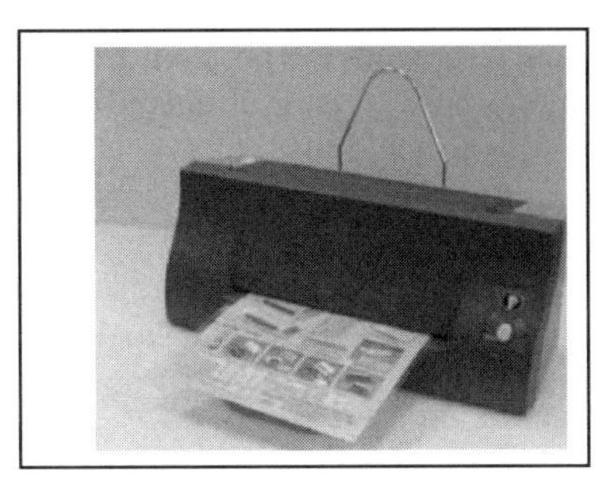

The resultant thermal binder and laminator product

Table 1 illustrates how the three methods contributed to understanding by confirming problems categories, identifying new problems categories and providing in-depth knowledge of when problems occurred, the product being designed and the communication between participants.

Table 1. Summary of problem categories identified by the three methods employed and process data and knowledge obtained

Problem Categories	Literature Review	Documents	Case Study 1
1. Time related		✓	✓
2. Information flow and communication	✓	✓	✓
3. Brief-specification		✓	✓
4. Project management	✓	✓	✓
5. Participation		✓	
6. Language (& culture)	✓	✓	✓
7. Political and economic	✓		
8. Product and quality	✓		
9. Technology			✓
10. Environment			✓
The stage of the project at which problems occurred	Unknown	Yes	Yes
Characterisation of the product designed	Few	Yes	Yes
Complete set of project documentation	No	No	Yes

To understand problems one must be able to identify them. However, when thinking about how to reduce problems, it is not enough simply to be able to identify problems. Instead it is helpful to be able to identify events that predict problems. Armed with such knowledge one is in a position to instigate an intervention strategy so as to prevent the problem from ever occurring. With this in mind, we traced problems back to their origins using both the CETRA documents and Case Study 1 records to yield a set of events that we defined as possible problems. In other words, in an ICDP a lot of events occur during the project, such as discussion, negotiation and decision making. Such events may not necessarily cause actual difficulty for the participants, but some may represent possible problems, i.e., events that are likely to result in actual problems. For example, when an agreement is made this could be targeted as a source of possible problems on the assumption that participants might disagree about or disregard such an agreement in the future. Possible problems were found in both synchronous (e.g., transcripts of videoconferencing sessions) and asynchronous communication (e.g., emails).

In total 57 possible problems were uncovered in Case Study 1 of which 41 were related to 40 actual problems and there were 10 actual problems that had no antecedent possible problem. This indicates two things: most actual problems (i.e. 80%) can be predicted and most predicted problems (i.e., 70%) result in actual problems when no preventative action is taken. Clearly, prevention is likely to be better than cure. Knowledge of the nature of possible and actual problems, the contexts in which they arise, and the relationships between them provided a foundation for the formulation of intervention strategies designed to prevent possible problems from becoming actual problems.

Developing intervention strategies

Having identified ICDP problems from the literature, the document study and Case Study 1, our attention turned to considering how these might be reduced, thereby enhancing the performance of such projects. As implied above, we proposed to achieve this by developing intervention strategies.

The literature relating to facilitation offered material for developing intervention strategies for an ICDP. Keltner's work (1989) was used as a reference for considering when, how often, what focus and how much control should be used when intervening. O'Hara-Devereaux and Johansen (1994) and Dubs and Hayne's (1992) research provided a framework for developing intervention strategies targeted on project stages, i.e., start-up, pre-meeting, during the meeting and post meeting. The intervention strategies were designed by hypothesising the likely causes of the problems identified via the methods described above. Given the probable cause of a problem, an intervention strategy was devised (where practical) to remedy it. Table 2, summarises the intervention strategies devised under broad categories, e.g., different strategies were employed for monitoring communication between participants to accommodate different communication media.

Case Study 1 produced 35 problems during meetings. A lot of the intervention strategies targeted for use at the pre-meeting stage were designed to prevent these from arising. However, 70% of the meeting problems had no antecedents, i.e., they could not be predicted from participants prior actions. The intervention strategies developed for the meeting stages therefore took two form; firstly, intervention by the facilitator to resolve communication/translation problems as they occurred and to quickly overcome technological breakdowns. The second form of intervention strategy focussed on the managerial issues and related to minute taking, recording the outcome of agreements, and noting actions.

Table 2. Summary of intervention strategies categories by meeting stage

Stages	Actions
Project start up	• Train the participants to use the communication media appropriately • Ensure that the product design brief is elaborated • Introduce the intervention strategies • Request information from the client for the designer • Provide a shared vocabulary for the project • Define the rules of feedback and response
Pre-meeting	• Develop and gather agenda items • Inform the participants • Monitor emails and contact with the participants
During the meeting	• Facilitate and resolve communication and translation issues • Overcome technological breakdowns • Make notes of any agreements • Remind the participants of agreements and actions
Post-meeting	• Record and distribute the minutes • Trace agreements and progress on their completion • Monitor emails and contact with the participants

Evaluation of intervention strategies

Although many of these intervention strategies reflect good management practice and therefore should be beneficial, it remained to be seen whether they were appropriate in a mediated-design context. Hence, we undertook another case study to assess the efficacy of the intervention strategies for:

1. preventing possible problems from becoming actual problems;
2. alleviating actual problems arising during the design process.

The needs for intervention suggested the adoption of an Action Research approach in which one researcher functioned as a facilitating member of the design team with a primary role of applying intervention strategies and assuaging actual problems as they emerged. Additionally, given and in-depth knowledge of breakdowns in ICDPs, the facilitator was required to refine or invent intervention strategies as necessary. The other participants in Case Study 2 comprised an UK-based design consultant and a manufacturer and mediator from Taiwan.

The client's company is a small enterprise in Northern Taiwan whose products include a variety of alarm clock movements. This cooperative design project was a milestone for the company, which wished to move from being a component supplier to a product manufacturer. The design project was to create a World-time alarm clock which would help users read the time in twenty-one major cites around the world by adjusting a dial ring on the clock which moved the hour hand to the correct time for the selected city. The proposed project life cycle was divided into six key stages, as per Case Study 1: namely design briefing, design analysis, concept development, concept refinement 1 and concept refinement 2. Case Study 2 employed the same communication infrastructure as that used in Case Study 1.

Preventing possible problems for becoming actual problems

It is tempting to compare the two case studies, particularly since the results favour improvement in the latter (cf. Table 3). However, we will resist this temptation, as it was never our intention to compare the case studies. Rather, the aim was to learn from the first case study and apply this learning in the second. The proportion of possible problems yielding actual problems is taken as the primary measure of learning and the effectiveness of intervention strategies. From Table 3, it is clear that only a small proportion of possible problems converted to actual problems.

Table 3. Possible and actual problems: Case Studies 1 and 2

	Actual Problems (APs)	Possible Problems (PPs)	APs associated with PPs	APs with no precursor PP
Case Study 1	50	57	40	10
Case Study 2	15	50	12	3

Of course, it is still possible to argue that one case study is insufficient evidence of the correlation between possible and actual problems. If there is no link between possible and actual problems then the proportion of actual to possible problems is meaningless. Furthermore, one might argue that even if there were a link, the small proportion of actual problems might be due factors other than the intervention strategies. Clearly, given the research approach taken and the data generated in the studies described above, such objections cannot be conclusively countered here.

However, notwithstanding the limited opportunity here for elaboration, we can say something about how intervention strategies addressed possible problems (although we cannot be certain that they prevented actual problems, as it is impossible to say whether related actual problems would have occurred in the absence of intervention).

One intervention strategy involved the elaboration of the design brief prior to the Design Briefing meeting. It is likely that the small number of requests for information relating to design requirements was due the enhanced design brief. In Case Study 2, there were a total of 33 design requirement discussions identified, i.e., possible problems. Here the intervention strategy was to explain design requirements and it is again likely that it contributed to the small number of related actual problems, i.e., six.

Translation intervention strategies addressing problems of "misunderstanding" and "inaccurate translation" probably contributed to the small number of such problems, i.e., three. In Case Study 2, inconsistencies of viewpoint did not occur, which may be attributed to the role of the intervention strategy that required the mediator and the client to reach a common understanding.

Hence, it is reasonable to argue that the intervention strategies addressed possible problems, and assuming a relation between possible and actual problems, probably account for the fact that so few of the former materialised as the latter.

Alleviating the impact of actual problems

Nevertheless, some possible problems resulted in actual problems. For example, during the Concept Refinement 1 meeting, the client and mediator promised to provide technical information to the designer relating to a design requirement. The intervention strategy was to record this in the minutes, to monitor asynchronous communication and to remind the participants before the deadline. Although active intervention was undertaken, the agreement was not completed by the agreed date and became an actual problem, i.e., 'failure to fulfil an agreement'. Following subsequent email exchange, the information (i.e., the wiring diagram) provided by the mediator did not completely match the designer's request. Hence, an actual problem of "lack of information" was recorded.

Hence, the intervention strategies employed could not prevent some possible problems becoming actual problems, particularly relating to language (e.g., lack of understanding, caused by linguistic problems due to cultural differences) and scheduling.

In both case studies, communication breakdowns proved most problematic in terms of actual problems, notwithstanding the intervention strategies employed in Case Study 2, such as explaining design requirements, redefining terminology and helping to resolve the impact of communication breakdowns, such as "lack of understanding".

Given time differences and geographical separation between distributed design team members, it is perhaps no surprising that the, in both case studies, the project fell behind the schedule, which then had to be renegotiated. The need for schedule change was difficult to predict as it depended on the number of concurrent projects on which the designer was engaged at any moment. Thus, intervention strategies focused on minimising the impact of schedule changes. Technological breakdowns also occurred and these had to be remedied immediately during the video session.

Summary

Overall, then, the results demonstrate that the intervention strategies helped the participants to achieve a greater level of managerial control and made a positive contribution to the ICDP by preventing and managing problems.

Conclusions

Prior research and the research described here have established that international distributed work adds layers of complexity to team design. We are beginning to build an understanding of the problems that may arise during ICDPs. Our research has contributed to this understanding, particularly in regard to small projects, by developing and refining existing taxonomy, by uncovering additional categories, and by establishing when and why problems arise during the life cycle of a project.

However, although it is useful to know what problems are likely to occur and when, we need techniques for identifying potential problems, so as to prevent actual problems from occurring in the first place, and for minimising, in action, the impact of actual problems that cannot be prevented or predicted. In tracing observed problems back to their causes, we have uncovered symptoms that provide reliable indicators of problems. Hence, we have developed techniques that enable team members to identify both possible and actual problems during the life of a project. The former are particularly useful as, given the symptoms of a problem we can take preventative action. Finally, we have devised and tested, using an Action Research approach, remedial strategies sensitive to problem taxonomy and project life cycle. The evaluation results indicate that such remedial intervention can make a positive contribution to the smooth progress of a project.

Both this and past research suggests a low level of project management skills in the design community. Some of the remedial strategies employed in this study can be regarded as good management practice. However, language and cultural problems, which are also implicated in many information flow and communication problems, are certainly exacerbated by international distributed work, and good practice has yet to be established. Furthermore, the level of project involvement required in our study in order to recognise possible problems, to implement known remedial strategies, to devise strategies on the fly, to monitor mediate and facilitate communication was intense. Indeed, this would seem to go beyond the call of duty of the conventional project manager and seems imply the need for project personnel with a specific brief of mediation and facilitation.

References

Aldersey-Williams, H. (1996). Design at distance: The new hybrids. *Design Management Journal*, 7, 43-47.

Cleland, D. I. (1994). Borderless project management. In D. I. Cleland and R. Gareis (Eds.), *Global Project Management Handbook*, New York: McGraw-Hill, pp. 1-3 - 1-15.

Dubs, S., and Hayne, S. C. (1992). Distributed facilitation: A concept whose time has come?, *CSCW '92*, Toronto, Canada: ACM, pp. 314-321.

Egginton, B. (1996). Multi-national consortium based project: improving the process. *International Journal of Project Management*, 14, 169-172.

Enshassi, A. (1994). The management style of multicultural construction managers in the Middle East. In D. I. Cleland and R. Gareis (Eds.), *Global Project Management Handbook*, New York: McGraw-Hill, pp. 24-1 - 24-25.

Hetland, P. W. (1994). Toward ultimate control of megaprojects in the North Sea. In D. I. Cleland and R. Gareis (Eds.), *Global Project Management Handbook*, New York: McGraw-Hill, pp. 30-1 - 30-21.

Keltner, J. W. (1989). Facilitation: Catalyst for group problem solving. *Management Communication Quarterly*, 3, 8-31.

Levcik, F., and Stankovsky, J. (1979). *Industrial Cooperation between East and West*: M. E. Sharpe.

O'Hara-Devereaux, M., and Johansen, R. (1994). *Globalwork: Bridging Distance, Culture, and Time*. San Francisco: Jossey-Bass.

Pandia, R. M. (1994). International projects: Opportunities and threats. In D. I. Cleland and R. Gareis (Eds.), *Global Project Management Handbook*, New York: McGraw-Hill, pp. 18-1 - 18-23.

Sakakibara, K. (1998). Global new product development: The case of IBM notebook computers. In M. Bruce and B. H. Jevnaker (Eds.), *Management of Design Alliances: Sustaining Competitive Advantage*, pp. 91-105. Chichester: John Wiley and Sons.

Schneider, A. (1995). Project management in international teams: Instruments for improving cooperation. *International Journal of Project Management*, 13, 247-251.

Siemieniuch, C., and Sinclair, M. (1999). Real-time collaboration in design engineering: An expensive fantasy or affordable reality? *Behaviour and Information Technology*, 18, 361-371.

How Engineering Designers Obtain Information

Ken Wallace and Saeema Ahmed, Engineering Design Centre,
Department of Engineering, University of Cambridge

Introduction

The industrial world is changing rapidly and engineering organisations need to adapt fast to remain in business. Some of the pressures faced by engineering managers include: (1) the trend towards globalisation and the reliance on IT; (2) the increasing commercial pressures which demand continual improvements in quality, shorter lead times and reduced costs; (3) the increasing complexity of both products and processes; (4) the problems of retaining and retrieving knowledge and experience; and (5) the need to move towards sustainable development.

Because the decisions made during the design process influence all downstream costs, the quality of an organisation's design process has a crucial impact on its long-term survival and profitability. The engineering design process can be described as a complex *information processing* activity, directed by the decisions made by the individuals in the design team. How engineering designers, both experienced and novice, obtain the information they require and how much time they spend doing so are questions of considerable importance when planning new information support systems.

Traditionally knowledge and experience have been retained in the heads of individuals who were regularly consulted throughout a design project. However, the opportunities to consult in-house experts are now diminishing, and may disappear altogether. It is therefore important to find new ways of storing knowledge and experience in order to access effectively previous solutions and the rationale behind them. Computer support systems are bound to be involved and, with almost limitless storage capacity on the horizon, they would, at first sight, appear to provide the solution. Storing information may not be a problem, but knowing *what* information to store, and how to structure and retrieve it, are key issues that need to be addressed. It is worth remembering how engineering design was done and information managed not so long ago.

In, say, the 1960s, an individual, frequently referred to as the Chief Designer, would guide and oversee the whole design process, which was normally undertaken in a single Design Office equipped with drawing boards. Design teams were co-located and supported by a small number of specialists. The product was defined by a set of drawings supported by appropriate reports and calculations, and every designer maintained a detailed logbook. A young engineer would often join a company as an apprentice and expect to remain with the same company until the retiring age – a career of up to 50 years. Many would stay in one particular functional area and become acknowledged experts in that technology.

When a new product was being designed, they would, as holders of the company's knowledge and experience, be the first to be consulted to answer information queries. Communication was frequent and informal within the team. The distance between the design and manufacturing departments was relatively small, and designers had 'hands on' knowledge of their products and the processes by which they were manufactured. The methods and tools were simple and comprehensible, with nearly all calculations being done using a slide rule. Mainframe computers were just beginning to be introduced into large companies. The knowledge aspects could be characterised as local, visible and stable. Knowledge was exchanged informally by personal contact, and handed down through Master/Apprentice relationships.

The situation is very different today, but great reliance is still placed on being able to access the knowledge and experience of individual experts. How would a design team cope if these experts were no longer available to consult? It is important to plan for this possibility, and to help with this two parallel approaches are suggested. The first is to undertake research to understand how information is currently obtained and the second is to speculate how new technology might change things in the future. It is important to achieve a balance between retaining what is good from present practice and adding to this the potential for improvement offered by new technology. It can be dangerous to embrace new technology too rapidly when current good practice, which has evolved over many years, is not fully understood. The purpose of this paper is to report on research into current practices rather than to speculate on the impact of new technology.

Two characteristics of the current situation are worth highlighting. The first is the transient nature of modern organisations and workforces, and the second is the rapid flow of knowledge and experience out of companies due to staff leaving, often through early retirement.

These days engineers tend to move jobs much more rapidly both within companies and between companies, and more staff are being employed on short-term contracts. Design teams are frequently distributed, and it is common for design and manufacture to take place in different countries. There is less personal loyalty to companies and the traditional functional experts are disappearing rapidly. In recent years the working life has become shorter due to more young people going on to higher education and older staff retiring younger. In some cases this has reduced a typical working life from 50 to 30 years, though there are signs that this trend may reverse in the future. As a consequence, new designs are frequently undertaken by less experienced staff, often referred to as novices, who have less access to traditional functional experts.

There is therefore an urgent need to retain knowledge and experience independently of individuals so that it can be accessed easily and novice designers can gain 'experience' more rapidly. Companies are therefore trying to establish novel information retrieval systems, e.g. lessons learned databases, and encouraging their staff to store their knowledge so that it can be shared. However, there may well be some reluctance on the part of individuals to do this. If the move towards transient employment and short-term contracts continues, and companies more frequently make people redundant in the pursuit of short-term profits, then individuals will wish to retain as much knowledge as they can as this makes them more readily employable.

A great deal of time is spent writing and storing reports. This is not a popular task and is usually done under pressure at the same time as work starts on the next project. There are two reasons for recording information: (1) to provide an audit trail in the event of some future failure and consequent liability; and (2) to store knowledge and information for future reuse. It is commonly assumed that frequent use is made of documented records when starting a new design, particularly by novice designers. Based on this assumption large sums are spent on storing documents electronically and providing improved access to them.

Two main questions are addressed: (1) How do designers currently obtain their information? and (2) What is the best way to help novice designers obtain appropriate information?

Traditional methods for gathering data on how designers act include conducting interviews and using questionnaires. However, both these approaches are retrospective and frequently fail to provide an accurate picture of what happens in practice. It is not that those responding to interview questions consciously try to deceive interviewers, it is simply that we all tend to provide the expected answers and to report the past in a rational and coherent way. Direct observation, although time-consuming, provides more reliable data.

Two observational studies undertaken in the same aerospace company are described briefly. The first study observed teams of four designers to see how they answered their information queries. The second study observed novice and experienced designers to see if there were differences in the way in which they approached their tasks and how they sought and used information.

Study 1

Marsh undertook the first study in an aerospace company in the UK, starting in 1993 (Marsh 1997). The aim of this study was a preliminary investigation into the capturing and structuring of design knowledge and experience. In the 1960s, the company had made a concerted effort to start recording its 'knowledge and experience' for future reference in a series of *technical reference documents*. It is reasonable to assume that this series of documents would be considered by designers to be particularly valuable and therefore referred to frequently. This turned out not to be the case.

Prior to starting the observational study, it seemed important to address the question: How frequently do designers refer to documents, including the technical reference series mentioned above? To answer this question, a survey, in collaboration with the University of Bath, was undertaken within the company (Marsh and Wallace 1995). The company maintained 31 series of documents, of which the technical reference series was just one. The results were normalised to estimate how frequently an 'average' designer referred to these documents. The results are summarised in Fig. 1. From this one can conclude that documents are referred to far less than one might expect. The two most frequently referred to series were: (1) procedures to be followed to meet specific requirements; and (2) materials selection. On average these were only referred to once every two weeks. The next two most frequently referred to series were standards manuals and the technical reference series mentioned above. On average these were referred to once every four weeks. It can be seen from Fig. 1 that over half the document series produced were hardly referred to at all.

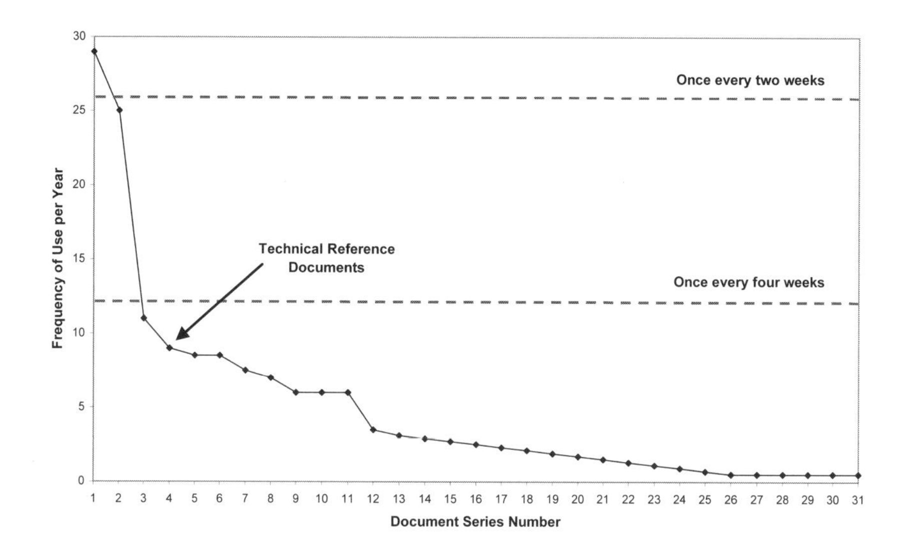

Fig. 1. Frequency of use of document series

It was possible that some documents were considered valuable but were only referred to infrequently. The survey was therefore extended to investigate the link between frequency of use and perceived usefulness. There was a clear correlation, i.e. the most frequently referred to series were considered to be the most valuable. Possible reasons why documents are not referred to include: (1) the document indexing is poor and the information cannot easily be retrieved; (2) the documents do not contain the required information or it is out-of-date; and (3) the information is in the documents, but not in an accessible form. These issues were addressed as part of the research, but one specific question was: How do designers currently obtain their information if they do not use documents?

To answer this question, an observational study was undertaken. Three groups of four designers were observed undertaking their normal design tasks. This yielded 51 person-days of data captured on a total of 1530 data capture sheets (Marsh 1997). The results were carefully analysed and some interesting results emerged. The main conclusion was that in nearly 90% of information requests, the designers contacted another person, see Fig. 2 wo people were therefore involved in each information search: the person asking the question and the person answering it. In 78% of the cases observed, the information was provided from memory. This finding is supported by Court who also found that, in general, designers made extensive use of their memories to retrieve information rather than searching in documents (Court et al. 1996).

From the data, it was simple to work out how much time on average a designer spent each day obtaining information and it turned out to be 26%, with the remaining 74% being divided between design time (67%) and meetings (9%). This division of time would appear reasonable if each of these activities occurred as a continuous period each day. However, information requests do not take place during one continuous period – they occur randomly throughout the day and cause significant fragmentation. From an analysis of the data, it was calculated that there was only a 50% chance of working for eight minutes without needing to obtain or give information. This conflicts with the requirement that creative design work needs intense, uninterrupted periods of concentration. The time spent searching for information is supported by other studies, e.g. the one by Rodgers who found that designers in a telecommunications company spent 20-30% of their time searching for information (Rodgers 1997).

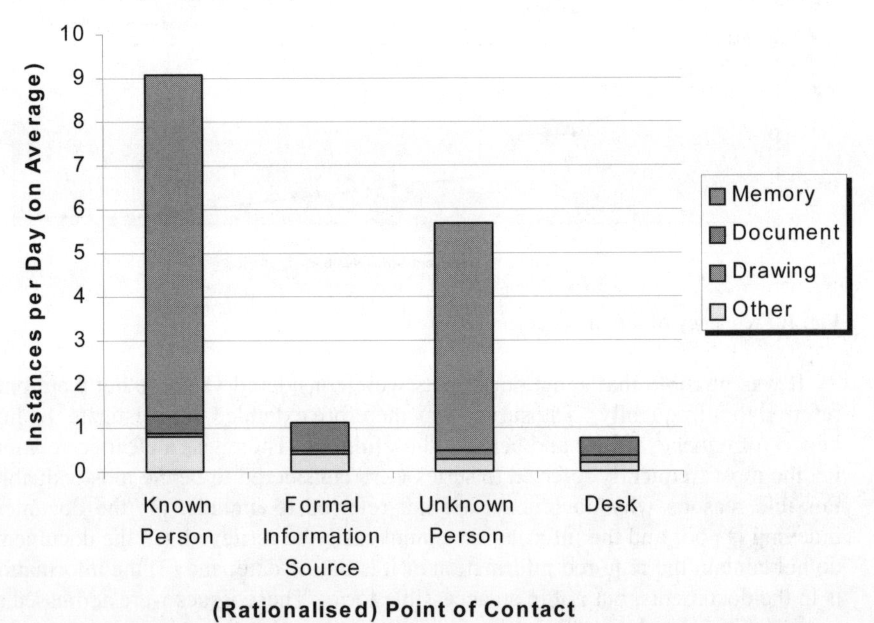

Fig. 2. Sources of information requests

It is interesting to consider why for 90% of information requests designers prefer to talk to another person. Three key factors appear to be speed, context and confidence. When undertaking some aspect of a design, one is often not sure what information one requires. It is therefore effective and efficient to have a 'dialogue' with an expert. This dialogue might be prompted by a particular question, but the person being asked usually asks for further information to set the question in context – something that computers cannot yet do. Once the context is established, then advice is given and in around 50% of the occasions observed the enquirer went away having had a completely different question answered. Even if one knows that the required information is available in a document, there is a tendency not to trust documents, believing that they are out-of-date. If one asks a trusted expert, one's confidence in the information being correct and up-to-date is much higher. The support gained from sharing a problem is also an important factor.

Some conclusions from the first study were:

- Documents were very seldom used as a source of design information
- For around 90% of information requests, designers contacted another person
- Approximately a quarter of each working day was spent requesting or providing information, and this significantly fragmented the working day.

Study 2

The arguments presented in the introduction suggest that increasingly design activities will be undertaken by less experienced or novice designers, and that there will be less opportunities to talk to more experienced designers. There is therefore a need to assist novice designers to gain experience more rapidly than in the past. This issue is partly addressed by the question: What is the best way to help novice designers obtain appropriate information?

It was clear from Study 1 that all designers, both novice and experienced, preferred to talk to another person when seeking information. A detailed analysis of the data showed, not surprisingly, that novice designers spent the most time consulting with their more experienced colleagues. Similar results have been reported from other studies. For example, Frankenberger, who undertook a protocol study in industry, noted: "Very often, the consultation of colleagues in the design process compensates for lack of experience". He also reported that the single greatest factor (30% of cases) attributed to wrong decisions and false analysis was the availability of information rather than the lack of experience (Frankenberger and Bade-Schaub 1996).

Before proposing new procedures and systems to support novice designers, it seemed wise to undertake a further observational study. Ahmed undertook this study in the same aerospace company in the UK, starting in 1997 (Ahmed 2001). The overall aim was to investigate the differences in the ways that experienced and novice designers approached design tasks. Part of this study addressed the following specific questions: (1) Can the strategies used by experienced designers be identified? and (2) Can these strategies assist novice designers with their information requests?

For this observational study six experienced and six novice designers were observed individually. The experienced designers had between 8 and 22 years of experience, and the novice designers between one and two-and-a-half years. They were drawn equally from two groups within the company: one group working on conceptual design, the other on detail design. The designers were observed for periods of between 90 and 120 minutes in their normal working environments working on their own tasks. Their tasks lasted much longer than this, so the observations only provided 'snapshots' of their working procedures. During the observations the designers were asked to 'think aloud' and the sessions were audio-recorded. Each observation was followed by an interview lasting 15-20 minutes in order to establish information about the background to each task and the experience of each designer.

A total of 20 hours of data was gathered and analysed. Undertaking observational studies such as this is extremely time consuming. For each hour of observation, a further 25 hours of additional time was required for planning, transcription and analysis. No categories for the data were determined prior to starting the analysis. Categories were generated as the transcripts were analysed a few lines at a time to identify all the activities being undertaken. The procedure was continued, with transcripts being reanalysed, until all the activities identified could be placed in a category. In total 22 categories were identified (Ahmed 2001). From this list, eight categories were identified as being associated with specific *strategies* adopted by experienced designers when tackling design tasks. Table 1 shows a comparison of designer behaviour for these eight selected categories and highlights the clear differences between novice and experienced designers.

In addition to observing the absence of clear strategies in novice designer behaviour, a significant finding was that novice designers were reluctant to approach experienced designers immediately when they had a query, preferring instead to save up a list of questions. Clearly social and personality factors play a role here, but it can be concluded that, because the novices were not aware of the strategies, they often did not know what questions to ask. This conclusion is supported by further work undertaken by Ahmed (Ahmed et al. 2003). It was also clear that the adoption of these strategies could be prompted by questions aimed at gathering missing information. From this emerged the important conclusion that when supporting the information needs of novice designers, it is necessary to provide *questions* as well as *answers*. This changes the basic approach to current information support systems, most of which simply attempt to supply information and data. It is also more likely that novice designers will feel confident when approaching experienced designers if they know that that the questions they are going to ask will match those that the experienced designers would expect to answer.

Table 1. Patterns observed for novice and experienced designers

Categories	Novice Designers						Experienced Designers					
Designer	N1	N2	N3	N4	N5	N6	E1	E2	E3	E4	E5	E6
Company experience (years)	0.3	0.7	1	1.5	2.5	1.5	8	9	18.5	11	19	32
Non-company experience (years)						1.5				8		
Consider issues	Low	Low	Low	Low	Medium	Medium	High	Low	Low	High	High	Medium
Aware of reason	Low	Low	Low	Low	Low	Medium	High	Low	Medium	High	High	Medium
Refer to past designs	Medium	Low	Low	Low	Medium	Low	Medium	High	Medium	High	Medium	Medium
Question is it worth pursuing	Low	Low	Low	Low	Low	Low	High	Low	High	High	Medium	Low
Question data	Low	Low	Low	Low	Low	Low	High	High	High	Low	High	Low
Keep options open	Low	Low	Low	Low	Low	Low	High	Low	High	Low	High	Low
Aware of trade-offs	Low	Low	Low	Low	Medium	Medium	Medium	Low	Low	Low	High	High
Aware of limitations	Low	Low	Medium	Low	Low	Low	Medium	Medium	Low	Medium	High	Low

Low occurrence　Medium occurrence　High occurrence

Table 2 summarises the eight strategies adopted by experienced designers and provides sample questions linked to each of the strategies. Ahmed has transformed these strategies and questions into the basis for support system called C-QuARK (Ahmed and Wallace 2001). This approach has been tested and the results were encouraging. They showed that when using C-QuARK, novice designers asked more relevant questions and began to adopt the strategies of more experienced designers.

Some conclusions from the second study were:

- Experienced designers adopted eight basic strategies when designing
- Novice designers were unaware of these strategies and therefore failed to ask the right questions
- The eight strategies could be phrased as generic questions and used as the basis of a support system.

Table 2. Eight strategies adopted by experienced designers and sample questions

	STRATEGY	DESCRIPTION	SAMPLE QUESTIONS
1	Consider issues	Assess relevant issues and determine the most important	What issues are relevant? Which are the most important?
2	Aware of reason	Recall reasons behind previous designs	What is the reason why this component is used? How does this component function?
3	Refer to past designs	Recall past designs from memory and other sources	Which designs are similar? Which designs were subjected to similar conditions?
4	Question is it worth pursuing	Evaluate current approach and alternatives for benefits	How much can I expect to achieve if I continue this approach? Is it worth pursuing?
5	Question data	Question all information and data received for accuracy and context	How accurate is this value? How much does accuracy matter?
6	Keep options open	Plan ahead and keep options open for future change	What should be considered further down the design process? Does this option limit later options?
7	Aware of trade-offs	Assess the relationships between issues	What other issues does this affect? What other systems does this affect?
8	Aware of limitations	Assess the limitations of the task and overcoming them	What should the current task expect to achieve? What should I be satisfied with?

Conclusions

There is considerable pressure to improve the design process and reduce the number of errors made. A key area to focus on is the effectiveness and efficiency with which engineering designers obtain the information they require. One approach being adopted is to improve electronic database support systems. There is certainly a role for such systems, but the results of the research undertaken at Cambridge indicate that such systems may not provide the complete answer.

The results of two observational studies were described. A clear conclusion from the first study was that for 90% of their information queries, designers preferred to talk to another person. It was also clear that novice designers needed to ask the most questions. This led to the second study during which the differences between the approaches adopted by novice and experienced designers were investigated. A clear conclusion from this study was that novice designers did not know what questions to ask and this made their information searching less effective than it could be.

It was suggested that the reason why designers liked to get their information from another person was that it was an effective and efficient approach when *searching* for information rather than simply *retrieving* information. Talking to someone allows the context surrounding an issue to be discussed and expanded so that the key questions are answered, and these are frequently not the questions originally asked. Information and advice from experts is also trusted more than information obtained from documents and databases, which are frequently believed to be out-of-date. The approval of an expert for the course of action being pursued increases confidence and leads to a feeling of shared responsibility.

These conclusions are worrying when, if the current trend continues, it is likely that fewer experts will be available to consult in the future and novice designers will have to rely more on computer-based information systems. It its therefore urgent that knowledge is captured and stored in such systems in a manner that will enable it to be easily retrieved and, once retrieved, trusted. An encouraging approach is to structure such support systems on the strategies adopted by experienced designers, i.e. these systems should suggest questions as well as provide answers.

It seems wise to aim to get computers to do what they do well and humans to do what they do well. Currently computers are very effective when one is simply retrieving information, but far less so when one is searching for information – particularly when one does not know exactly what one is looking for.

Acknowledgements

This research was funded by the Engineering and Physical Sciences Research Council (EPRSC) and the BAE SYSTEMS/Rolls-Royce University Technology Partnership for Design. The authors acknowledge the support for this research from Dr Michael Moss from Rolls-Royce plc.

References

Ahmed S (2001) Understanding the use and reuse of experience in engineering design. Ph.D. thesis, Cambridge University Engineering Department.

Ahmed S, Wallace KM (2001) Developing a support system for novice designers in industry. In: International Conference on Engineering Design (ICED 01), Design Management – Process Information, IMechE, pp 75-82

Ahmed S, Wallace KM, Blessing LTM (2003) Understanding the differences between how novice designers and experienced designers approach design tasks. Research in Engineering Design, vol 14 (1), pp 1-11

Court A, Culley S, McMahon C (1996) Information access diagrams: a technique for analysing the usage of design information. Journal of Engineering Design, vol 7 (1), pp 55-75

Frankenberger E, Badke-Schaub P (1996) Influences on engineering design in industry. Report, TH Darmstadt, Germany

Marsh JR (1997) The capture and structure of design experience. Ph.D. thesis, Cambridge University Engineering Department

Marsh JR, Wallace KM (1995) Integrity of design information. In: International Conference on Engineering Design (ICED 95), Prague, 4, pp 1449-1454

Rodgers PA (1997) Capture and retrieval of design information: an investigation into the information needs of British Telecom designers. Cambridge University Engineering Department Report, CUED/C-EDC/TR5

Interaction between individuals: Summary of Discussion

Herbert Birkhofer and Judith Jänsch, University of Technology, Darmstadt

1. A shared mental model

Considering the multitude and variety of factors that determine interaction between individuals, it is sensible to establish a common understanding of the issue as a starting point for a detailed discussion. This common understanding can be achieved by a shared mental model of interaction between individuals, i.e. a system with related elements (Fig. 6). This model, which is based on the holistic view in Fig. 5, can also be used for emphasizing independent objects of research and facilitating the demarcation of research areas and their objectives.

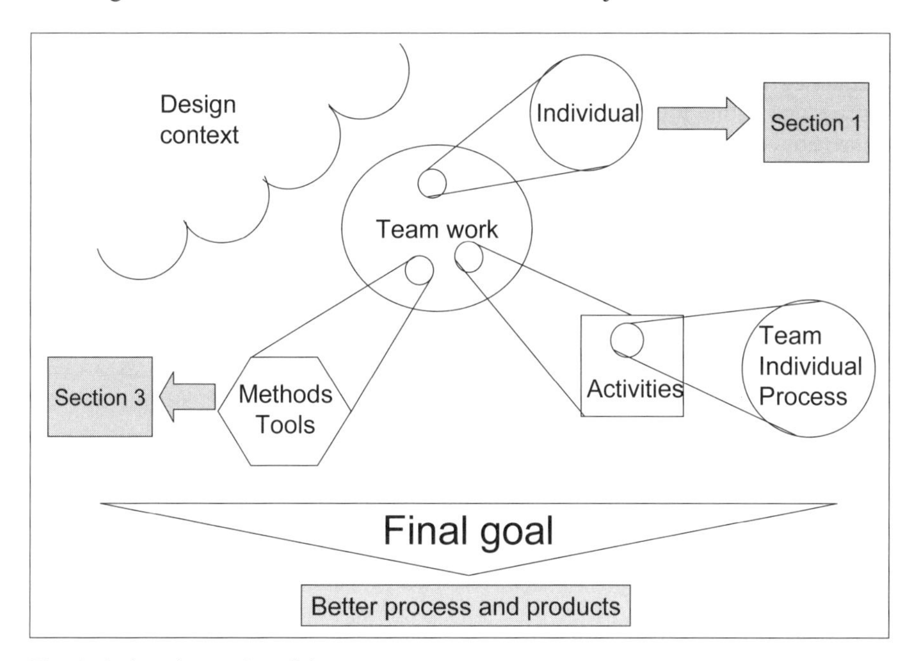

Fig. 1: A shared mental model

The model consists of the design context and the teamwork which represent interaction between individuals. Teamwork by itself includes individuals carrying out activities supported by methods & tools. Thus, the interaction between individuals basically takes place within design teams. For this reason, the focus of the following considerations is on teamwork and group processes, respectively. Teamwork in a design context should be oriented to achieve better design processes and better products for the market.

2. Starting questions

By means of the model in Fig. 1, it was first found possible to form three main questions, which should be regarded as key questions in upcoming investigations:

- What are the activities in group processes?

- What are characteristics from teams and individuals which generate a successful teamwork?

- What are the requirements on a team for the successful use of methods and tools?

It seems impossible to answer these questions comprehensively and conclusively as the contributions of chapter 2 provide a survey of a wide field of research from various viewpoints like asking questions, mental imagery, obtaining information, etc. But it was deemed attainable and useful to reduce the analysis once again to include only the success factors of teamwork according to the model mentioned in Fig. 6.The following chapter contains the findings, proposals and discussions within this research area, regarding success factors of teamwork mentioned above.

2.1 Findings regarding the activities in group processes

Concerning Fig. 1, success factors for activities within group processes can be separated into factors which concern team activities, individual activities and process activities. For each of them one can formulate activities in detail, which are crucial for the success of teamwork.
First, the team activities refer basically to group-forming actions:

- selecting a leader (his/her competence and faculties, recognition within the team, skills in moderation and presentation)
- selecting members (their competence and qualification, willingness to share teamwork, interdisciplinary their interdisciplinary what?)
- identifying experts (their specific competence, willingness to share their expertise with others)
- …

Second, there should be team activities to shape the team. These activities support the team members in improving their orientation within the team and in understanding the ongoing design process and its needs:

- motivating the team (motivation is heavily influenced by the design task and the design situation, and needs room for the development of a positive group climate and sufficient time for discussing design variants mentioned by team members)

- monitoring the team (teamwork tends to develop its own dynamics and it is essential to define and visualize the balance between sophisticated discussions about quite specific details and an abstract view on the issue neglecting important characteristics)
- control team (team members from different hierarchical levels, with different rhetorical abilities and different attitudes have to be integrated in a harmoniously acting team, thinking ahead and contributing to each other)
- ...

Third, there are team activities which ensure processes within the team. These activities make teamwork more transparent for the team members. As a result they have a better understanding of their individual contribution to the whole process.

- planning the process (should be done at the very beginning of project work as a guideline for all members, members should participate in the planning process by contributing their own concepts and improving their motivation)
- reflecting on the process (reflection asked by team members or team leader may avoid digressions and repetition in design work)
- ...

Beside the team activities there are individual activities which are necessary for successful processes within teams:

- attend to requirements (understanding the need of requirements as a key-factor for good decisions and a commitment of every team member to the requirements
- making decisions (a most crucial activity for the success of design work and the product designed)
- communication (deficits in communication abilities are often mentioned as a key cause of an insufficient group climate and demotivation)
- ...

Activities which support the design process within its progress may be embodied in the methodical procedure. It is important for team members to recognize their own work in order to have a good command of structuring design processes and the use of action-based guidelines. This supports the implementation of design methods in the working process:

- clarify tasks
- introduce task
- generate requirements
- visualize requirements
- verify requirements
- develop solutions
- attend to requirements

- generate concept/embed/detail
- evaluate
- select
- ...

On the one hand, it is obvious that convincing success in teamwork can only be achieved if activities are carried out in accordance with the design context and its limitations. On the other hand, there is a need to create an appropriate design context for the teamwork itself. The interplay between design context in general and the framework for successful teamwork might be a challenging subject for future research.

2.2 Findings regarding the characteristics of teams and individuals

Success factors for teamwork are without any doubt caused by the individual and group characteristics. Individual characteristics are provided by the personalities of each team member. According to the expertise of the workshop participants the following characteristics of individuals are considered desirable in team members:

- self-confidence (based on successful design work in the past, but not exaggerated/excessive self-confidence)
- motivation (heavily influenced by design context and the history of the individual's own work)
- expertise/knowledge (a major subject for successful design work, requires a positive attitude to life long learning)
- stress resistance (important in actual design work with its challenge for multitasking and simultaneous engineering)
- social competence (influences group climate and supports or hinders knowledge acquisition by means of communication)
- ability to communicate (besides the power of information technology, one has to conclude that most information in actual design work is acquired by communicating with other designers or with people inside or outside the company)
- tolerance for ambiguity (decisions in design work could often not be made with an exact 'yes' or 'no'; the variety of criteria with vague assessment and influences of social values call for a remarkable amount of open-mind ness in designers)
- …

Next to individual characteristics, the characteristics of the group are decisive for the success of the group-process:

- trust (in the honesty of other members and in the backing of the team leader and the company in general)
- reliability (every member has to carry out his tasks punctually and obtain good results)

- team role flexibility (changing the role within the team may be a stimulating element and can avoid boring discussions and unsuccessful work with rising demotivation)
- communication (key factor for the success of teamwork; should be orientated in the goal and carried out with constructive criticism)
- coordination (a main task for the team leader, who should do it in accordance with the team members; good coordination requires the delegation of tasks to the "right" people and justifying decisions in an appropriate manner)
- chemistry (a kind of soft factor concerning the perceptions, mental images and concepts of every member; might be based on experiences from a long time ago)
- shared knowledge (should be teamwork to avoid misunderstanding and useless discussions)
- team size (should be defined according to the size of the task and its complexity)

2.3 Requirements on the team regarding the successful use of methods and tools for group activities

It was a clear statement within the expert-team that, under no circumstances, an appropriate or even successful use of methods and tools is achieved by any one person's performance alone. An incomplete set of requirements for the successful use of methods might illustrate this statement. Methods should

- be flexible and adaptable
- be easy to learn and use
- have a clear benefit/effort ratio
- be dedicated to a concrete field of application (goal, restriction, limitations, ...)
- identify performance and critical factors
- support convergent-divergent thinking
- support communication with the goal of consensus finding
- support the visualization of mental ideas and images
- be able to structure and group topics, ideas, solution, problems, ...
- critically reflect team work, results and solutions
- ...

There is no doubt that these requirements aim at a common and flexible toolkit for designers. Compared to the success factors of individual and team characteristics mentioned above, a large area of potential research in analyzing and using the contexture of methods and tools was recognized. For example, good trainers adapt methods to a specific design situation using intuition and experience, whereas novices often completely fail, primarily due to an insufficient awareness of a design situation. A possible goal for the future design of methods, therefore, might be the enrichment of methods with hints for successful use and limitations which must be considered.

3. General findings and criticisms

Considering the interaction process more comprehensively and referring to the shared mental model (fig. 1) there have been some findings and criticisms:

1. Designers who work in teams require shared mental models for a better understanding and efficient teamwork. To achieve this, they must present and discuss their views on the design task and try to reach a common understanding of the specific design context. This could be accomplished at the very beginning of design work with the design step "clarification of the task" and during the design process with the step "status analysis". Methods and techniques supporting the elaboration and the sharing of mental models are available and should be consciously used.

2. A further goal of education and training should be to improve one's ability to produce external representations and to properly express the content of internal ones. Here the designer's education plays an important role, as well as the training of designers in understanding specific external representations, like specialist's languages, terminology, symbols, etc..

3. Important interaction types besides information transfer should be known by designers and used specifically. Especially positive group-dynamic effects can substantially influence teamwork. Negative influences, such as killer-phrases in creativity techniques should be deliberately avoided. On the other hand, designers should be aware that teamwork tends to level individual performance and greatly smooth out diversities.

4. However, one should keep in mind that different views on one item/issue and different backgrounds and mental models are an important source for new ideas and comprehensive views. The goal should not be to "create" uniform designers with identical mental models. Rather, there should be a common understanding of and an agreement on specific methods, external representations and terms, in order to make interaction more effective and efficient.

5. A key qualification for designers interacting in teams is the ability to transform mental representations into appropriate external ones. Designers should have a good command of external representations and their proper uses. On the one hand, the variety and performance of external representations is a main source of creativity. On the other hand, problems could arise within the interaction process when external representations are used which are not understood by the partner or the change of representations occurs too fast.

4. Proposals

Teamwork in a holistic sense was discussed and success factors extracted. Table 1 gives advice to build up and support a good design team. The table presents a guideline for activities and requirements within a team to achieve the following effects.

success factors	**⇒ effects**
selecting “right” leader	⇒ better project management
selecting team members	⇒ team “chemistry”
organization structure	⇒ effectiveness and efficiency of team
shared motivation	⇒ better cooperation
shared and individual knowledge	⇒ more specific solution
ability to communicate and stress resistance	⇒ improved crisis management
sharing ideas	⇒ meditated design “state”
external motivation	⇒ team “chemistry” and group climate
good network	⇒ shorter processing
additional experts or sub-teams	⇒ higher overall competence and quality of solutions
physical environment	⇒ motivation, supporting working conditions
Small distances	⇒ direct communication, more verbal interaction
clear responsibilities	⇒ higher efficiency and higher motivation
team’s concentration on the main task	⇒ better progress and higher quality
…	⇒ …

Table 11: Success factors with effects related to teamwork

All success factors produce effects on the team and design processes and/or the quality of the product. Many discussions in the “Bild & Begriff”-sessions were dedicated to these chains of effects that arise from a single argument or occasion and cover the entire teamwork and the product development process like an avalanche.

5. Statements

A lot of research was done within the area of teamwork and interaction in design and published in psychological and design-methodology literature. For use in design practice, training courses and guidelines for achieving and improving special characteristics exist. The statements mentioned below give a short overview of the need for future research in interaction and teamwork.

- There is the need for techniques and methods detecting events which cause problems.

- It would be helpful, in terms of knowledge management, to develop techniques to externalize private experiences.
- Also, a real-time feedback about designers' performance is necessary.
- Designers prefer to gather information from other people. What are the reasons for doing it and how can we professionalize knowledge transfer with databases and internet applications?
- The optimal composition of team leader and members is crucial for good team work, but how can we quickly define fitting criteria for selecting a team leader?
- There are two distinct processes in problem-solving groups. Can we give recommendations as to which one we prefer and how it should be adapted to a given design situation?
- There are strong differences between individual design and group design. What are the characteristics and benefits of each work type and how do they depend on communication?
- What are the specific characteristics of successful designers and successful design teams altogether?
- Do these characteristics differ from the common characteristics of "good" individuals and "good" teams?

6. Conclusion

Current design research in teams seems to be a multifaceted puzzle, in which researchers are looking for different pieces, elaborating them well and in detail, but neglecting the whole view. The questions might be asked why methods and tools are not used more frequently or even with enthusiasm, and why guidelines for teamwork often don't result in a substantial improvement in team interaction. It was one result of the discussion that the answer could be reduced to the (too) narrow view of researchers on the complexity of interaction processes between designers.

A second main topic within the workshops was the role of the design context and the organization and how methods are integrated into organizations. Currently, there is an adequate understanding of methods but too little effort in establishing and maintaining them in companies. The reason for this lack could have its roots in the insufficient analysis of design context and the elaboration of sufficient flexible support.

In summary, design research has to in some way be aware of the tendency towards the conscious or unconscious neglect of real world problems and has to boost the development of guidelines, methods and tools as appropriate solutions and their transfer in design practice.

Topic III: Methods, tools and prerequisites

Günter Höhne and Torsten Brix, Ilmenau, Technische Universität

Introduction

Design methods and tools support the designer in their creative work and determine the effectivity of the design process. Development, application and training of them are essential tasks for scientists and teachers in Engineering design. Summarizing the outcomes of the workshop "Bild & Begriff" in the last 8 years (Reports from 1994 to 2001) three issues are significant for developing methods and tools and their application in Engineering Design:

- "Cognitive outsourcing" (allocation of functions between humans and computers) and mental load,
- Computer aided creativity – possibilities and limitations,
- Training design methods.

This topic should be addressed under recognition of the practical work of designers, teaching engineering design and the scientific dialog and cooperation between engineers and psychologists at this symposium and in the future.

"Cognitive Outsourcing" – Types and Function

The term "Cognitive Outsourcing" was created and used in workshops "Bild & Begriff" 1999 and 2000. The limited mental capacity of the man requires external help during cognitive operations especially during designing new products. This support covers external representations of ideas or images (Bilder) and concepts (Begriffe). It was found that external representations have to carry the following functions:

- communication between partners and with technical devices (e.g. computers),
- organizing and support of mental models,
- modelling and manipulation of facts and relationships,
- storing of contents.

Cognitive outsourcing is characterized by interaction between internal and external representations and the relationship between the both modalities of representations image and concept (Fig. 1). For description, representation and modelling of technical products are used external representations of different levels of abstraction.

Modality	**Image**	**Concept**
Internal	Figurative	Textual

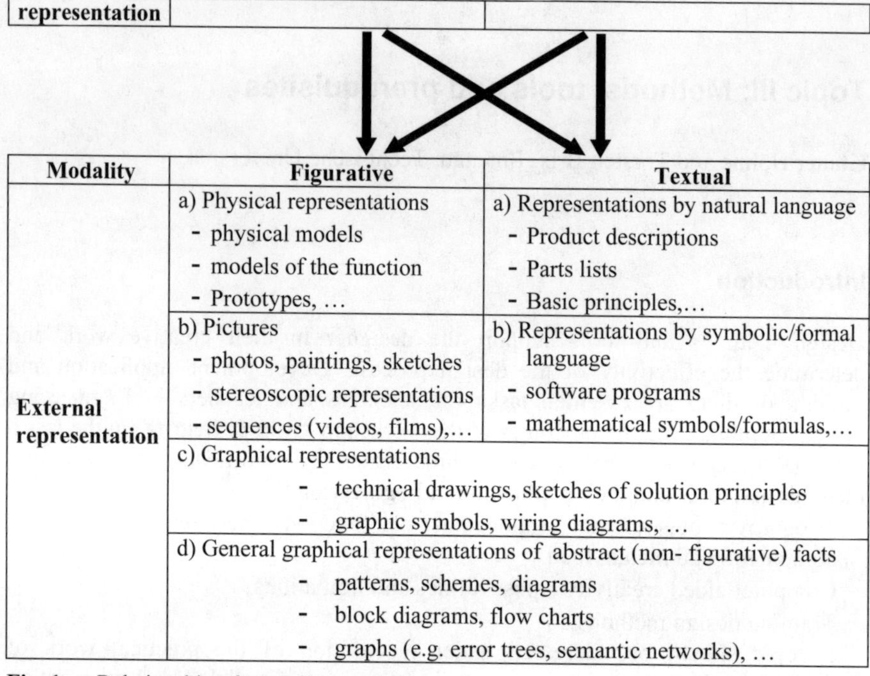

representation		

Modality	Figurative	Textual
External representation	a) Physical representations - physical models - models of the function - Prototypes, …	a) Representations by natural language - Product descriptions - Parts lists - Basic principles,…
	b) Pictures - photos, paintings, sketches - stereoscopic representations - sequences (videos, films),…	b) Representations by symbolic/formal language - software programs - mathematical symbols/formulas,…
	c) Graphical representations - technical drawings, sketches of solution principles - graphic symbols, wiring diagrams, …	
	d) General graphical representations of abstract (non- figurative) facts - patterns, schemes, diagrams - block diagrams, flow charts - graphs (e.g. error trees, semantic networks), …	

Fig. 1. Relationship of Modalities and types of representations of technical products.

This representations are essential components of design methods and tools. The optimisation of the change between modalities using adequate types of external representations is one important objective for the development of methods and tools (*Andreasen*, *Chandrasekaran*, *Höhne and Brix*, *van der Lugt*)

Computer Aided Creativity – possibilities and limits

The current situation in the industry is characterized by a wide application of CAD-Systems in the design process. But also sophisticated CAD Systems do not support the conceptual design phase. Therefore there are a need and efforts to develop and to introduce CAI-Systems. However we have not enough knowledge about the cognitive process to control creativity. Cognitive outsourcing and computer tools are focused to support the following operations:

- Generation of a great number of ideas (*Chandrasekaran*),
- Evenly spreading the idea generation to cover the solution field,
- Supply of information (*Tomiyama*),
- Selection of solutions (*Chandrasekaran*).

It should be investigated which kind of external representations are able to initiate and support intuition and imagination in the early states of the design process. Also embodiment design it need creative work which is characterized by 3D-imaging of product shapes and their manipulations in the space (*Lindemann and*

Pache). Virtual prototyping, Virtual Reality and Augmented Reality (*Schönfelder*) are technologies for this applications. However the effect of them for synthesis tasks is not enough investigated.

Teaching and Training design methods

For Engineering Design are available a large number of methods, tools and utilities (Fig. 2).

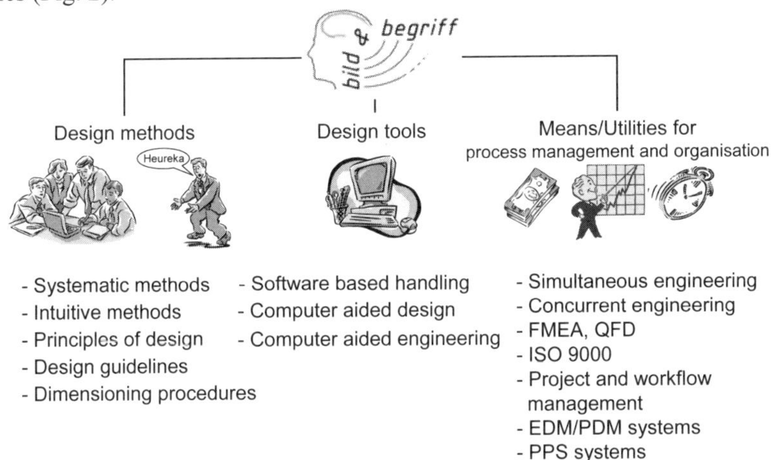

Fig. 2. Design methods, tools and utilities.

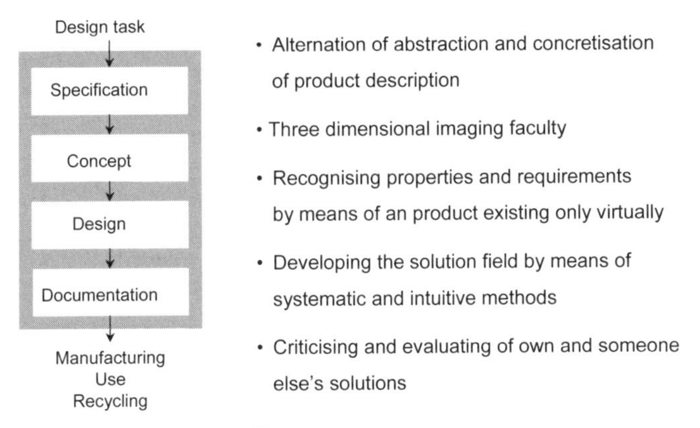

Fig. 3. Design process and soft skills.

The methods and tools are means to support and control the designers behaviour during the problem solving as well as the internal representation in form of a mindset (*Andreasen*). Important for training design methods is to understand the methodological kernel of the method and to use it by adaptation to the actual problem. The function of sketching (*Lindemann and Pache*, *van der Lugt*), supply and treatment of knowledge (*Tomiyama*), target specification by interviewing users (*von der Weth*) and changing of modalities (*Höhne and Brix*) are essential issues of teaching and training methods. Soft skills as shown in Fig. 3 must be trained during the designers education.

List of Contributions

The contribution within this chapter dealing with following questions:
- Which benefits have external representations?
- How can computer tools as a special type of "cognitive outsourcing" influence the creativity in the design process?
- How can be improved the design process by means of methods and tools?
- How should be run the design process (local, globally)?
- How should be trained/taught designers and students?
- Which skills an methods are particularly suitable for design?

For the orientation the papers about methods and tools are classified in Table 1 in relation to the main topics of the stream III.

Table 1. Contribution and their key aspects in relation to the main topics (CO-cognitive outsourcing, CAC-computer aided creativity, TDM-training design methods).

Contribution	CO	CAC	TDM
Mogens Myrup Andreasen Method Application and Designer's Mindset	x		x
Balakrishnan Chandrasekaran Design Problem Solving: Strands of my Research	x	x	
Günter Höhne and Torsten Brix "Cognitive outsourcing" in the Conceptual Design Phase	x	x	x
Udo Lindemann and Martin Pache Sketching in 3D - Future Tools for Conceptual Design	x	x	
Ralf Schönfelder VR/AR-applications, limitations and research in industrial environment	x	x	
Tetsuo Tomiyama Knowledge deployment: How to use design knowledge			x
Remko van der Lugt Reconsidering the Divergent Thinking Guidelines for Design Idea Generation Activity	x		x
Rüdiger von der Weth Designers and users: an unhappy love affair?			x

The short comments of the individual papers below should give a first impression of their contents.

- *Mogens Myrup Andreasen* concentrate on mindset as a mental framework. The mindset is the knowledge, experience, attitude, values etc. which constitute and influence the mental constructs which determine the result of use of the method and tool. The designer's procedural mode of applying a tool has been investigated concerning different DFX and Mechatronic tools. The mindset consists of several interrelated elements like interpretation of the task, the context, understanding the theory behind the method, mastering proper use of the method, and ability to judge the appropriateness and results of the method. The toolmaker should have a proper understanding of the user's mindset and how to build it up, as a mental framework leading to an effective execution of a tool.

- The main research interest of *Balakrishnan Chandrasekaran* is cognitive architecture for intelligent agents focused on specific tasks that bring out the nature of thinking, and design problem solving. It was sought to identify what is common to all design activity, whether it is design of mechanical artefacts, electronics systems, or software. This contribution of problem solving comprehends design task analysis, a formalization of the concept of "function", external representations in design and exploration of large design spaces.
- *Günter Höhne and Torsten Brix* review various forms of external representation for engineering models in the design process. At the conceptional phase of engineering design the switching of modes between image and concept is of particular significance. Analysis of technical solutions requires step-by-step abstraction so that features can be evaluated. In the field of computer-assisted tool development, there is a need for cross-phase, consistent product models to be created which link (in both directions) the descriptions of the configuration, the principle and the function.
- *Udo Lindemann and Matin Pache* represent the prototype of a 3 dimensional free-hand sketcher, which combine the advantages form the "classical" free-hand sketch and modern 3D-CAD-systems. It is a tool for conceptual design. The 3D lines drawn by means of an interactive pen are visualised by a stereoscopic impression.
- *Ralf Schönfelder* in a first part illustrate a snapshot of today's Virtual Reality (and also Augmented Reality) technology. After great expectations in the beginning of the nineties the VR made with VR technology has become a valuable component of the designers environment with successful industrial applications. In the second part problems of immersive modeling are discussed if it aims at a deployment in industrial design. Two different application types are distinguished: a) the use of more conceptual immersive modeling for sketching purposes and b) the use of more precise immersive modeling in CAD environments.
- *Tetsuo Tomiyama* explore how to effectively and efficiently use which knowledge during design is a critical issue within design research, based on an assumption that design is a largely knowledge-centered activity. Knowledge deployment needs the development of mechanisms for systematizing, structuring, and integrating in design. So far, research into design knowledge has been focusing on such topics as representation, reasoning, modeling, learning, discovery, data mining, management, reuse and sharing. Knowledge deployment is also crucial from the viewpoint of creative design, because innovate design results from innovative combination of known, existing knowledge, rather than from innovative, new knowledge.

- *Remko van der Lugt* present a cyclical model of idea generation that uses sketching as the primary mode of representation. Designers produce -more or less- indeterminate idea sketches and then interpret their meaning for further idea generation activity. This light, inquisitive form of analytical thinking appears to be conducive to idea generation when sketching is involved. The brainstorming guidelines for divergent thinking are built on that notion, which makes the brainstorming guidelines somewhat conflicting with idea generation activity that involves sketching. The paper provides a theoretical reflection of these four guidelines and proposes an alternative set of guidelines that emphasizes a swift interpretive idea generation cycle.
- Many studies about designers´ activities (e.g. in engineering and software design) show that especially in the early stages of the design process (clarification of the task, conceptual design) usability aspects are neglected. This situation is criticized by many engineers. Concepts like „usability engineering“ show the importance and the necessity for improvements in this field. This can happen in several ways. (a) Methods for designers should be developed and evaluated which integrate the user´s perspective in designers thinking and work process. (b) Users should participate in the design process not only by the formulation of requirements and by the testing prototypes and products. He should also be involved in other stages of the design processes e.g. conceptualisation of first ideas. *Rüdiger von der Weth* introduce two studies, following the two strategies described above. (a) The first study decribes the evaluation of a method which shall broaden the designers focus on more aspects then form and function (b) the second study describes a method of „participative“ software design which involves users in early stages of the design process and should enable them to make knowledge explicit which is useful for designers to improve usability of their products.

Improving Design Methods' Usability by a Mindset Approach

Mogens Myrup Andreasen, Department of Mechanical Engineering, Section for Engineering Design and Product Development, Technical University of Denmark,

1. Introduction

The designer's procedural mode of applying a tool has been investigated in our research projects concerning different DFX and Mechatronic tools. An important part of the mental framework leading to the execution of a tool is the so-called *mindset* of the designers. The mindset consists of several interrelated elements like interpretation of the task, the context, understanding the theory behind the method, mastering proper use of the method, and ability to judge the appropriateness and results of the method. We have observed, that this mindset is often absent from those who use the tool in practice or has a weak form compared to what is expected by the toolmaker.

The paper raises the question of how we can fit the method to the user. On the one hand the toolmaker should have a proper understanding of the user's mindset and how to build it up. On the other hand we should understand user's mindset in general and the conditions for tool implementation and use. The paper shows mindset-related findings and proposals for models, which support the part of a mindset, which may be called theory understanding.

2. Design methods development

In the following I will assume, that design methods are created based upon theory and empirical studies, from which the method is derived and designed. The method is seen as a system of rules for acting, often supported by a tool. Executing the method is a progression in time.

Normally research leading to new methods and tools consists of two interrelated activities: the research activity clarifying phenomena; and a design activity (which is not scientific!) creating the method and tool. Most method-related research is planned only for the tool proposal, not for the proper empirical studies of the method in practice, leading to an actual proof, see (Blessing 2002).

In his research on acquisition of product development tools in industry, Claudiano Sales de Araujo (2001) investigates the life-cycle of tools. He states, *"In general, the work of most tool developers, promoters and marketers is based on three fundamental assumptions:*

1. That a need in the context of product development (PD) exists for the new tool being developed/ promoted/sold.

↑ Three assumptions on the development/use of design methods for PD.

1. That the new tool can be implemented in the PD organisation (that is, practitioners can learn how to use it, etc,).
2. That the use of the tool can ultimately deliver benefits for the company's PD organisation, and consequently to the company as a whole".

The first assumption is true; the rapidly changing world creates new conditions for business and industry, leading to new needs for tools. The second assumption is less obvious and *"apparently based in simplistic optimism from the side of developers together with lack of understanding the nature of PD tools implementation in practice"*, (Araujo 2001). Implementation of tools generally needs a tremendous effort and the present low level of adoption of tools by industry, (Araujo et al. 1996) is a rather problematic issue of implementation, (Norell 1996).

The third assumption is also problematic, increasing suspicion has grown that companies adapting and implementing PD tools are not collecting the good results that they expected, (Barkan 1994).

3. Findings concerning methods

A survey of tools utilisation in UK, (Araujo et al. 1996) shows that the benefits achieved by the companies adapting new PD tools are perceived by practitioners as being much lower than they generally expected. *"Part of the problem is said to be intrinsically related to the poor quality of the tools delivered by the researchers, which are seen by practitioners as being*

- Too theoretical in nature
- Too complicated to understand
- Presented in a "strange" kind of language
- Difficult to implement
- Difficult to use
- Problems to evaluate the results attained by its use"

These findings are in accordance with several other findings concerning use of methods, especially the introduction of CAD systems in industry. Araujo sees it as a central problem, that what is delivered to the public is often the result of a person's unique interpretation of what he saw or experienced, see Fig. 1. The interpretation and the design method made by the researcher is based upon a highly personal view and imagination. And the researcher's "materialisation" of the method into an often sparse (called "short and precise") description is transformed into a mode of operation by the user, influenced by many uncontrolled factors. This leads to poor quality.

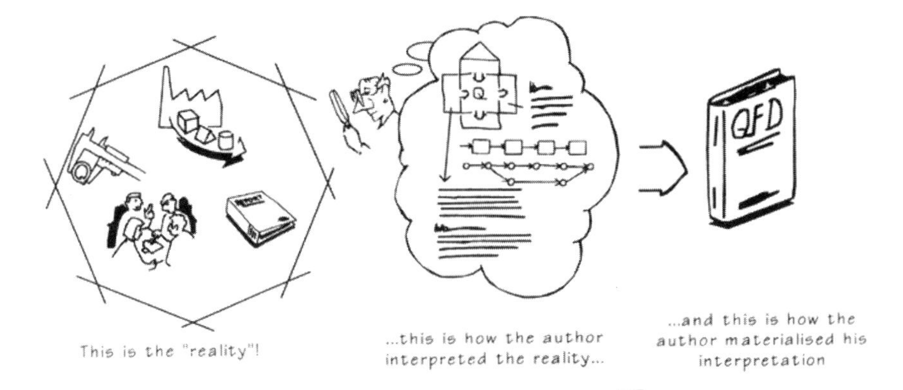

This is the "reality"!
...this is how the author interpreted the reality...
...and this is how the author materialised his interpretation
QFD

Fig. 1. Establishment of new tools: Whose interpretation? (Araujo 2001).

"Because the introduction of new tools will always involve changes in the modus operandi of the PD operation, and the people within it, the implementation effort is always a major bottleneck", says Araujo (2001) who also points out the likely factors accounting for the differences in the implementation practice; including:

- The perceived impact on the organisation from the new tool.
- The perceived degree of relevance of the new tool.
- The person or group that carry out the implementation (personal goals, experience, resources, support etc).
- The attitude to the tools of the persons carrying the implementation.
- The perceived cost or risk of a potential failure of the implementation.

A central observation is that the implementation is often left to the user. The respondents to a survey made on the use of tools in Danish industry, Araujo (2001), believe the following factors to be important to implementation failures:

- Negative acceptance and lack of support from top management.
- Poor information on the tool and lack of clear instructions.
- Lack of investigations into fitness, usefulness and benefits.
- Lack of necessary skills for using the tool.

The description of a method and the related tool(s) is normally a very sparse part of what actually happens when user is practising or executing the method. Araujo (2001) has visualised the operators procedural mode as shown in Fig. 2. The operator creates "the mental constructs", (Jayaratna 1994) as an interpretation of the task, the execution of the method fitted into a mini-plan on "how" to execute, and where in the procedural mode to apply the method. The actual execution is done under influence of the way the task is understood, interpretation of the reality against which the method is used, the user's knowledge about the method itself and necessary framework of knowledge related to the use of the method, and under influence of motivation, personality etc.

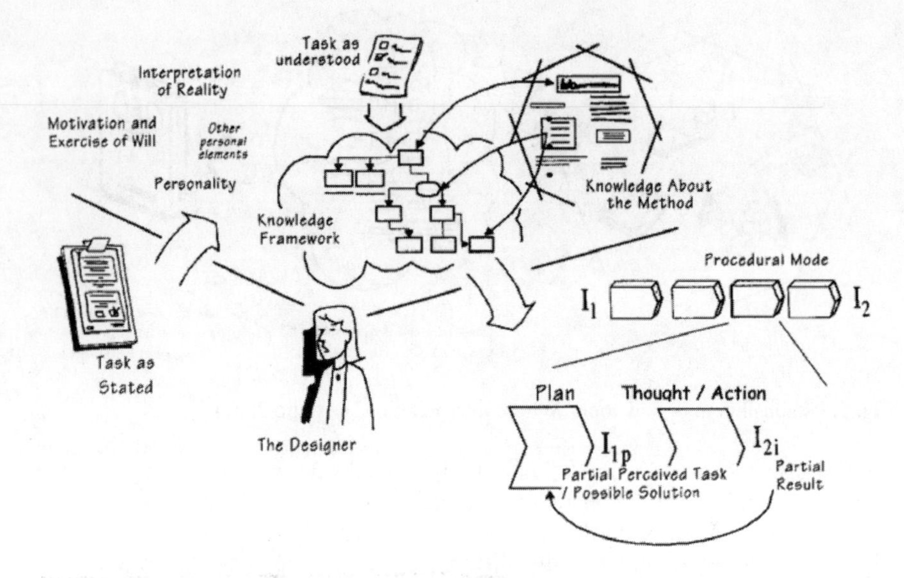

Fig. 2. Understanding the operator's procedural mode, (Araujo 2001)

4. Mindset, the operators understanding

As educators at universities and in industry, we are concerned about the part of Araujo's model of the operator's procedural mode, which may be seen as a consequence of the operator's *mindset*. How is the proper mindset, which may be seen as the uncontrolled and partly uncontrollable part of a method's application, created?

As mentioned above the mental activity is influenced by several factors and the method's execution leads to a mental construct, the actual mindset, consisting of the following interrelated elements:

- interpretation and understanding of the task
- interpretation of reality and context, judging appropriateness and timeliness of using the method
- understanding the theory or phenomenon on which the method is used, translating it's concepts into the method's language and models, and judging the validity of the method and it's actual use.
- imaging the use of the method, based upon experience and skill
- understanding the result of the method's use and proper action to be taken in the procedural mode of designing

interrelated elements leading to difficulties in method usage and adaptation.

5. Mindset and theory understanding: observations

A central part of a mindset is the understanding of the actual "mechanism of the tool", i.e. the theory behind the tool which make it work, and the concepts and models related to the tool.

It can be assumed, that this understanding influences the output of a method's application decisively. In our group of design researchers at the university we have made observations concerning the existence, nature and quality of mindsets and theory understanding in the following projects:

1. In relation to our research on Design for X (DFX), we observed, that the designers were not aware of the strong influences the product's design has on operations and conditions in other functional areas and the product's life conditions. A new *Theory of dispositions,* (Olesen 1992) was created: The product's design characteristics and some characteristics of the operation's equipment, methods and performance relates in dispositional patterns. These patterns are the core of any DFX, and therefore they are an important part of the mindset related to DFX-methods.
2. Researching *Design for Quality,* (Mørup 1993), we found, that engineering designers in industry had confused and contradictory understandings of the concept of quality. This fact is very surprising, because quality-supporting methods like ISO 9000 and QFD are generally seen as the foundations of product development. Based upon our research we proposed new concepts of quality:

   – separating the innovative part of quality (customer value) from the supportive (producability, assemblability etc.) part of quality.
   – distinguishing between the qualities Q ("big Q") related to the product and it's aspects of usability, operation, reliability etc., and the qualities q ("little q") related to the ease of establishing the Q qualities in the operations.

This new quality mindset was eagerly adopted by industry because it leads to new, productive dialogue and handling of quality and new competitive ideas.

3. *Conceptualisation* is a design area rich in methods and approaches. Most authors of textbooks see conceptualisation as an activity following after the crystallisation of an idea and goal-specification of the ideal solution. This approach blurs the fact, that a product concept must be a combination of two ideas: The "idea *with* the product" (purpose, task, value) and the "idea *in* the product" (principles, technology, operation), (Hansen and Andreasen 2002). This new mindset seems to be very productive in search for innovation concepts, judged early from our teaching experiences.
4. *Mechatronic Design* was empirically investigated by McAloone. He shows, that one of the barriers to "equal opportunity" in projects (i.e. the equal chance for both mechanical, electronic and software engineers to have a say " in which slice of the cake" should be solved in their own competence area), could partly be solved by taking a common starting point, guided and steered by creativity techniques. McAloone sees this symptom as a missing theory or first of all a

missing productive mindset for creating mechatronic product concepts, (Andreasen and McAloone 2001).

These examples on empirical research into mindsets are related to a very basic question on the nature of design theories and methods, to be treated in the next section.

6. Design theories: truths or productive ideas?

As pointed out in the chapter: "The role of Artefact Theories in Design" (Andreasen 1998) in Grabowsky et al: "Universal Design Theory", design theories have their origin in two very different sources, roughly speaking, namely the nature of human thinking and the nature of things. Many authors from the design methodology area have proposed artefact theories (mainly for mechanical products) and methods for designing products. These theories propose structural characteristics and relations between structural characteristics and behavioural properties and functions.

It is a well-known fact that there exist several proposals for concepts and theories; my own presented in the mentioned source, belonging to the WDK-school, is such a proposal. This means that these artefact-based theories hardly are real theories or may be seen as truth, but productive (hopefully) thinking-patterns or languages, by which the designer will be supported in his/her designing. The theories cannot be proved or falsified, but their productivity when used as methods can be evaluated.

In the section above we saw the necessity and appropriateness to create mindsets, i.e. understanding of basic concepts and identified situations when such mindsets was not present. But what happens when methods are used with weak mindsets and misunderstandings? In our education and in industrial projects we have observed that good results often are obtained even if false concepts or false reasoning is used. The explanation is, we believe, that the methods are soft and actually do not control the reasoning; therefore the designers deepening into the matters and his/her common sense leads to results.

7. Discussion

The problem treated here is the establishment of proper mindsets in relation to the execution of methods. Generally proper implementation of methods, including the building up of the necessary understanding, is disregarded, resulting in a poor reputation and low dissemination of tools.

Our research shows the following characteristics of the methods and their application:

- Methods are often described in an insufficient way for creating proper understanding (mindset).
- Available methods have low interest in industry and a poor reputation.

- The necessary mindset is often not present in situations, where methods are applied.
- Even if methods are used improperly they may lead to good results due to common sense, i.e. the methods are soft.

So the situation concerning the use of methods is very unsatisfactory. Several single elements contribute hereto, meaning that the repair of the situation can be brought about in several ways:

The transfer from a theory to a method is, as previously mentioned, a design step and several alternatives may be developed. A central point is to balance the correctness or precision against the ease of operation or the deepness of knowledge necessary for using the method. One example is the use of precise standard basic functions for establishing a function structure of a machine, balanced against the much more flexible use of functional units described by any verbal label the user may choose.

The softness of methods, which shows its consequences when methods are used wrongly, but apparently can lead to meaningful results, is a parameter in this transfer or method design situation. Contrary to the situation when a craftsman uses a physical tool we are not able to register how a design method is used unless we find subtle ways of performing empirical research.

The *performance of a method* is primarily a question about the value of the result balanced against the cost of applying the method, as we know it from the powerful Design for Manufacture methods, which are rather simple to use. But the performance also depends upon the users skill and the fit to the actual problem, which may lead to a high probability of unsuccessful applications. The methods' fit to the problem is the question of the methods broad-spectredness or flexibility. We know, for instance, that the appraised functional reasoning used in the mechanical area fails to perform in electronics and software design, because the concept of function has not been developed here.

Finally the performance of a method has a clarification dimension, i.e. to what degree does the preparatory work or the method itself give such insight into the matters that the robustness of the result can be estimated. Many methods for analysis are of such a nature, that the goodness of the concept to be analysed decisively determines the result, but the methods do not give any hints to strengthening the concept.

The *appropriateness of the user's mindset* seems to be a very strong precondition for the proper application. One solution to the mindset problem seems to be the development of pairs of methods and mindset, meaning that both the method and mindset are established on a theory, and the design of the method and the pedagogic of building up the mindset are created for efficiency of the use of the method. The use of metaphors, visual models and well-chosen words could be elements of the pedagogic, but also proper training might be successful.

Summing up, we see that researchers and toolmakers are confronted with a challenging situation. Until now the general approach has been that designing should be seen as a chain of methods, linked together in a computer system (a designer's workbench). But from our experiences the main problem is to create a well-structured framework of proper understanding, just like the craftsman's tools need a skilled hand for being useful.

The research on design methodology is mainly focused upon the development of methods, but only few research projects are planned for a proper testing and evaluation of the methods. Therefore we need new research approaches for telling us the nature and role of mindsets and tools and telling us about the performance of user and method when the method is applied.

We see it as a challenge to develop such a framework of mindsets, based upon a theory of technical systems, which link the concepts to each other in a logical structure, but at the same time leading to productive mindsets.

8. Improvements in methods' usability

The examples on mindset observations in section 5 focussed on that part of a mindset, which is reflecting the proper understanding of the theory or the theoretical mechanisms built into the methods. The examples were followed by a proposal for a new articulation of the theory, in form of a better representation or a more operational explanation of the theory. We have used, as mentioned above, metaphors (big Q, little q), visual models (a sugar-top model for the hierarchy of qualities of product ranked after importance) and new words (like disposition for the chain effect or propagation of decisions in designing).

Our efforts for teaching design methodics in education and industry show promising results, due to this pedagogical "styling" of the mindset.

Summing up, we see that researchers and toolmakers are confronted with a challenging situation. Until now the general approach has been that designing should be seen as a chain of methods, linked together in a computer system (a designer's workbench). But from our experiences the main problem is to create a well-structured framework of proper understanding, just like the craftsman's tools need a skilled hand for being useful.

The research on design methodology is mainly focused upon the development of methods, but only few research projects are planned for a proper testing and evaluation of the methods. Therefore we need new research approaches for telling us the nature and role of mindsets and tools and telling us about the performance of user and method when the method is applied.

We see it as a challenge to develop such a framework of mindsets, based upon a theory of technical systems, which link the concepts to each other in a logical structure, but at the same time leading to productive mindsets.

9. Conclusion

It is impossible to buy a violin and get one proper tone out of it, if you have never played a violin before. But we expect, that a sparse verbal description of a method shall make us a master in utilising the method. The development of proper mindsets (and of course necessary tool skills) seems to be a decisive factor for the success of the use of a tool.

Our research has added to the established insight, that the use of design methods in practice is troublesome, sparse and erroneous. We have observed, that the kernel of the mindset, namely the proper understanding of the method or phenomenon behind the method, seems to be a central cause for the problematic.

Our first proposals for improved, visualised or metaphoric articulation of mindsets has shown industrial acceptance and improved use of methodics.

10. Acknowledgements

The author is in debt to his PhD-students, firstly Claudiano Sales de Araujo, who's thesis has been broadly utilised for structuring the paper, and to other students creating contributions to our understanding of mindsets, and to his colleague, Assoc. Prof. Tim McAloone who's results concerning mindset are brought together in this paper. The author would also like to thank Dr. Rüdiger von der Weth, HfT Stuttgart, for valuable critique and advice.

References

Andreasen MM (1998) The Role of Artefact Theories in Design. In: Grabowski et al (ed) Universal Design Theory. Shaker Verlag Aachen, ISBN 3-8265-4265-7, pp 67-72.

Andreasen MM, McAloone T (2001) Joining three heads - Experiences from Mechatronic Projects. In: Meerkamm H (ed) 12 Symposium, Design for X, Neukirchen 11-12 October 2001. Lehrstuhl für Konstruktionstechnik, Friedrich-Alexander-Universität Erlangen-Nürnberg.

Araujo CS (2001) Acquisition of Product Development Tools in Industry: - A Theoretical Contribution, PhD-Thesis. Technical University of Denmark.

Araujo CS, Benedetto-Netto H, Campello AC, Segre FM, Wright IC (1996) The utilization of Product Development Methods: A Survey of UK industry. Journal of Engineering Design 7,3, pp 265-77.

Barkan P (1994) Benefits and Limitations of Structured Methodologies in Product Design. In: Eastman SDaC (ed) Management of Design: Engineering and Management Perspectives. Kluwer Academic Publications, Massachusetts.

Blessing L (2002) What is this thing called Design Research?. In: Annals of 2002 Int'l CIRP Design Seminar, 16-18 May 2002. Hongkong.

Hansen CT, Andreasen MM (2002) The Content and Nature of a Design Concept. In: Boelskifte P et al (eds) proceedings of NordDesign 2002. 14-16 August 2002, Trondheim, Norway. Dept. of Technical Design, NTNU.

Jayaratna N (1994) Understanding and Evaluating Methodologies. NIMSAD-A Systematic Framework. McGraw-Hill, Berkshire, England.

Mørup M (1993) Design for Quality. Doctoral thesis. Technical University of Denmark, ISBN 87-90130-00-6.

Norell M (1996) Competitive Industrial Product Development Processes - a Multidisciplinary Knowledge Area. In Kleimola MKaM (ed) NordDesign'96. Helsinki University of Technology, Faculty of Mechanical Engineering, Espoo, Finland.

Olesen J (1992) Concurrent Development in Manufacturing – based upon dispositional mechanisms. Doctoral thesis. Technical University of Denmark, ISBN 87-89867-12-2.

Design Problem Solving: Strands of My Research[1]

B. Chandrasekaran, LAIR, Department of Computer & Information Science,
The Ohio State University, USA

Four Strands

My main research interest is cognitive architecture for intelligent agents. I have always driven my research by focusing on specific tasks that bring out the nature of thinking, and design problem solving has been one such task area. I have sought to identify what is *common to all design activity*, whether it is design of mechanical artifacts, electronics systems, or software. What unifies all of my work on design is my interest in logical, ontological and computational bases of the generic design task. My work on design can be thought of in terms of strands of research that relate to one another.

The first strand is my interest in modeling expertise in design and analyzing the *task structure of design*, results of which are summarized in a widely cited paper (Chandrasekaran 1990). The task structure of design that I developed was intended to bring out the generic character of design problem solving, the common structure across domains. This work was influential in the community that works in the intersection between engineering design and Artificial Intelligence. The second stream focused on another common element, a *device ontology*, again independent of the domain. Notions such as function, structure and behavior pervade all design activity. Over a decade of research my colleagues and I developed a framework called *Functional Representation* (FR) that accounted for how one understands the relation between the structure of a system and its function (Chandrasekaran 1994), (Chandrasekaran and Josephson 2000).

The third stream is a technology that my colleagues and I recently built to assist engineers to explore vastly larger design spaces than has been possible so far, especially to handle the multi-criterial nature of design (Josephson, Chandrasekaran et al. 1998). As a practical tool, we think that the technology has a great deal of potential in design. The fourth stream is relatively recent, and arises out of my interest in diagrammatic and other perceptual reasoning. In a recent paper (Chandrasekaran 1999), I place external visual representations in design in the larger contexts of perceptual representations and problem solving in general. I outline a framework that

[1] Prepared through participation in the Advanced Decision Architectures Collaborative Technology Alliance sponsored by the U.S. Army Research Laboratory under Cooperative Agreement DAAD19-01-2-0009.

brings together external and internal representations, and conceptual, perceptual and kinesthetic modalities. I make use of a task analysis of design to identify the roles these various representations might play in design. In the remainder of the paper, I summarize the results of the first three streams. The stream on perceptual representation is omitted, partly because of it is the least mature of the four strands, and also due to space limitations.

Task Structure of Design

A task structure is a description of the task, proposed methods for it, the internal and external subtasks, knowledge required for the methods, and any control strategies for the method. Thus, the task analysis provides a clear road map for knowledge acquisition. The task structure for design proposed in (Chandrasekaran 1990) is based on the idea that design activity proceeds by use of a general class of methods called *propose-critique-modify* methods. That is, the designer starts by mapping from the specifications, or subsets of the specifications, to proposed solutions or solution templates, which would need to be critiqued and modified or refined. Modification or refining are themselves design tasks, with a similar task structure, so the task structure has a recursive nature to it. The table below, taken from (Chandrasekaran 1990), summarizes the task structure.

Note that there is no suggestion that there is a preferred method for any of the subtask, let alone for design as a whole. What method is to be used depends on the desired properties of the solution or the solution process, and availability of the knowledge the method requires. Different types of methods can be combined. In particular, the task structure also makes clear how AI-like methods and other algorithmic or numeric methods can be flexibly combined, much as human designers alternate between problem solving in their heads and formal calculations. Further, the task structure brings out how the design task has as subtasks other types of reasoning. For example, the Critique subtask may use methods identical to diagnosis, and Verify subtask may use simulation and other predictive methods. The view elaborated here is that there is a generic vocabulary of tasks and methods that are part of design and that design problems in different domains simply differ in the mixture of subtasks and methods. Expertise, that is, methods, and knowledge and control strategies for them, emerge over a period in different domains to help tractably solve the task in a given domain. Thus, the key to understanding real-world design computationally is not in a uniform algorithm for design but in the structure of the task, showing how the tasks, methods, subtasks, and domain knowledge are related.

Table 1. . A Task Structure of Design

TASK	METHODS	SUBTASKS
Design	Propose, Critique,Modify family (PCM)	Propose, Verify, Critique, Modify
Propose	Decomposition methods (incl. Design Plans) and Transformation methods	Specification generation for subproblems
		Solution of subproblems generated by decomposition (another set of Design-tasks)
		Composition of subproblem solutions
	Case-based methods	
	Global constraint-satisfaction methods Numerical optimization methods Numerical or Symbolic constraint propagation methods	Match and retrieve similar case
Specification generation for subproblems	Constraint propagation incl. constraint posting	Simulation to decide how constraints Propagate
Composition of subproblem solutions	Configuration methods	Simulation for prediction behavior of candidate configurations
Verify	Domain-specific calculations or simulation	
Critique	Qualitative simulation, Consolidation Visual simulation Causal behavioral analysis techniques to assign responsibility	
	Dependency-analysis techniques	
Modify	Hill-climbing-like methods which incrementally improve parameters	
	Dependency-based changes	
	Function-to-structure mapping knowledge	Design new function. Recompose with candidate design
	Add new functions	

A Framework for Functions

One common thread that runs through all design are the notions of function (design cannot start until the customer has specified what he wants the artifact for), structure (design solutions are described in terms of components with specific parametric values to be composed in a specific way), and behavior (a design achieves its function behaves it behaves in a certain way under certain conditions). By the way, the same ontology of function, structure and behavior is also central to diagnostic problem solving. Thus, this ontology is for the world of *devices*.

The last two decades have seen a flowering of work on reasoning about physical systems, specifically, on the notion of function. For example, (Iwasaki and Chandrasekaran 1992) uses qualitative simulation to verify design functions. The various investigations of device function have mostly lacked a unified technical vision. Different intuitions about functions are pursued in different contexts and application domains. A framework for functions should, in our opinion, satisfy the following desiderata. It should apply to intended functions of human-designed devices, and to functions or roles in natural systems. It should apply to functions of both static and dynamic objects, whereas almost all of the work on reasoning about objects and their functions has focused on functions that are defined in terms of state changes of objects, e.g., electronic circuits, buzzers, gears, and so on. However, the notion of function applies just as well to static objects, e.g., support beams and windows. Finally it should apply to functions of both abstract and physical objects. Even though most work has been done for physical objects, one can speak of functions of modules in software, and of steps in plans. We present a brief summary of our framework for function, presented in more detail in (Chandrasekaran and Josephson 2000) which discusses object and behavior composition, abstraction of structure and behavior and a variety of related issues.

The Design Task

Let E be an environment and let G be a predicate defined for E. Let a cognitive agent have a goal to have G be true in E. This sets up a design task: to specify an object O, and specify a way to embed O in E, such that when O is so embedded, G is caused to be true.

Traditional definitions of the design task focus on the need to specify the object, e.g., to provide a list of components from some component library and a way of composing them. Our definition additionally requires that a way of embedding O in E be specified; the design task is not complete until the designer specifies a *mode of deployment* of O. The mode of deployment makes the connection between the properties and structure of O and the achievement of G in E. Specifying the mode of

deployment becomes necessary if G is defined without any commitment to the properties or structure of O.

Often, device functions are specified in terms of the device itself. Sometimes, however, it may be more useful not to make any reference to the structure. Consider the example of a buzzer. In the literature, the function is typically stated as "when the switch is pressed, a sound is made," which makes reference to the switch, a part of the structure of the device. However, suppose a buzzer has not been invented yet, but there is a need. The need could be expressed by stating what we want to happen in the environment:

"no sound in the environment $\longrightarrow$ buzzing sound in the environment"

By isolating the function definition from any reference to the structure, we are leaving it open for the designer to come up with a very different object to achieve the function. Perhaps one design would achieve the function when it is twisted, another when it is blown on, and so on. In our current formalization, we are focusing on the concept of function independent of the structure of a device.

Ontology: an object, in an environment, viewed from a perspective

The world is composed of objects in causal interaction with each other. The primitive representational notion for us is that of *an object, in an environment, viewed from a perspective.* Representationally, the basic elements are:

<object> in <view>
<object properties>
<generic environmental properties in
potential causal relation with object>
<property relations>

An object in the real world has an open-ended number of properties: science can discover new properties or relationships between existing properties, and one can define new properties from old properties. A *view* is a specific modeling stance; it selects certain properties of the object for representation. The view also implicitly specifies the classes of external objects with which an object can be in causal interaction.

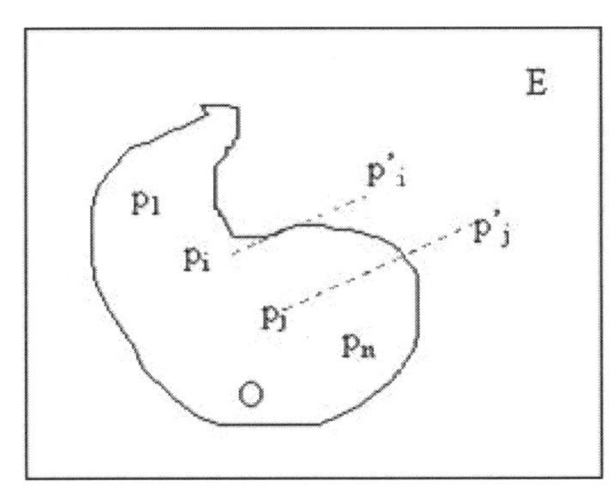

Fig. 1. An object O in a generic environment E

The central idea is illustrated in Fig. 1. An object interacts with its environment because some of its properties either affect or are affected by the properties of objects in the environment. When an electrical wire comes in contact with an electrical terminal of an object in its environment, depending upon which of the voltages is the independent variable, the voltage of one of the terminals causes the voltage of the other terminal to have the same value. The terminals are simply special cases where the property is localized to a physical location, but a more general way of talking about causal interaction between objects is by means of the properties that causally interact. When we wish to describe the object's potential interactions in some generality, the environment is described as a type. Finally, a set of *property relations* is given that represent the modeler's causal understanding of the object. The relations state all the causal relations between the properties, both the object's and environmental ones, believed by the modeler to be relevant. The property relations can be in any form: continuous, discrete, qualitative, etc. As mentioned earlier, we focus on the sense of function as effect on the environment.

Function of an Object as Effect on its Environment

Definition. Function. Let G be a formula defined over properties of interest in an environment E. Let us consider the environment plus an object O. If O (by virtue of certain of its properties) causes G to be true in E, we say that D performs, has, or achieves the *function* (or *role)* G.

A description of how O is used to achieve the function (or serve the role, etc.) G has three parts:

1. Functional formula expressing G: what predicate of the environment will be true, under what conditions.
2. Description of properties: what properties of O are used in achieving G.

3. Mode of deployment: what property relations (using what properties of the environment) determine the causal interactions between the object and its environment. This is commonly given by specifying the types of connections between an object and objects in its environment.

Example. Pump: The properties of interest in the environment are the quantities of water, $Q_1(t)$ and $Q_2(t)$, at time t, in locations L_1 and L_2 respectively. Let G be the formula corresponding to $Q_1(t_0) - Q_1(t_f) = Q_2(t_f) - Q_2(t_0) = K > 0$. That is, a positive quantity of water is moved from L_1 to L_2 from the initial instant to final instant. For simplicity let us call this formula, Pump(K, L_1, L_2).

Note that while G is described as a function of object O, both preconditions and effects in the specification of G are defined exclusively in terms of properties outside of O. The function of an object is the effect it has on its environment, not its behavior in isolation. In the Pump example, the formula Pump(K, L_1, L_2) describes an effect on the environment. If an object is introduced that causes the formula to be true, we will say that the object "plays the role of a Pump" or "has a Pump function." A particular pump, P, say a reciprocating pump that uses a piston to repeatedly move equal units of water, has relevant properties of having an inlet port $Port_1$ and an outlet port $Port_2$, and is deployed by having $Port_1$ connected to L_1 and $Port_2$ connected to L_2 so that (water at $Port_1$) = (water at L_1) and (water at $Port_2$) = (water at L_2).

Applying the definition with appropriate locations L_1 and L_2, we can also say that the heart has a Pump function in the body. The definition of function is neutral with respect to whether the cause-effect description is intended or is described after the fact.

Architecture for Exploring Large Design Spaces

Design is a Multicriterial Problem.

Design problems are typically multicriterial, i.e., each design candidate is subject to assessment along several, potentially conflicting, criteria. The Seeker-Filter-Viewer architecture (Josephson, Chandrasekaran et al. 1998) enables the designer to avoid making mathematically convenient, but often inappropriate, assumptions, such as weights to convert the problem into a univariate optimization problem.

The Seeker

The Seeker generates or acquires alternative designs, and evaluates them according to multiple criteria. The technology includes CFRL, a compositional modeling and simulation language, to generate and evaluate designs in many different domains. In general, however, generating the alternative designs and evaluating them may involve domain-specific techniques and algorithms.

The Filter

The evaluated designs are passed on to the *Dominance Filter*: Design candidate A is said to *dominate* candidate B if A is superior or equal to B in every criterion of evaluation and strictly superior for at least one criterion. Dominated designs are removed without risk: there is another candidate that is at least as good in the surviving set, which are *Pareto optimal,* and in which there is no way to improving on any criterion without giving up something along another criterion. The dominance filter is very effective and scalable. In an experiment in the domain of hybrid electric vehicle (HEV) design, about 1.8 million design candidates were generated, and these were evaluated on four criteria. 1078 candidates survived, giving a survivor percentage of 0.06! The designer has the assurance of having examined close to 2 million designs in the design space, while only needing to look in some detail at about 1000 survivors. The filter *scales* very well.

The Viewer

Almost any member of the Pareto-optimal survivor set is a good design. However, designers have additional tacit preferences that can help them select from the survivor set. These are trade-off preferences, i.e., a sense of how much of performance on one criterion is worth trading off for how much of performance on another criterion. The Viewer permits the designer to view these tradeoffs in multiple trade-off plots and select subsets interactively. The different trade-off plots are linked such that selections in one plot are reflected in other diagrams.

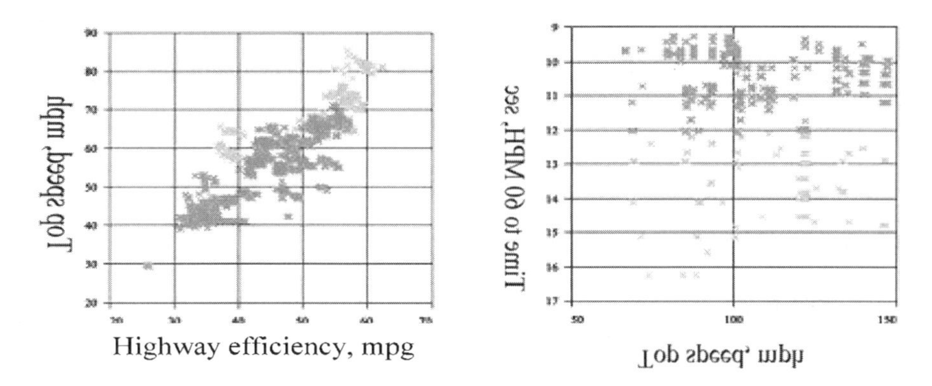

Fig. 2. Cross-linked trade-off plots in the Viewer

Fig. 2 illustrates a pair of cross-linked trade-off plots of the 1078 survivors in our experiment in the HEV domain. The figure is a selection from a picture of the Viewer display, in which all the design candidates in the survivor set are seen in six trade-off scatter plots. (The actual display is in color, but in the figure, different colors appear as different shades of gray). The designer in this case first focused on the "Time to 60 mph" criterion (display on right), and selected those that take less than 12 seconds – the selected candidates appear darker in comparison to the rest. The same candidates appear in a darker shade in the other trade-off plots as well. This gives the designer a chance to visually examine whether the extrema in other plots are acceptable. In this case, he sees that within the selected subset there are plenty of car designs that have desirable efficiency and top speed values. He now makes the decision to keep the selection, i.e., discard the candidates not in the set. This choice is immediately reflected in all the scatter diagrams. He proceeds in this way, by making a series of narrowing selections, focusing attention on different trade-off diagrams.

The architecture has several implications for the practice of design.

- *Change in Perspective from Traditional Optimization of linearly weighted criteria.* Decision makers find weighting an unnatural requirement, and our Viewer discussion shows why: the weights are highly nonlinear over the performance space.
  - Change in Perspective from Traditional Design Practice of incremental improvements to exploring the entire decision space.
  - *Understanding the Design Space.* The Viewer can also be used to give the designer a deeper understanding of the design space, by enabling him to see how the criteria values and the design specification interact.

- *Validating the Simulators.* Often, the Viewer brings out modeling errors in the simulator: anomalous performance in certain regions often pointed to modeling assumptions that were not valid over the entire space.
- Allocation of responsibility between the machine and the human. The computer and the human do what each is best at.

Concluding Remarks

I have been struck by how often in different gatherings of design theorists from different design areas, I hear the same wishlist expressed. One of them always is the need to understand design across disciplines. I have also been struck by how little known to many participants in such gatherings is work in AI that has gone on for over two decades with exactly the same goals: an account of the generic phenomena – knowledge and inference – that are involved in design across domains. This is not to deny two ways in which design differs in different domains. First, clearly, there will be domain-specific device terminology and procedures: temperature, thrust, gear, momentum conservation, etc. are terms that occur in one domain, while currents, voltages and Kirchoff's Laws will be relevant for another domain. Second, certain methods in the task structure would be more applicable in one domain than another. These domain-specific terms and methods are specializations of something more generic, and it is important for design theorists to understand design at the generic level as well. This paper is an attempt to bring to the engineering design audience work that exemplifies this concern with generality.

References

Chandrasekaran, B. (1990). Design Problem Solving: A task analysis. AI Magazine. 11: 59-71.

Chandrasekaran, B. (1994). "Functional representations: A brief historical perspective." Applied Artificial Intelligence 8: 173-197.

Chandrasekaran, B. (1999). Multimodal Perceptual Representations and Design Problem Solving. Visual And Spatial Reasoning In Design: Computational And Cognitive Approaches, MIT, Cambridge, USA, Key Centre of Design Computing and Cognition, University of Sydney, Australia.

Chandrasekaran, B. and J. R. Josephson (2000). "Function in Device Representation." Engineering with Computers 16: 162-177.

Iwasaki, Y. and B. Chandrasekaran (1992). Design Verification Through Function- and Behavior-Oriented Representation: Bridging the gap between function and behavior. Artificial Intelligence in Design '92. J. S. Gero, Kluwer Academic Publishers: 597-616.

Josephson, J. R., B. Chandrasekaran, et al. (1998). An Architecture for Exploring Large Design Spaces. National Conf on AI (AAAI-98), AAAI Press/The MIT Press.

Cognitive Outsourcing in the Conceptual Phase of the Design Process

Günter Höhne and Torsten Brix, Ilmenau, Technische Universität

1 Introduction

Investigations of how the design process actually proceeds have shown (B&B Reports 1994-2001) that engineering design can be made both more efficient and better at fulfilling the original purpose if the design process flow is based on order and clearness. This applies to the description of the objects being designed at the various stages of their development; it applies, likewise, to the formulating of rules, regulations and critical paths with which it will be necessary to support a methodical approach, and it also applies to their use. Objects being designed are to be represented in a variety of external forms, but, frequently, the full possibilities of such representations are not exploit. For the concrete stages of the design process, technical drawings with their laws and norms are one well-established and useful type of representation. Another is geometrical modelling, whether with two-dimensional or three-dimensional CAD systems. However, there is no set of representational and other support tools to assist in the earliest stages of engineering design. In hopes of closing this gap to some extent, the paper concentrates on cognitive outsourcing (see introduction of the chapter) and helps for conceptual design, with the aim of offering suitable methods and computer-assisted tools which may support the designers conceptual work.

2 Modes and possibilities of cognitive outsourcing in the design process

While there are important distinctions among the various procedures described in the literature on design methodology they all entail a comprehensive structure with a certain flow being imposed upon the design process. There are various differences as to terminology applied and demarcations drawn between individual stages or steps, but apart from these, the procedures offered typically vary as to their use of "thinking aids" to assist in the solving of parts of the problem, affecting the designer's imagining of what he or she is to design – and thus also the external representations. It is the earliest design stages where such imagination is most relevant.

In the "Bild und Begriff" (Image and Concept) workshops (B&B Reports 1994-2001) there were lectures and discussions on the problem of the appropriate form of representation to be used in the early stages of engineering design – with particular reference to the alternation between conceptual and graphic ideas when technical products are being projected. Both concept and image may exist not only as "internal" but also as "external" representations – externally expressed, they take many different forms, and it may be by either the manual or the computer-assisted method that they are produced and/or modified (Table 1).

Table 1. Examples of external representations of an Oldham coupling at different levels of abstraction

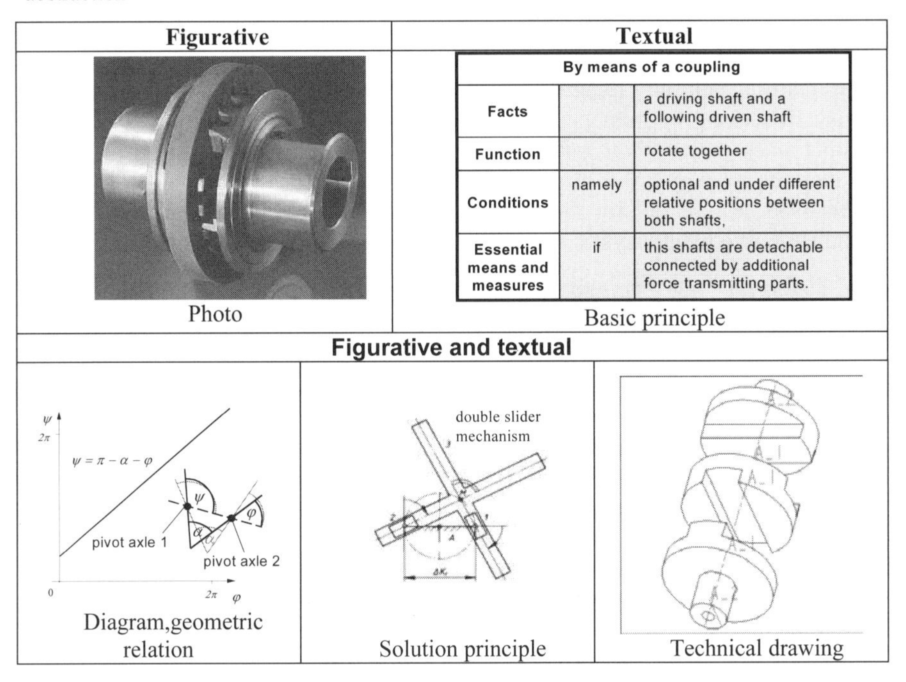

Figurative	Textual
Photo	Basic principle

By means of a coupling		
Facts		a driving shaft and a following driven shaft
Function		rotate together
Conditions	namely	optional and under different relative positions between both shafts,
Essential means and measures	if	this shafts are detachable connected by additional force transmitting parts.

Figurative and textual		
Diagram,geometric relation	Solution principle	Technical drawing

Minuting the process step by step and state by state as the design advances is absolutely essential. These minutes are firstly necessary as a support to the engineer's problem-solving thought processes, helping to make the mental models become clearer and more checkable; secondly, they are an aid to communication. Developers of new products tend predominantly to "nail down" their initial thoughts, before ever the technical object exists, in image form. The three functions performed by such external representation are summarised by Müller (Müller 1994) as

- Communication
- Recording
- Support (technical and manual).

The thought processes to determine a yet non-existent product is the main task in the design process. For this we found two main purposes, which are consonant with the three given above:

1. Descriptions of a product before it exists are a necessary means, affecting all involved in the product's life-cycle, of fixing the stage of thinking reached in the designer's work. The transfer and storage of this information are each subject to such demands as those of being unequivocal, reproducible, and low in redundancy, and thus possess the characteristics of a classic sender-receiver communication channel. It should be noted that graphic representations are not only necessary as the basis for communication between the people involved, but are also useful to the original thinker as a means of reviewing what has been imagined. Internally, computers employ a quite different type of data structure, which only requires conversion to the graphic form if the purpose is visual checking by humans; the CAD-CNC-CAQ data string could, of course, function entirely without images.
2. There is no doubt that the designer uses his image of whatever is being designed as a model for the as yet non-existent object, and manipulates that model in two main ways:

   – One is *analysis* – here the designer confronts the design with figures for size, etc. (which are also capable of depiction), so as the check out how it might fulfil particular specifications. Analysis also helps in the comparison, judgement and ranking of alternative solutions. Also, it is the starting point for abstract conceptualisation.
   – The second way the designer manipulates his image (or model) is *synthesis*. Here, a number of familiar elements are together configured into more complex structures. Sketches are an indispensable support to synthesis. However, there are additional types of manipulation which are assisted by the use of an image as the initial model: it permits abstract imaginings to be given concrete components, it permits the structure first found to be varied, and it permits dimensioning (i.e. measurements synthesis).

The forms of external representation shown in Table 1 take the strain off the developer's thought processes, particularly to dealing with complex scenarios in the course of design. They are indispensable as a means of increasing the efficiency of the process, whatever means (manual or computer-assisted) used to produce or modify or exploit them. When analysis and synthesis, those two vital procedures, interact together during the technical problem-solving process, these external forms are used in contrasting ways, as shown in the two following sections.

3 Abstraction, analysis, optimisation

From the initial commission onward, the developer is confronted at every stage in engineering design with need to analyse, rank, judge, evaluate, and improve either what is "given", or what is "found" by means of synthesis. This is the cycle of problem-solving which must be followed at every design stage, and for it, abstraction is the prerequisite.

To make real objects or sequences capable of scientific or technical manipulation, simplified representations which match the purpose of the moment are required, and abstraction (B&B Reports 1994-2001) is the means by which concepts and models are achieved. During the design process, abstraction serves

- to highlight the essentials in the preliminaries to problem-solving,
- to discover the common features of potential design solutions (and thus act as a means of classification),
- to simplify relationships while models are being constructed and calculations made.

For this sort of representation, generalisation, isolation and idealisation of the technical facts may all be required in due measure. The degree of abstraction will also be of relevance, for it will significantly influence the efficiency of the attempt to design in a systematic way. If the commission (the technical function specified) is formulated very generally, there will usually be very many potential solutions generated. If the commission is specified in too great a detail, it is even possible that no solution might be produced. The level of abstraction selected is thus a means of regulating of how well the intention will be fulfilled by the design procedure adopted. However, it is still largely experience and intuition which assist in determining the level of abstraction to choose. This is illustrated below with some examples.

To take one instance: if one needs to judge whether a function is fulfilled, it is only necessary to take account of those characteristics of the product as projected which are relevant to that function and contained in symbolic representation of the product's principles. Fig. 1 shows a situation where the technical principle according to Fig. 1(b) is not permissible in the description of the configuration as in Fig. 1(a), though it does describe the bearing of a mirror. The arrangement of the axis around which the mirror and thus the reflective surface tilts is functionally important for modification of the breadth of cut when the beam has undergone diffraction and for the setting of the mirror dimensions. The technical principle must, therefore, be selected according to Fig. 1(c).

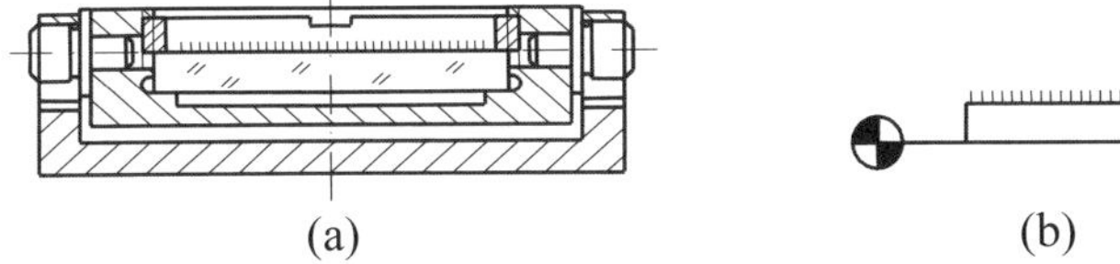

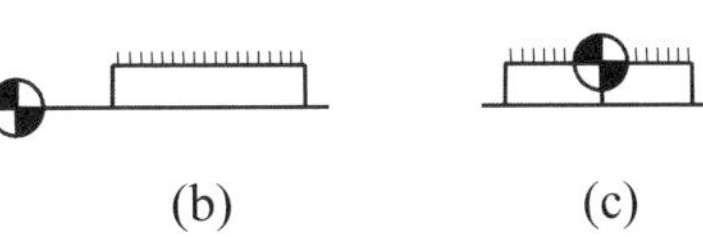

(a) (b) (c)

Fig. 1. Bearing of a mirror to deflect a beam of light: (a) description of the complete configuration; (b) and (c) sketches of the principle involved

Even so, symbolic representation of technical principles is not always an easy matter, as is shown by comparing the two examples in Fig. 2(a) and 2(b). In each case, a solution principle of a slider mechanism free from backlash is represented. However, Fig. 2(a) contains far too much information describing the function, and this detracts from the ease and speed with which it can be interpreted.

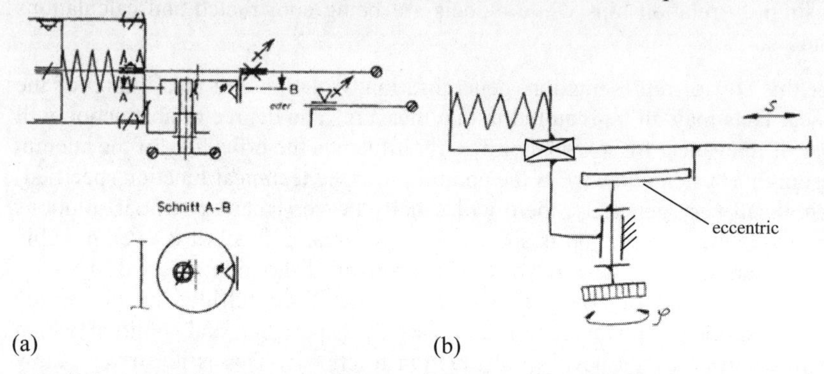

Fig. 2. Solution principle of a slider movement free from backlash driven by an eccentric: (a) represented according to (Lippardt 1996) and (b) represented in a version reduced to structural features of relevance to the function

It is only possible to evaluate the manufacturability of a configuration generated in the course of technical design when the manufacturing process is known. The detail of the configuration shown in Fig. 3(a) is thus favourable if the manufacturing process will be lathe work or ceramic pressing, but not if it will be die-casting. In contrast, the version in Fig. 3(b) is a configuration which is fine for die-casting but wrong for procedures such as ceramic pressing.

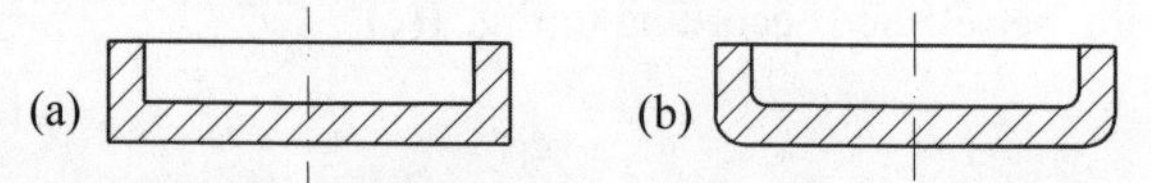

Fig. 3. Design variants of a rotationally symmetrical part: (a) without roundings (e.g. manufacturable by ceramic pressing) and (b) with roundings (e.g. nonmanufacturable by ceramic pressing).

The instances given in Fig. 1 and Fig. 3 show that technical drawings which relates to function and manufacture may still be short on information. It is possible nowadays to provide users with computer support while they are learning and practising with this vital task of abstraction (Höhne et al. 2001).

How function-oriented reduction of the description of the configuration is carried out is shown in Fig. 4. Here, all parts of the object are represented in the illustration without reference to their significance for function, manufacture or similar situations. So that the function can be recognised, the links the product will use for the exchange of inputs and outputs with its environment must be entered. Only when it is clear that the function flow goes to the tilting from the knurled-head screw with its scale do those elements of the structure become apparent on which the functions really depend. Abstraction to the technical principle will then proceed in the following methodological steps:

- Elimination of elements not vital to the function (e.g. fittings for attachment purposes),
- Reduction of the remaining components to formal elements vital to functioning
- Symbolic representation of the technical principle.

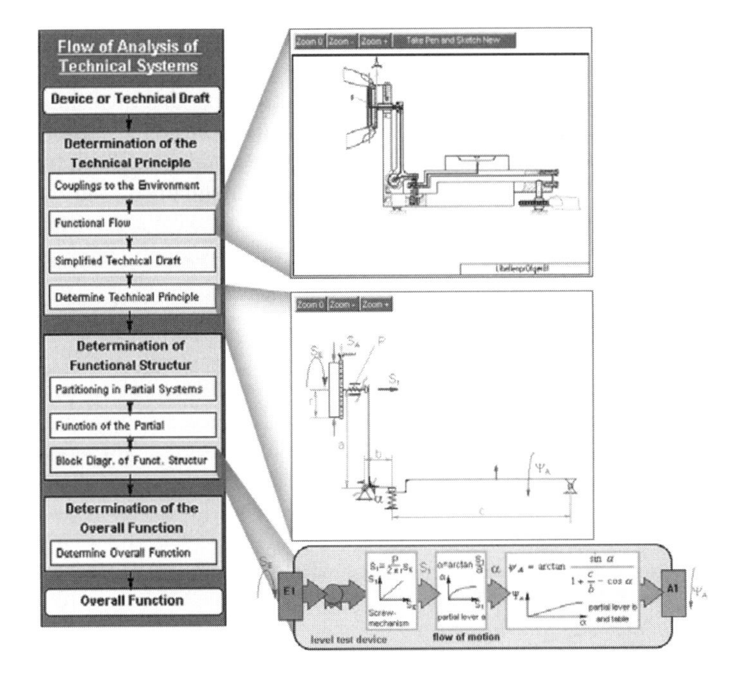

Fig. 4. Procedure for function-oriented abstraction of a given technical drawing (Screenshot of a web-based program for self-studies, Höhne et al. 2000)

The functionally important structural parameters required for the setting of dimensions and for the mathematical description of the function are very easy to determine using this sort of abstraction. It is then possible clearly to represent the relationships determined in a function structure. The configuration detail and physical positioning relationships for the elements describing the structure are, on the whole, ignored. Instead, their functions and their links are entered. Carrying out the abstraction process in this way is associated with simple and clear visual representation.

Technical principles and function structures constitute useful models for the computer-assisted analysis, simulation and dimensioning of engineered products during the early part of the design process. Constraint-based modelling is one example of a suitable tool (Brix et al. 1999).

The use of constraint networks for computer-assisted external representation (in the first principles stage, for example) is actually similar to the procedure for the layer construction used for mechanisms, vector plans in the mechanics field, and other modelling methods based on the geometry of a product. The user is given a tool which, with help from the constraints method and used interactively, will generate first principles structures of different levels of complexity if supplied with pre-defined symbols and variable parameters such as angles or distances apart (Brix et al. 2001). This activity resembles that in a feature-based modelling approach, as the symbols made available to the designer have already been provided with the relevant geometrical and functional features. This quantitative description of the principles makes it possible to carry out analysis, simulation and optimisation runs for such things as simulations of mobility and kinematic analysis (see Fig. 5), analysis of degree of freedom or of error, tolerance optimisation and so on. Likewise, it makes it possible to check whether the features specified in the commission have been maintained. The quality thus achieved far exceeds that of pencil and paper for the sketching of objects being designed and will often significantly benefit the designer's thinking, actually encouraging creativity and inspiration.

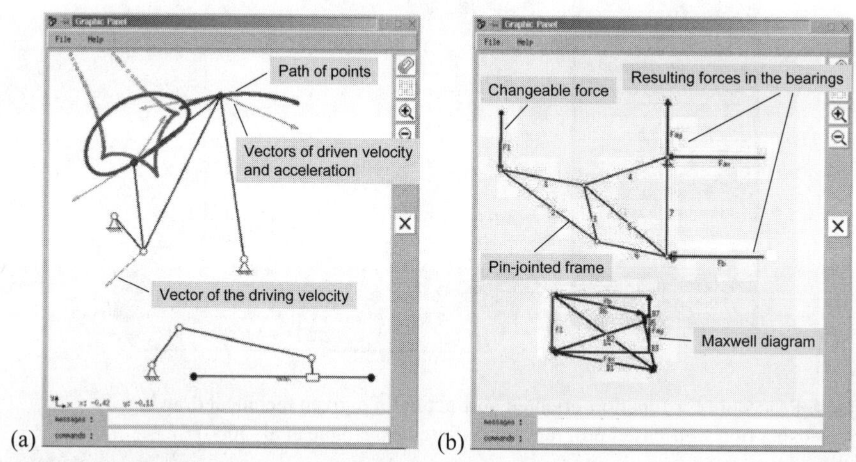

Fig. 5. Design system for modelling and analysis of solution principles (Brix et al. 2001): (a) kinematic analysis of mechanism and (b) force analysis on a pin-jointed frame

There have been studies of the order in which a design comes into being which show that the designer usually starts with some initial concrete, mainly graphic, ideas of possible solutions, or of small details which form part of a solution, for the problem to be solved. Then the designer frequently remains at this concrete level and attempts to modify the approach until it is good enough to produce an adequate means of fulfilling the specifications.

However, if innovation is the order of the day, it is better for the thinking to be channelled into a generalised description of species of solution. This calls for conceptual abstraction to extend the range of potential solutions. Abstraction multiplies the number of options. Furthermore, it is likely to produce a solution which is actually closer to the best one possible.

The task might be, for instance, the fixing of a component and the first idea might be to use bolts. However, if one generalises the task into the form whereby "fixing" is replaced by "preventing the present degree of freedom being available after joining", the basic principle according to Table 2 is the result.

Table 2. Basic principle for mechanical connections

By means of a mechanical connection		
Facts		two or more parts
Function		have a fixed relative position
Conditions	namely	under influences of forces and moments
	if	the parts
Essential means		1. have connectable surfaces, which are
and measures		2. brought in a physical context and
		3. locked together, so that degrees of freedom nonexistent

In Fig. 6, all imaginable ways of making the mechanical connections required in engineering design are depicted. Abstraction brings with it, in addition, a basis for systemising a known range of solutions, and, in the case of development work, for opening up a range of solutions yet to be determined.

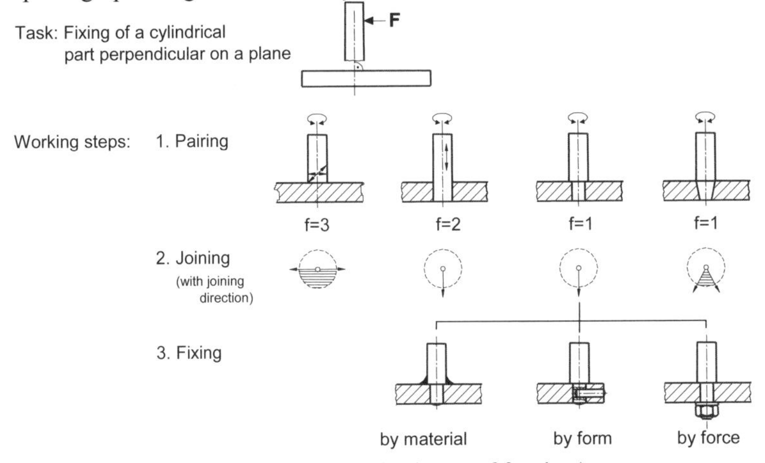

Fig. 6. Design of mechanical connections (f – degree of freedom)

Table 3. Combination table for a linear positioning system

Function element	Variants	Selection of possible solution
1. Transducer	1.1 Electric motor	
	1.2 Manual drive	x
	1.3 Spring drive	
	1.4 Gravitation drive	

2. Coupling	2.1 Flexible coupling 2.2 Sliding coupling 2.3 Rigid coupling	x
3. Transforming gear	3.1 Gear/rack 3.2 Screw gear 3.3 Belt drive 3.4 Cam mechanism 3.5 Slider crank	x
4. Guide	4.1 Sliding guide 4.2 Roller guide 4.3 Spring guide	x

There is yet another method, the working out of new solutions by combination (also termed the morphological box method). A combination table for a linear positioning system with the possibility of combining 180 options is shown in Table 3.

While this task was being worked on, EEG measurements were taken to quantify the cognitive effort required in moving between the graphic and conceptual modes (Höhne et al. 1996; Schack and Krause 1995). The actions required are shown in Fig. 8, which reveals that the search for ideas is associated with more cognitive effort than is the making concrete of the idea as long as the latter takes place using images in the combination process. The general trend is that identification (the concept) is more taxing than is sketching (the image). Signs and symbols are associated with less cognitive effort. A suitable set of alternative or complex solutions is shown in Fig. 7.

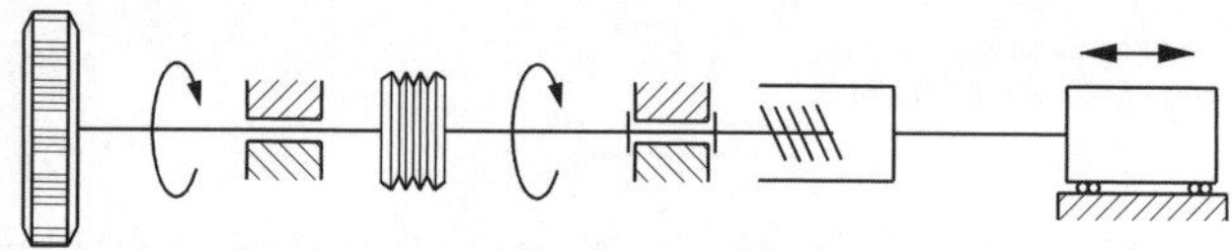

Fig. 7. Solution principle for a combination of function elements from Table 4 (marked with x)

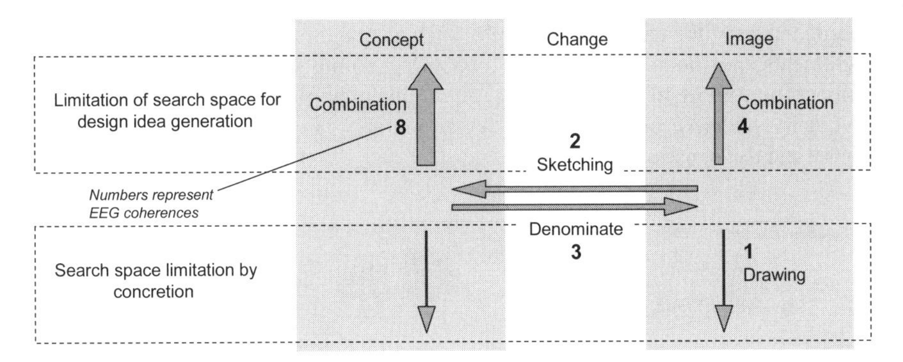

Fig. 8. Thinking activity during the search for engineering solutions and their representation. The number of EEG coherences found is interpreted as a measure of the cognitive effort applied.

The investigations brought results for mode in relation to effort which led to the following hypotheses, on which it would be interesting to base further investigations:

- The effort required for decoding may be reduced if the modes (image/concept) for the external and the internal representation are compatible, i.e. correspond, with one another. If an image is expected, an image should be offered – and vice versa (cf. Birkhofer 1994).
- Extending the search area in pursuit of ideas (by changing the mode from image to concept) will necessitate higher cognitive effort. Combining ideas in the concept mode is more taxing than in the image mode. In whichever mode, it is more demanding to extend the search area in pursuit of further ideas than it is to limit the search area by making the ideas more concrete. Cognitive training is necessary to help reduce effort required. Verification of these hypotheses would require measurement of the thought processes taking place in the designer's head, using a process which offers images.
- From the psychological point of view, it may be possible to stimulate the development of new ideas by changing the mode between concept and image. The stimulus may be provided by the form the environment takes. The information is provided as images, whereas the (verbal) exchange of information takes place within groups. This may mean that the (internal) shift from one mode to the other can be induced by the way the environment is (externally) set up.
- It is also possible to use computer assistance in making a technical solution more concrete (Höhne and Brix 2003). The example in Fig. 9 shows the transition from the engineering principle to the rough configuration. The type of principle found to be best, that of the crank, is what determines the scheme of configuration specifications for the transition to the rough representation. This rough configuration contains the specifications from the nature of the task, e.g. track features for the drive element, kinematic resistance features, forces and moments to be transmitted, velocities, and any values which can be

deduced from the technical principle, such as ranges of movement for particular elements, positional requirements, coupling points and, potentially, planes of operation important to functioning, directions of application of force, speed and velocity patterns, and permitted deviations such as tolerances in length, free play and deformation.

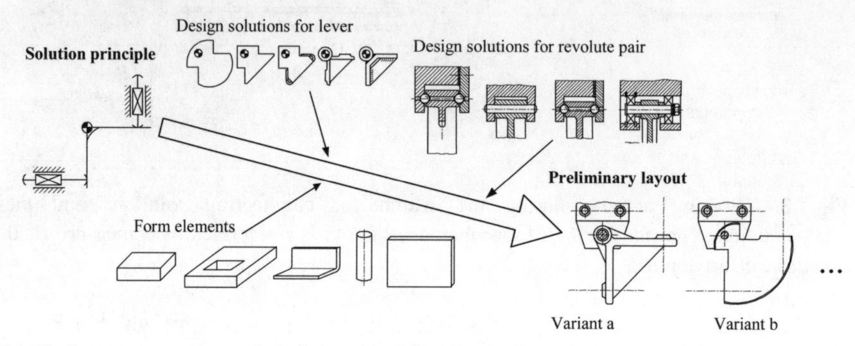

Fig. 9. Lever as an example of the transition from the solution principle to the rough design.

When these specifications have been assembled, a dialogue can ensue for the design of the possible forms. The principle will provide the constraints for the form. If these constraints are impossible to keep within, because of new specifications added, it will immediately be clear that a change of principle is necessary. A bi-directional link between the models of the various levels of abstraction will make it possible to work in more than one phase at once.

It must be said that this sort of cross-phase linkage is not at all, or hardly at all, computer-supported, which will mean additional cognitive effort on the part of the designer. If the product has a modular structure, the process of synthesis may, effectively, be that of configuration. The design process for such products starts with the function structure. Individual sub-functions can be the responsibility of various pre-prepared modules. A module should be available in different levels of description and different quantitative variants to perform functions within a certain range of parameters. The combination of the modules will produce the desired product variants (Höhne 1997; Riitahuhta and Pulkkinen 2001). How the configuration proceeds will depend on the type of function structure. Products with a given function structure can be configured directly in steps with the following order: parameter specification, choice of components, layout generation. An alternative would be to combine sub-structures for the configuration. If a two-(or multiple-)co-ordinate positioning system is required, it can be configured from the simple, one-coordinate system by a serial arrangement (Fig. 10).

(a) W_{el} | x-position → linear actuator → s → linear guide → s

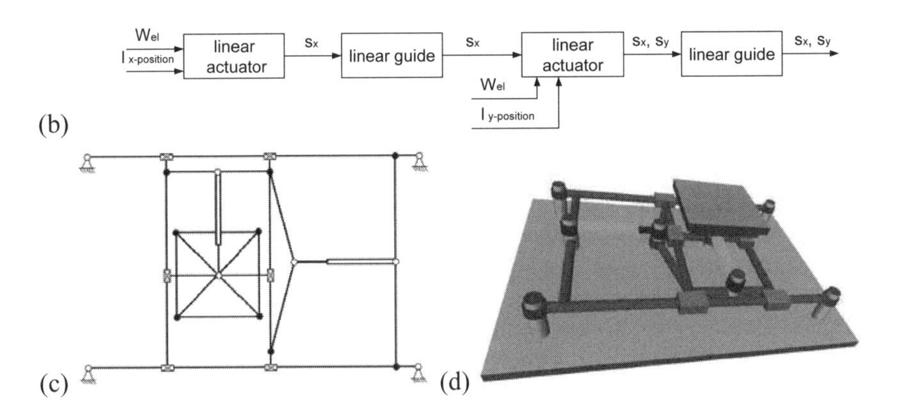

Fig. 10. Configuration of a two-co-ordinate positioning system by combination of two linear units: (a) given functional sub-structure; (b) functional structure; (c) solution principle; (d) preliminary layout/ embodiment design

4 Conclusion

The paper reviews the various forms of external representation for engineering models in the design process. At the conception stage in engineering design, the switching of modes between image and concept is of particular significance. Analysis of technical solutions requires step-by-step abstraction so that features can be evaluated. It also requires models to be developed to enable the behaviour of the product to be simulated and subjected to calculation. Targeted conceptual abstraction in the search for new solutions will lead to an extension of the search area. In comparison with the subsequent procedures for making solutions more concrete, for which computer support is possible, search area extension requires higher cognitive effort.

People developing new products need to be able flexibly to use the available external forms of representation, matching them to the current problem and objectives, and they need to be able certainly to execute the transitions between them. In the field of computer-assisted tool development, there is a need for cross-phase, consistent product models to be created which link (in both directions) the descriptions of the configuration, the principle and the function. Such models will mainly serve to free the designer from stereotypical work and other tasks which can be converted to algorithms, thereby improving the efficiency of design engineering and the likelihood of its fulfilling its purpose.

References

B&B Reports (1994-2001). Reports of the workshops on “Bild und Begriff”.

Birkhofer (1994). Bild und Begriff im Ablauf einer realen Produktentwicklung. In: Report of the 1st Workshop on "Bild und Begriff ", Munich, Germany, 14-23.
Brix, T., Brüderlin, B., Döring, U., Höhne, G. (1999). Using a Geometric Constraint Solver for Analysis in the Early Phases of Product Modeling. In: Proceedings of the 12th ICED, Munich, Germany, 1869-1872.
Brix, T., Brüderlin, B., Döring, U., Höhne, G. (2001). Feature- and constraint-based design of solution principles. In: Proceedings of the 13th ICED, Glasgow, Scotland, 613-620.
Frankenberger, E. (1997). Arbeitsteilige Produktentwicklung. s.1, no.291, VDI, Düsseldorf.
Lippardt, S. (1996). Der Mangel an guten Bildern: Untersuchung der anschaulichen manuellen Darstellung des Produkts in den frühen Phasen des Konstruktionsprozesses. In: Report of the 7th Workshop on "Bild und Begriff", Quedlinburg, Germany, 27-37.
Müller, J. (1994). Einleitung. In: Report of the 1st Workshop on "Bild und Begriff", Munich, Germany, 6-13.
Höhne, G., Sperlich, H., Krause, W., and Sommerfeld, E. (1996). Untersuchungen des Modalitätswechsels zwischen Bild und Begriff. In: Report of the 3rd Workshop on „Bild und Begriff", Dresden, Germany, 115-130.
Höhne, G. (1997). Configuration of Modular Products by Means of Mechatronic Components. In: Proceedings of the European Conference in Focused Aspects of Mechatronics, Ilmenau, Germany, 121-130.
Höhne, G., Lotter, E., Chilian, G., and Henkel, V. (2000). Teaching Software and Teleteaching in Engineering Design. In: Proceedings of the International Workshop on "Education in Engineering Design", Pilsen, Czechia.
Höhne, G. et al. (2001). Using Multimedia to Transfer Knowledge in the Teaching of Engineering Design. In: Proc. of the 13th ICED (4), Glasgow, Scotland, 411–418.
Höhne, G. and Brix, T. (2003). Function-oriented Configuration of Products by Means of Feature and Constraint-based Modelling. In: Proceedings of the 6th Workshop on Product Structuring, Copenhagen, Denmark, 2003.
Riitahuhta, A. and Pulkkinen, A. eds. (2001). Design for Configuration. A Debate based on the 5th WDK Workshop on Product Structuring, Springer-Verlag, Berlin.
Schack, B. and Krause, W. (1995). Dynamic Power and Coherence Analysis of Ultra Short-Term Cognitive Process–A Methodical Study. In:Brain Topography, Vol.8/2, 127-136.

Sketching in 3D
What should Future Tools for Conceptual Design look like?

Martin Pache and Udo Lindemann,
Product Development, Technical University Munich

Introduction

At the institute of product development of the Technical University of Munich, a workshop was held to bring together designers and CAD software developers. At the workshop, a designer, who worked for an automotive supplier for clutches, complained that CAD restricted him in being creative. On this complaint, the representative of a company that distributed a major CAD-system responded that "for a supplier, restrictions from automotive manufacturers are so severe, that there is no room for creativity, anyway." This statement may reveal a potential misunderstanding according to creativity. The designer may not be able to invent a new principle for clutches, but does he therefore not need to be creative?

According to Ehrlenspiel (1995), creativity is "the ability of humans to create ideas, concepts, products and combinations which are new in essential aspects and have been unknown to the processor so far". In our opinion, the quote stated before results from the misconception that creativity only takes place in design from scratch. Strong restrictions may exclude the development of new functional principle and may necessitate the reuse of a highly optimised working principle. Still, the adaptation of a known solution may require new partial solutions on a concrete level and therefore demand creativity in particular.

So creativity can be necessary in any stage of the design process and therefore needs to be supported whenever it occurs. Especially CAD-systems, often regarded as omnipotent tool for individual design, should support the emergence of new solutions. According to surveys (Pache et al. 1999) designers often sketch particularly as a preparation for CAD work and also sketch besides to CAD. Designers even draw within plots of actual CAD-models. Since more than 90% of the designers report "developing solutions" (besides "supporting communication") as intention for using sketches, it may be presumed that CAD-systems lack functionality in properly supporting solution development.

In this paper a tool for conceptual design – the so-called "3D-Sketcher" – is presented that aims to combine the advantages form the "classical" free-hand sketch and modern 3D-CAD-systems. In the following, the theoretical background of the 3D-sketcher's functional principle is described. Afterwards the findings, observation, assumptions and hypotheses regarding the sketching process are presented, which lead to the concept of the 3D-sketcher. The concept is characterized in principle, followed by the description of the actual configuration of the 3D-sketcher and its further development.

Theoretical Background

Ideally, a designer should elaborate and imagine all possible solutions for a design task within short time and compare them with each other. Because of the limitation of human mental capacity (often labeled as "working memory") (Miller 1956), this is certainly impossible for usual design tasks. External representations can decisively support the designer in developing and handling an extensive search space. They can be quickly accessible data storages which reduce the information processing load. Due to the externalization of information, an overload of mental resources and an interference of mental working processes can be prevented or at least, restricted (Hacker et al. 1998). Surveys among designers from industry (Pache et al. 1999) have shown that CAD and sketches are the most common representational media in engineering design processes.

Doerner (1999) states a procedure, which may be used to create new solutions. The so-called "ASAKRAM"-procedure is used for problems of composition, that includes sub-processes of abstraction and concretion (he names those processes "finding superordinate and subordinate concepts"). Doerner (1975) defines abstraction as "omitting inessential attributes respectively accentuating essential attributes". More generally, abstraction is the generation of terms, respectively elements, which represent a certain variety of subordinated terms, respectively elements. The "ASAKRAM"-procedure may be conducted by a "dialogue with oneself" in which the designer switches between different levels of abstraction and complexity. Thereby he can create combinations that are new to him as a whole.

Akin to the dialogue described by Doerner, Schoen (1983) describes designers having reflective conversations with their own sketches, in which they find useful clues for further procedure. Arnheim (1977) resumes basic mechanisms of visual perception that facilitate such effects. He even assigns "intelligence" to those mechanisms due to the cognitive functions they fulfil. Adapting Ehrlenspiel's definition for creativity, those mechanisms might as well be regarded as creative.

Observation, Assumptions and Hypotheses as Basis for the 3D-Sketcher

The idea for the concept of the 3D-sketcher arose during the work on a research project that analysed the process of sketching for conceptual design. Therefore the outline of this project is briefly presented at first, followed by those observations, assumptions and hypotheses, which laid ground for the creation of the 3D-sketcher. In this article, these insights shall not be verified.

120 design experiments were carried out with 45 engineering design students and 15 professional designers, each processing two conceptual design tasks that

differed in their complexity. No “thinking aloud” was demanded from the proband and the choice of tools was free, except that no CAD-system was available.

The probands were asked to design the conceptual solution of a device for laser welding of sheet metals. Shape and dimension of the sheet metal, as well as the laser and a linear motor for actuation were given. The proband had to develop a device that makes the laser move with defined speed, distance and angle along the sheet metal. Since the metal was shaped rather complex, some of the main problems were the guidance of the laser along the sheet metal's shape and the transmission of the linear actuation from the motor to the laser (see Fig. 1).

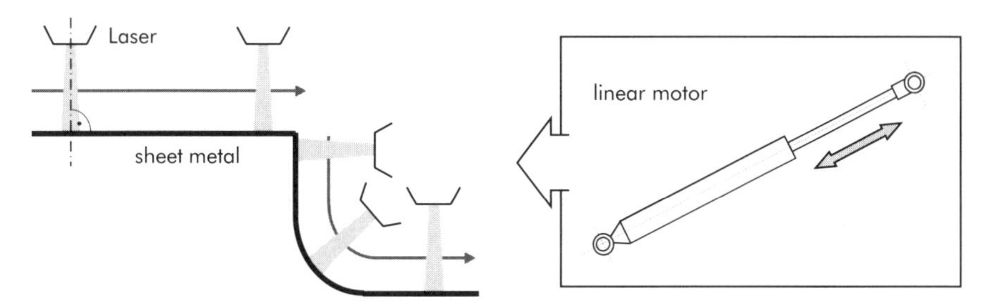

Fig. 1. Design task to develop a device for laser welding of sheet metals

Information modalities in sketches

The sketch shown in Fig. 2 is the early stage of a conceptual solution for the laser device that was developed by a student. The information within this sketch is very complex, therefore we will refrain from explaining its technical content. Instead we will discuss the modality of information within this sketch in detail.

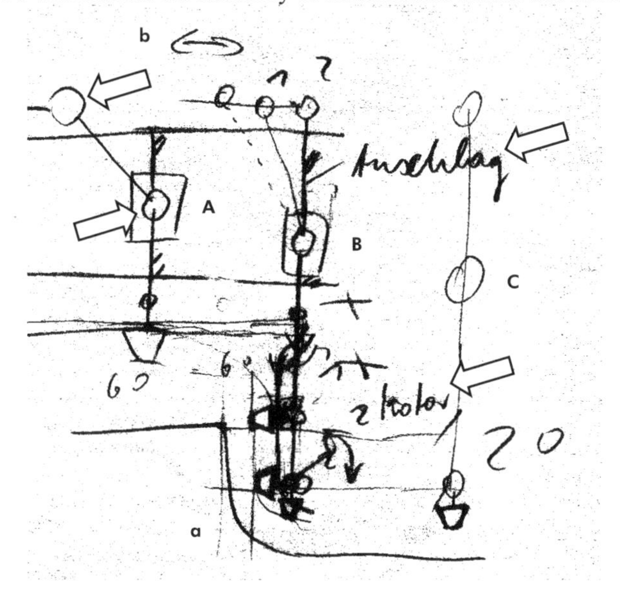

Fig. 2. . Student's sketch of a conceptual solution for the laser welding device

The student used (mainly) thick lines to represent the shape of technical components and he used thin lines to represent ancillary elements that indicate that certain elements are on the same horizontal or vertical level (a in Fig. 2). However, the representation of components is not limited to the use of depicted shapes, since words were used for these purposes as well. These words do not only label geometrical elements, but represent these components entirely. In this sketch "Anschlag" (= limit stop), as well as "2. Motor" represent components that are not defined geometrically yet. The student also used symbols, such as circles to represent joints. Similar to textual predefinitions, these circles do not further specify the type of joint (words and circles are labelled by white arrows in Fig. 2).

Additionally to those elements representing technical components, other elements are used to represent processes, such as motions. Technical components are drawn repeatedly to represent different states of a motion (an example is labelled in Fig. 2 by A, B and C). Symbols, such as arrows, indicate that certain components move in a certain direction (b in Fig.. 2 indicates a back-and-forth motion).

A sketch can provide the representation of technical components (along with the representation of their functional attributes) and technical processes. For these purposes, geometrical elements may be used, as well as symbols and words. Through the contextual meaning of symbols and especially of words, a designer can take advantage of the variety of different levels of abstraction that are provided by language. In this regard, the sketch can provide memory relief for the designer's dialogue with himself (as described by Doerner), providing him with external representations of the words in his mind. Still, regarded this way, a sketch is a "simple" memory relief, fairly similar to a text.

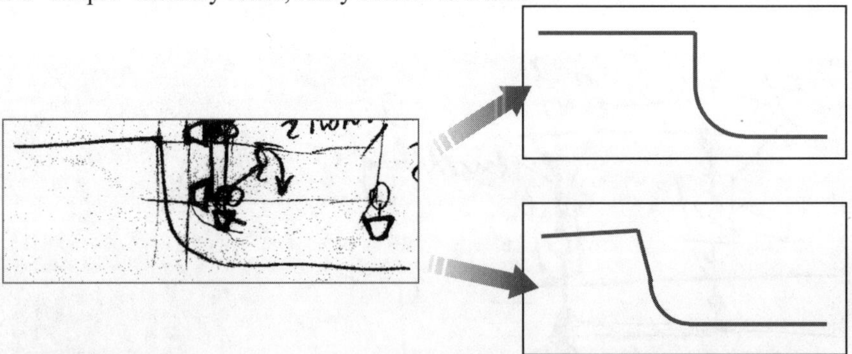

Fig. 3. Two concrete shapes that may be represented by an ambiguous shape in a design sketch

The shape of the technical elements represented in the sketch can not always be clearly recognized from the way they are depicted. Fig. 3 shows a sector of the sketch, in which the shape of the sheet metal (that is given in the task as an unchangeable boundary condition) is depicted. Although the student meant to represent the shape on the upper right side in Fig. 3, it might as well be interpreted as the representation of the shape shown on the lower right side. The shape of a component that was drawn roughly and quickly can be depicted in an ambiguous way. Therefore, such shapes may represent a variety of specific shapes. The recognition of this kind of predefinition depends on the contextual knowledge of the viewer.

This kind of ambiguous predefinition described before may be seen as abstract elements in certain aspects. It might be possible that such predefinition is used by a designer as a representation of a variety of possible shapes. Keeping in mind Doerner's definition of abstraction (see Chapter 2), this quality is an important characteristic of abstract, superordinated elements in general. Although in this case, the attributes that are omitted (respectively accentuated) can hardly be named by words, the effect may be just the same. Visually perceiving such an ambiguous predefinition allows perceiving the shape at once and still "exploring" the variety of specific shapes represented by it. Even if this ambiguity respectively abstraction has not been intended by the sketching person, the underlying variety of concrete shapes may be a source of inspiration for new solutions.

Re-interpretation of conceptual sketches

The following design episode has taken place early within a design experiment. A Student started from the desired direction of motion for the laser, which was given by the shape of the sheet metal as described in the formulation of the task: He sketched the sheet metal (see Fig. 4, state 1: the lower line represents the sheet metal) and then decided to achieve the motion of the laser by two tracks, which were shaped just like the sheet metal (each track represented by a double-line). The laser was fixed to a "cart" with two wheels, each running in one of these tracks. Thereby, the laser perforce had to move along the shape of the tracks. This solution is comparable with the principle of a monorail or a roller coaster. The cart with the laser was drawn twice to point out its motion.

Still, this approach did not cover the actuation of the cart. The student then visualized the actuation problem by sketching speed vectors to represent the crab's motion. The vector triangle was attached to the crab pointing in the direction of the motion (see Fig. 4, state 2).

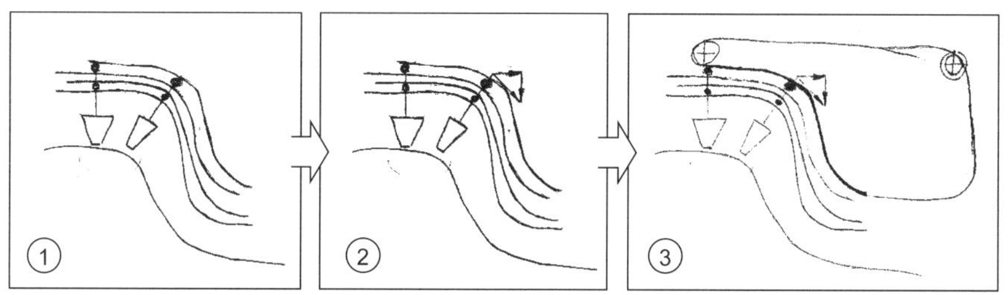

Fig. 4. Different states of the sketch when a new idea arose to realise a defined motion of the laser

After regarding the sketch (for a repeated time), he drew one of the track curves (the one that was closest to the vector of the crab) again right over the existing one and thereby intensified it, but did not alter its shape. Without removing the pencil from the sheet, he drew beyond the existing shape and completed the curve to a loop (see Fig. 4, state 3). He added pulleys, represented by simple circles, to some corners of the loop and attached the laser directly to the loop. Therefore, by adding few lines, the student altered the solution principle from a crab running in tracks into a circulating belt with the laser attached to it. Within the described sequence, the design concept was changed significantly. So what might have happened within these minutes?

As an explanation approach it may be assumed that the speed vector of the crab and the track curve have been attended together and have been correlated. Each element has been drawn independently and without a direct, contextual relation to each other. Still, the student may have referred this vector to the track and interpreted the combination of both elements as a curved shape that moves "in itself", which may be a belt or a chain. Furthermore, the assumption that such process has taken place is corroborated by the fact that from all curves representing the tracks, the one that almost touches the vector has been chosen for modification (although all curves have the same shape).

This explaination approach (for details see Pache et al. 2001) corresponds with the unexpected discoveries by Suwa et al. (1999) and the backtalk of self-generated sketches by Goldschmidt (1999). Still, in this case, the discovered relation has altered the meaning of the sketch radically. The track curve, a simple line, has geometric and symbolic character. Its shape roughly determines the geometry of the track. The curve gains a symbolical meaning because of the representation of the crab and only therefore, it is recognised as a track. In this regard, the curve again gains a completely new contextual meaning when correlated with the vector. The fact that the representation of the belt was developed directly out of the track curve by completing the loop (not by drawing a new sketch that represents the new solution), shows that this process is possible, although it cannot be verified in this experimental setting.

The 3D-Sketcher

On the basis of the insights into the use of free-hand sketches, several requirements on future sketching tools were worked out. These requirements aim to enable the occurrence of the mechanisms that were assumed in the preceding paragraph:

1. A future sketching tool should enable the designer to use shapes, words and symbols and put them into a spatial-relational context
2. It should be possible to represent technical components as well as processes.
3. The designer should be able to combine geometrical and symbolical meaning in one element (e. g. a curved arrow to indicate a complex motion)

4. It should be possible to depict ambiguous shapes that provide a perceptible variety of concrete shapes.
5. In all other aspects, the sketching tool should abide by the paper-pencil-metaphor as close as possible.

Beyond these demands from the insights into sketching, there are requirements that result from the subsequent CAD-processes and data management processes.

1. Since usually the products designed in mechanical engineering are spatial objects, it might be desirable to achieve a sketching process that especially supports the creation of 3D-objects.
2. Furthermore, a digital outcome from the sketching process would enable the integration of such result into digital product data management.
3. Manipulation of the digital sketch should be possible in all those ways that are provided by digital tools for representation (e. g. rotating, shifting, up and down scaling). This holds for the sketch as a whole as well as for elements of a sketch.

The 3D-Sketcher aims to fulfil all these requirements within one intuitively usable tool by the help of Virtual Reality. It enables the user to create a 3-dimensional free-hand line by motion of some input device through design space. The 3D-line is represented within the design space, depicting the device's motion line at the very location where the motion has taken place. It is displayed in real-time in relation to the device's motion and consists of small cylindrical increments forming a "pixel line" that is floating in space. The 3D-sketch that emerges can be rotated and shifted in space by some (additional) 3D-input device. Fig. 5 shows a 3D-sketch that was created by means of the 3D-sketcher. It shows a stand for presentation of the tool at an Industrial Exhibition.

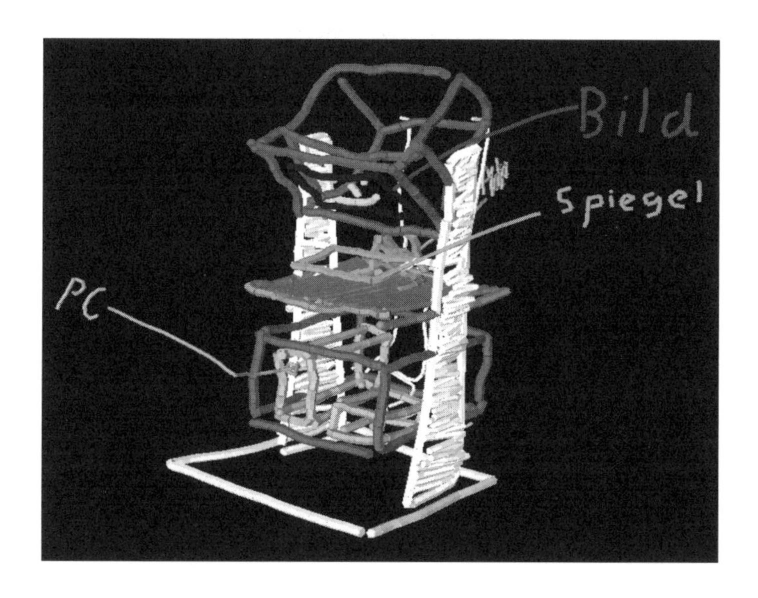

Fig. 5. 3D-sketch of a stand for presenting the 3D-sketching tool

A first prototype shows a low-cost solution, which is integratable into the common working environment of a designer. Basis of the prototype is a PC-based high-quality graphics workstation (Fujitsu-Siemens Celsius 460) with a 21-inch CRT monitor. The stereo view is realized by using an infrared emitter and shutter glasses. Software-base for the prototype of the 3D-sketching tool is World Tool Kit 9 by Sense8. This software is run on Windows NT 4.0.

As an input device the "Phantom Desktop" was used, which is basically a pencil-like grasp that is connected to the main device by a system of levers. The grasp can be moved in space and is localized according to 6 degrees of freedom by sensors that are integrated in the joints between the levers. The integration of the Phantom Desktop was realized with Ghost SDK, a developer kit that is shipped with the Phantom. The grasp of the Phantom holds a button that can be used to activate the drawing mode of the 3D-sketcher. Thereby, the 3D-line is created by moving the grasp and simultaneously pressing the button. In order to achieve the imagination that the 3D-line emerges at the very tip of the pencil-like grasp, the configuration shown in Fig. 6 was chosen. The user watches the display via a semi-transparent mirror that is placed horizontally over the worktable. Thereby the displayed 3D-sketch appears to be floating underneath the mirror. At this location, the Phantom is placed, which can be seen as well due to the semi-transparency of the mirror. Thereby, the action-space and perception-space merge underneath the mirror.

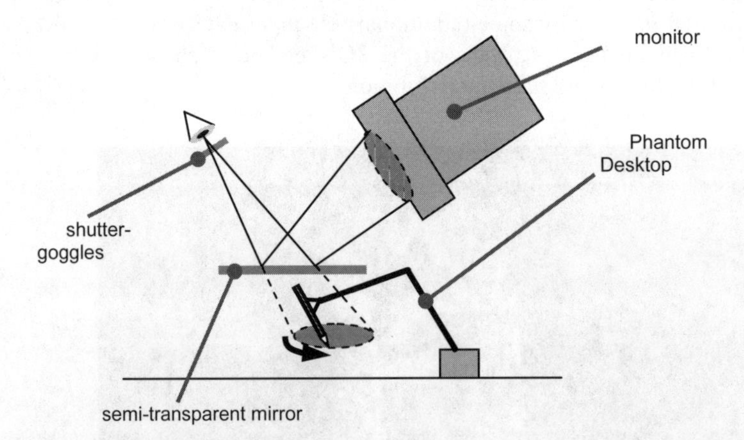

Fig. 6. The 3D sketching tool

The sketch can be moved in space either by a separate space-mouse or also by the Phantom Desktop. Since this is an input device with 6 degrees of freedom, shifting and rotating can be done in a very intuitive way. In order to do so, the drawing mode has to be switched into the moving mode by the help of the keyboard.

For input as well as for output, several VR-devices are possible. Basically any kind of input-tool needs to be localized in space. Besides grasps that are connected to a system of levers (as in the case of the Phantom), any pencil-like input-tool may be suitable that is localized by any (non-mechanical) tracking-system. Moreover, a data-glove may be considered as an input-device, which would turn the fingertip into a 3D-pencil as soon as a certain gesture (like pointing the index finger at a certain point in space) is carried out. Shifting and rotating the sketch may be done by the motion of the hand when a grasping-like gesture is carried out. Working with two data-gloves (one for sketching and one for moving the sketch) may be possible as well. For output a large variety of devices exists as well: head-mounted-displays, simple desktop-VR working environments, projection tables and caves can be considered.

Besides the variation of input and output devices in future several other characteristics respectively features of the 3D-sketcher will be systematically varied and analyzed regarding their benefit: representation of the sketch could be adjusted to the user's angle of view. Thereby, the viewer can regard the 3D-sketch from different angles just by moving his head, which probably meets natural habits. In the actual prototype this is basically possible already, but the magnetic tracking system interferes with the magnetic field of the monitor, and therefore the position of the displayed objects is unstable. A non-magnetic tracking system would be more useful in this case.

Further features for manipulation of the sketch will be developed, which are for example duplication, rotation and shifting of single lines (respectively other elements) of the sketch. Therefore the data representation of the "pixel cloud" needs to contain information about the semantic relation between the pixels (e. g. which pixels form a line or a certain symbol). Such information would also be helpful for integration of the 3D-sketcher into existing development processes and IT environments. This concerns particularly the linkage to existing CAD systems or even the integration of the sketcher into CAD. As a medium-term goal, it should be possible to concretize 3D-sketches within a CAD system and to "load" already existing CAD-models into the 3D-sketching environment. In the long run, all sketching features could be fully integrated into a CAD-environment, which is also "truly" 3-dimensional by the help of Virtual Reality. The process of step-by-step concretizing a vague 3D-sketch into specific geometry of a CAD model could be supported by means of software assistants. Thereby the 3D-sketcher could be an essential element of next generation CAD systems.

Summary

During several studies, the authors analyzed the use of free-hand sketches by experiments and surveys. Thereby several observations were made regarding hypothetical mechanisms of creativity in conceptual design. Based on these observations, several requirements on future sketching tools were formulated, as for example to provide the user with the variable line being the infinitesimal element for presentation. Furthermore, requirements were formulated regarding today's processes and tools that are typically used within design processes. Thereby, the most important requirement, besides being digital in general, was the 3-dimensionality of a future sketch.

Based on this consideration, the concept of the 3D-sketcher is derived, which enables the designer to draw lines within a 3-dimensional design space. By the help of Virtual Reality, the designer can create a 3D-sketch that consists of lines that appear to be floating in space. The lines are generated by moving an input device within the design space, whereby the line evolves in real-time at the tip of the device. A prototype of this system has been developed at the institute of product development at the Technical University of Munich, which realizes this basic principle along with some additional features for manipulation of the evolving 3D-sketch.

In future, additional functionality will be added to the software and further devices for in- and output will be implemented and evaluated. In the long-run, the objective will be to fully integrate the 3D-sketcher into future CAD-systems.

References

Ehrlenspiel K (1995) Integrierte Produktentwicklung. Hanser, München

Pache M, Weißhahn G, Roemer A, Lindemann U, Hacker W (1999) Effort-saving modeling in early stages of the design process. In: Lindemann U, Birkhofer H, Meerkamm H, Vajna S (eds) Proceedings of ICED 1999, Vol. 3. Technische Universität München, pp 679-684

Miller GA (1956) The Magical Number Seven. The Psychological Review 63: 81-97

Hacker W, Sachse P, Schroda F (1998) Design thinking – possible ways to successful solutions in product development. In: Frankenberger E, Badke-Schaub P, Birkhofer H (eds) Designers, The key to successful product development. Springer, Berlin Heidelberg New York, pp 205-216.

Doerner D (1999) Bauplan für eine Seele. Rowohlt, Reinbek

Doerner D (1975) Problemlösen als Informationsverarbeitung. Kohlhammer, Stuttgart

Schoen DA (1983) The Reflective Practicioner. Basic Books, New York

Arnheim R (1977) Anschauliches Denken. Du Mont, Köln

Pache M, Roemer A, Lindemann U, Hacker W (2001) Re-Interpretation of Conceptual Design Sketches in Mechanical Engineering. In: Proceedings of DETC'01 (ASME 2001 Design Engineering Technical Conferences), ASME, Pittsburgh, on CD-Rom

Suwa M, Gero J, Purcell T (1999) Unexpected discoveries: How designers discover hidden features in sketches. In: Gero J, Tversky B (eds) Visual and Spatial Reasoning in Design. Key Centre of Design Computing and Cognition, University of Sydney, pp 145-162

Goldschmidt G (1999) The backtalk of self-generated sketches. In: Gero J, Tversky B (eds) Visual and Spatial Reasoning in Design. Key Centre of Design Computing and Cognition, University of Sydney, pp 163-184

VR/AR – Applications, Limitations and Research in the Industrial Environment

Ralph Schönfelder,
Virtual Reality Competence Center DaimlerChrysler Research, Ulm, Germany

Introduction

When people are asked about their definition of VR, they might say something like "VR is being within artificial worlds or dreams", "VR is the technical creation of real experiences" or "VR is interacting naturally with non-real entities". Those quotes are not very precise, but they already contain the "three I", introduced by (Burdea and Coiffet 1994) for a definition: immersion, which is the attempt to provide the VR user with feedback that helps to create the feeling of "being there" (presence); interaction, which integrates the user even more into the virtual environment offering him the power to change and use it; imagination, that is the ability of human beings to believe something is real even if its representation is somewhat unreal. Another, more precise and technical definition would be "VR is a human-machine interface, which allows for perceiving an artificially generated environment as a reality with multiple senses involved in that process.", which is based on (Hennig 1997).

In the context of VR the terms Mixed Reality (MR) and Augmented Reality (AR) are commonly used. To best describe them, (Milgram and Kishino 1994) offers an explanation which is summarized in the depicted graph: while MR describes the whole range between reality and virtuality, AR consists of more real than virtual components. (Regenbrecht et al. 2002) defines AR as follows: "Augmented Reality (AR) attempts to enrich a user's real environment by adding spatially aligned virtual objects (3D models, 2D textures, textual annotations, etc) to it. The goal is to create the impression that the virtual objects are part of the real environment."

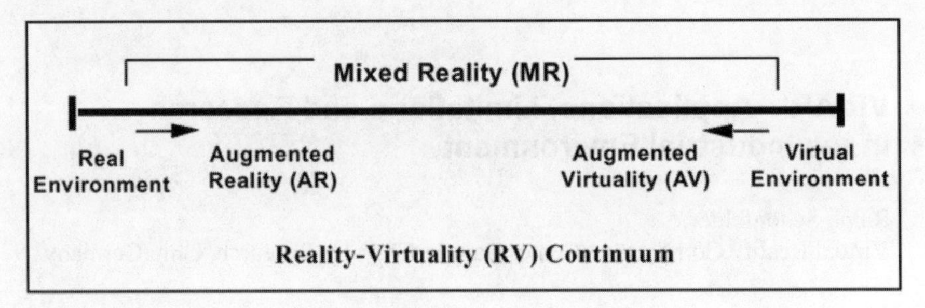

Fig. 1. Reality-Virtuality Continuum (Milgram and Kishino 1994)

Technologies

The main effort of VR/AR research has gone into the development of new technologies. The focus has been on immersive display systems (optical, audio, haptic, tactile), tracking systems, interaction devices and VR Systems (Frameworks, Connectivity and Software). (Hennig 1997) offers a more detailed overview.

Immersive Display Systems

Today's optical display systems can be classified roughly into: a) the ones attached to the head of the user (head mounted displays (HMD) or head mounted projective displays (HMPD)) and b) the ones more or less stationary in the world.

The first group usually provides stereoscopic views by offering each eye a separate image, produced by LCDs, Lasers or projection systems (Hua et al. 2003). The major drawbacks of these systems are usually a bad field of view, low resolution and the somewhat heavy glasses. Also, when used in AR, the image of the real world has to be either recorded by cameras and displayed in the HMD, or a see-through mode has to be used with the need of aligning the virtual objects correctly with the see-through real world.

The second group (e.g. walls, CAVEs, Workbenches) is usually driven by different projection systems like LCD/CRT/DLP projectors, allowing for active (that is the temporal separation of the left and right eye image) or passive stereo (that is the separation of the left and right eyes images through either polarization or wavelength separation/transformation). Alternatively, CRTs, LCDs, plasma or other displays can be used, although these designs usually do not offer the possibility of stereo.

Though optical display systems are a key part of VR/AR installations, the use of audio, haptic, or tactile feedback is becoming more common. These modalities can add great value to intuitive interaction and immersion.

Tracking Systems

The task of tracking is crucial when it comes to the need of showing the user a virtual or augmented reality: head movements have to be translated into equivalent movements of the viewpoint, interaction devices need to have their virtual representations follow the real transformations or even a virtual half-world has to match a real half-world.

Many different tracking technologies have been developed, every one of them has its advantages and drawbacks. Mechanical tracking is very accurate but does not allow for fast, momentum-free movements, magnetic/capacitive tracking does not require large installations, has a good update rate but is easily subject to distortions, inertial tracking has the drift problem, optical tracking has good potential, but mostly requires markers and computationally intense algorithms (yielding a low update rate, when no expensive hardware is used), acoustic tracking is susceptible to room geometry and GPS is not very accurate and only available outdoors.

To overcome all those limitations hybrid approaches are widely used. The resulting precision and update rates are well enough but to the cost of a higher software and hardware complexity.

Interaction Devices

With the capability to experience the third dimension through more immersive displays the need to control it becomes important. The 2D desktop interaction devices and metaphors are not suitable anymore to all tasks in the 3D domain. Here many of different instances have been generated, from which none can claim to be the one that fits all needs, applications and installations. Just like in real world, it seems that numerous different tools are required.

Additionally, when compared to the traditional desktop computer interfaces, the often free-hand operated tools of the VR/AR domain cannot match the exactness and possible usage duration, but they allow for simultaneous manipulation of more degrees of freedom.

VR Systems

A typical VR system consists of a rendering kernel, a layer that performs certain animation/simulation tasks, a layer which contains the interaction logic, and a layer that allows the attachment of the required devices. Actual efforts lie in the incorporation of certain software and hardware trends like distribution, modularisation, scripting, runtime-configurability, wireless transmissions, plug&work etc. Those attempts mostly aim at the ease of VR/AR installations, which are still very complex and time intensive.

Applications

Today most non-research applications of VR/AR technology can be found in industrial environments of the military, aerospatial, automotive and entertainment sectors. The typical installation is rather space intensive, expensive due to specialized hard- and software and its maintenance, complex in its setup and difficult use. Those factors have inhibited a deployment of VR/AR in the consumer market so far, but with technology evolving (especially in the area of computer graphics hardware), the numbers of affordable home entertainment VR/AR technologies will most likely rise soon.

With VR/AR being seen as a tool in the working environment, one can identify even more problems: In many cases the available technology does not match the requirements of ergonomics or the easiness of natural tools. Thus, users have a much harder time in learning and working with VR/AR tools, leading sometimes to low acceptance of the system. Due to the research status of several involved aspects the cost/benefit ratio is difficult to estimate for the decision-makers. A great deal of persuasion work is sometimes needed to make them enter the "adventure" of using this new technology.

Nevertheless, certain applications present an ongoing proof for the potential of VR/AR, supplying its environment with valuable advantages. Here is a short list of installations developed at DaimlerChrysler research:

- The process of creating a new car design involves several creation-validation cycles. Traditionally, validation has been performed using e.g. wooden models, with their creation taking a great deal of time. By allowing designers to validate the car models in a large scale immersive environment, this model generation time is greatly reduced. That leads to a significantly shorter production time for new car types, which is essential in the competitive automotive sector (May and Stahs 1997).
- When modelling new parts for a car, not only their aesthetics or functionality need to be in the mind of the engineer. He also has to think about the maintainability. Thus, the question whether a service person will be able to remove a part without getting stuck with his hands or tools needs to be answered. This is simulated using a model of the car and its components, being displayed in an immersive environment. All geometry is collision enabled, that is, if two parts collide, a feedback is transmitted to the user. By allowing near-natural 3D interaction through a haptical or tactile feedback device the engineer receives similar feedback the service person would get while trying to remove the part (Buck and Schömer 1998).
- Long intercontinental flights usually strain their passengers because of little space in the aircraft's passenger cabins. The more important is the supply of fresh air to the individual, which has to be considered in the design phase of such a space. Because airflow is a 4-dimensional phenomenon (3 spatial dimensions + time), an evaluation using a 2D monitor is somewhat suboptimal. For this scenario the following solution was found: A part of the cabin with the passenger seats is being looked at through a see-through enabled HMD.

Airflow is now added into the scene in the form of volumetric "clouds". Also, other plane parts and passengers are digitally included, allowing the engineer a walk-through and a validation of the simulated airflow (Regenbrecht and Jacobsen 2002).
- Reflecting the improvements in comfort and safety, new cars and trucks contain a large number of wires. Since those wires take space away it is important to find length-optimised pathways through the inner spaces of the vehicles. This can be simulated by using a real part of e.g. a truck and attaching virtual wires to it, seen through a HMD (Baratoff and Regenbrecht 2003).

Immersive Modelling

In the design process modelling plays an important role. The application of VR/AR technology in modelling seems to be apparently useful, but a closer look shows there is no easy solution. In this part we try to identify reasons for this difficulty and discuss opportunities to overcome the existing problems.

We define immersive modelling as the task of shaping or composing spatial objects with the use of immersive technologies. This definition takes into account that modelling is not always the direct or indirect manipulation of just geometric data. Especially in the sketching phases of the design process, modelling can have a very conceptual nature. To the contrary, exact modelling is the task of aligning geometrical properties to a specific target state, an activity that does not allow for aberrations.

There are several approaches to using immersive modelling for sketching purposes (immersive sketching). Nearly all of them use highly immersive environments like CAVEs, Powerwalls, etc. (Bimber et al. 2000; Stork 2003; Deisinger 2002), often coupled with free-hand interaction techniques. We observe that those setups have not found their way into industrial every-day processes. The main reason might be the low availability of that type of "sketchbook". While paper, whiteboards and napkins are accessible at nearly any time, immersive sketching applications are too large and costly to fulfil that requirement. Maybe a good solution based on portable tablet computers may be accepted by the design community. Also, we believe that modelling through direct manual interaction requires physical restrictions or aids that allow to sustain intermediate states without application of physical force.

The advantage of an immersive sketching system might not even be the possibility of handling and interacting with 3D sketches rather than 2D projection drawings of 3D objects. When immersive technology succeeds to create the feeling of presence, the designer is more capable of evaluating the sketch in relation to "their" environment. New, helpful ideas might evolve in this process. Also, the conceptual character of a sketch could be possibly improved: When assuming that the object to be designed has a certain use, immersive technology (reaching out to emulate reality) enables one to catch the "use concept" better than plain paper. To give an example, the sketch of a tool could include the designers idea on how to use that tool. An immersivly experienced 3D animation of that "use concept" including the designer's avatar might be able to teach the tool user in a better way than a handbook can.

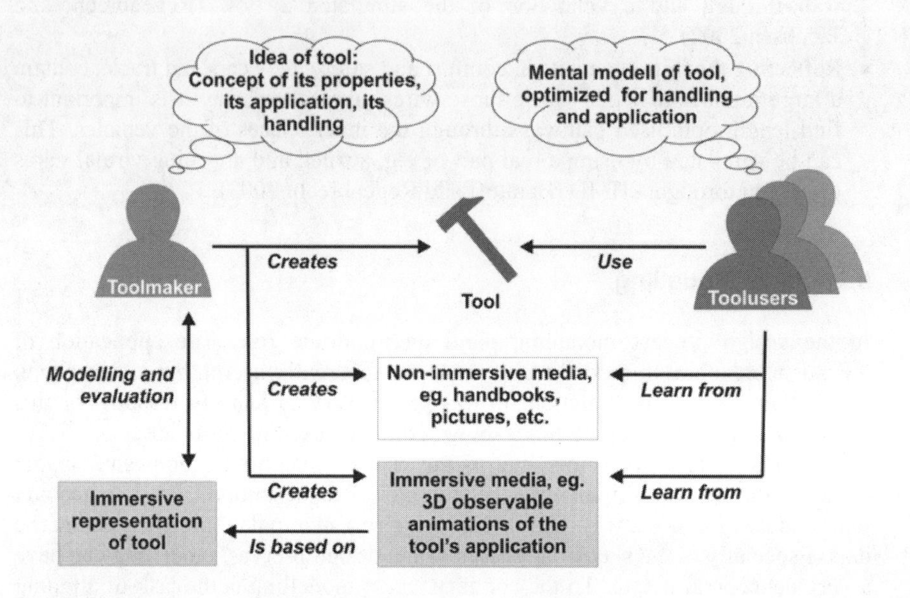

Fig. 2. Example of Immersive Technology's Role in Tool Design

Our second application of immersive modelling aims at today's CAD systems. Here interaction techniques sometimes need to be very precise, and observation is performed mostly in 2D. How can immersive technology ameliorate the process of CAD design?

First, there can be various tasks and situations identified, where a better understanding of the spatial model is crucial and would lead to an increase of performance. The option of enabling a stereoscopic view of the model is in our opinion not enough: There should be established certain, more natural navigation methods to actually get a spatial impression of the virtual object. Even, if the used display is just a CRT, the left-right movement of the user's head translated into different viewing angles onto the model creates a more intuitive way of understanding the geometric properties, compared to rotation of the object through keyboard or mouse interaction.

Second, in certain situations it can be more effective to give up the constraints of 2D interaction for the sake of gaining efficiency: When exactness is not required, the simultaneous manipulation of all three spatial dimensions enables more conceptual modelling. E.g. the attachment of one object's surface to another can be performed in this way, reducing the interaction required from 2+ steps to just one step.

Third, the vast amounts of functionality of today's CAD systems require special techniques to access and understand those methods. Here immersive technologies can a) help speeding up method selection and b) show the user the concepts and the lead-through of a certain method in an understandable way.

Outlook

With the recent increase of graphic rendering power in consumer PCs VR/AR technology has become more affordable for smaller businesses and even home users. These customers will create demand for new applications. This will, in return, lead to improvements in the available hard- and software, creating a positive feedback-cycle. Thus we can expect to see more and more VR/AR technology in several areas of life. Because most of these developments are rather technology-driven than need-driven, it is not clear which ones will succeed and which ones will not. Immersive modelling is no exception here, with its usefulness and usability still to be proven.

References

Baratoff G, Regenbrecht H (2003). (submitted) Developing and Applying AR Technology in Design, Production, Service, and Training. International Journal of Advanced Manufacturing Technology.

Bimber O, Encarnação LM, Stork A (2000). A Multi-Layered Architecture for Sketch-based Interaction within Three-dimensional Virtual Environments. Computers and Graphics - The international Journal of Systems and Applications in Computer Graphics 24, 851-67.

Buck M , Schömer E (1998). Interactive Rigid Body Manipulation with Obstacle Contacts. Journal of Visualizytion and Computer Animation 9, 243-57.

Burdea G, Coiffet P (1994). Virtual Reality Technology ed. anonymous: John Wiley & Sons, Inc.

Deisinger J (2002). Entwicklung eines hybriden Modelliersystems zur immersiven konzeptionellen Formgestaltung [Development of a hybrid modelling system for immersive conceptual form-styling]. In IPA-IAO Forschung und Praxis, ed. anonymous, Stuttgart: University of Stuttgart.

Hennig A (1997). Die andere Wirklichkeit [The Other Reality] ed. anonymous: Addison-Wesley.

Hua H, Brown LD, Gao C, Ahuja N (2003). A New Collaborative Infrastructure: SCAPE. In IEEE VR 2003, ed. anonymous, pp. 171-9, Los Angeles: IEEE.

May F, Stahs T (1997). Virtual Reality für Entwicklungs- und Produktionsprozesse bei Daimler-Benz [Virtual Reality for Development- and Production Processes at Daimler-Benz]. In Fachgespräch: Virtual Reality Anwendungen in der industriellen Produktion, ed. anonymous, Paderborn: Heinz Nixdorf Institut.

Milgram P, Kishino F (1994). A Taxonomy of Mixed Reality Visual Displays. IECE Transactions on Information and Systems (Special Issue Networked Reality) E77-D, 1321-9.

Regenbrecht H, Jacobsen S (2002). Augmentation of Volumetric Data in an Airplane Cabin. In ISMAR, ed. anonymous, Darmstadt, Germany: IEEE.

Regenbrecht H, Wagner M, Baratoff G (2002). MagicMeeting: A Collaborative Tangible Augmented Reality System. Virtual Reality 6, 151-66.

Stork A (2003). SketchAR: collaborative, immersive free form modelling in virtual and mixed realities. Computer Graphik topics 15.

Knowledge Deployment: How to Use Design Knowledge

Tetsuo Tomiyama, Faculty of Mechanical Engineering and Marine Technology, Delft University of Technology, Netherlands

Introduction

The research has two backgrounds. One is the recognition that design can be regarded a largely knowledge-based activity. Historically, research in design knowledge began with knowledge representation and reasoning. Design knowledge representation looked at knowledge about design objects and design processes. While the former is obvious, design process knowledge is all about how to design and is related to design reasoning that deals with how to use design object knowledge. In contrast, design reasoning was mostly concerned about how to computationally arrive at conclusions, but not how to use which knowledge, which is the central question of knowledge deployment. As theories and technologies to deal with knowledge made progresses, research focuses gradually extended to such topics as modeling, acquisition and capturing, learning, discovery, data mining, maintenance, management, reuse and sharing. These addressed, however, little about how to make knowledge well ready for use.

The other background is a finding that innovative design results mostly from innovative combination of existing knowledge, but not solely from innovativeness of knowledge itself. In other words, a new combination of knowledge sources that have never been combined brings added-value. Often, the individual knowledge that was combined was not necessarily innovative; it can even be trivial. From this point of view, we can easily expect that design knowledge must be appropriately ready for combination, as a consequence of knowledge deployment.

This paper focuses on knowledge deployment, in particular, of design knowledge and discusses its research issues. As stated above, design can be regarded a largely knowledge-centered activity. This suggests that how to effectively and efficiently use which knowledge during design is a critical issue within design research. To be able to use effectively and efficiently knowledge, design knowledge must be appropriately deployed. Knowledge deployment needs a meta-level point of view to understand and to develop a mechanism for systematizing, structuring, and integrating design knowledge. This paper first discusses what knowledge deployment is. It then describes each of systematizing, structuring, and integrating design knowledge referring to my group's research results. In this context, the paper gives an overview of those elements of knowledge deployment and suggests future research directions.

What is knowledge deployment?

It is crucial for designers to be able to use effectively and efficiently design knowledge. The designers must be provided with in the form ready for effective and efficient applications. This is the knowledge deployment that is an activity from a meta-level point of view to understand and to develop a mechanism for systematizing, structuring, and integrating design knowledge.

A simple example of knowledge deployment can be depicted as follows. While such home appliances as refrigerators, vacuum cleaners, and washing machines were invented many years ago and need not necessarily be further developed, we can still see many "new designs" in the market every year. Perhaps, this reflects not technological development but rather new development of requirements of consumers (or the market). In other words, for these machines needs-oriented development is more active than seeds-oriented development. It is hard to believe that innovative, new knowledge had to be discovered and applied to design these "innovative" designs. This stands in a sharp contrast to cutting-edge technological areas (for example, at the beginning of the 21st century, they could be molecular scale machines) in which new design always comes from new knowledge.

In these examples, then, how these innovative or creative designs come? The most reasonable answer could be that innovate design results from innovative combination of known, existing knowledge, rather than from innovative, new knowledge. Of course, this cannot be proven and remains a postulation at this stage; it is, in fact, a good research subject.

We know that good design reflects considerations from many aspects. To do so, various knowledge theories should be integrated. Concurrent engineering and DfX (Design for X) address exactly this knowledge integration issue. The author's group has been looking at knowledge integration issues of various design object models, and developed a concept called a metamodel (Tomiyama et al. 1990; Yoshikawa et al. 1994). This is another example of knowledge integration.

Therefore, knowledge deployment should mean, selecting from a variety of knowledge, to find and to make knowledge ready for a right combination that can generate added-value in an integrated manner. Fig. 1 depicts the knowledge deployment process consisting of knowledge systematization, knowledge structuring, and knowledge integration. We will discuss these below.

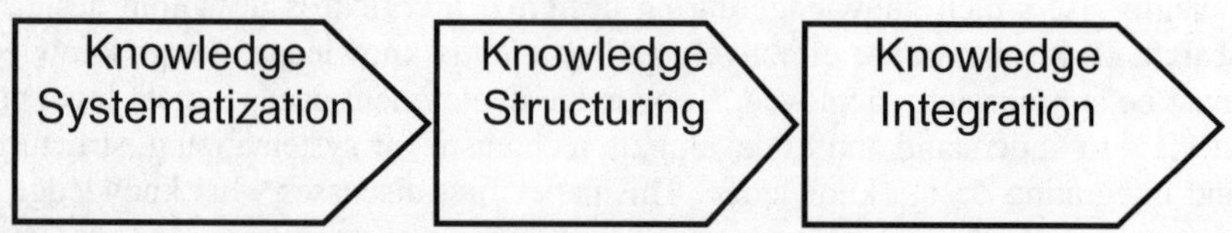

Fig. 1. Knowledge Deployment

Knowledge systematization

To correctly deploy knowledge, it is necessary first to have a good systematized source of knowledge. Knowledge systematization is nothing but a scientific process to establish a theory (Fig. 2). It begins with observation of phenomena, subsequently followed by processes to set up a view and a background theory, to articulate observations, to extract facts from observations, to codify facts, to build a hypothesis (or to choose a theory), to reason about the hypothesis, to test the hypothesis and reasoning results against the observations, and to verify the hypothesis (Tomiyama et al. 1992, 1996, 2000). After this knowledge systematization, we obtain a well-formalized theory that consists of axioms **A** (i.e., hypothesis becomes axioms after verification), facts **F**, theorems **Th** (i.e., reasoning results derived from axioms and facts), and lastly inference rules σ (usually *modus ponens* is used). A theory is, therefore, formulated as follows.

$$\mathbf{A} \cup \mathbf{F} \vdash_{\sigma} \mathbf{Th} \qquad (1)$$

The symbol |– means the right hand side can be logically derivable from the left hand side. Here note that each of **A**, **F**, **Th**, and σ is a set of logical formulae.

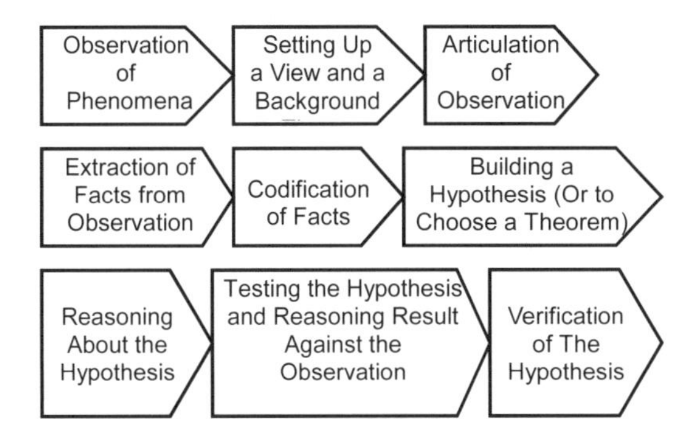

Fig. 2. Knowledge Systematization

As a result of knowledge systematization, knowledge is now modeled and represented to allow reasoning. These processes are acquisition and capturing. Sometimes, knowledge can also be learned or discovered (by learning, knowledge discovery, or data mining). Once systematized knowledge must be maintained, managed, and eventually reused and shared by many others. Since these concepts can also be found in product life cycles, we can call these processes knowledge life cycles.

Knowledge structuring

As a consequence of knowledge systematization, we now have a collection of theories. These theories are basically independent. However, it is often the case that there are some relationships among theories.

A theory consists of axioms, facts theorems, and inference rules. Because usually we follow the deductive system that employs *modus ponens* as the inference rule, theorems are secondary to axioms and facts, and they can be deduced from axioms and facts. Facts are usually given through observation (in many branches of science) and can be considered outside the theory. Axioms indeed define things that can be derived, which means that the primary structural element of a theory is axioms.

Second, there are concepts defined by the terms that appear in axioms and facts. For instance, when a theory is represented in predicate logic, predicates and terms signify fundamental concepts in this theory. However, obviously theorems include only those predicates and terms appeared in those axioms and facts. Therefore, a theory has a vocabulary that consists of those terms and that defines the theory's target domain.

These two are important structural elements of a theory; the former is structural relationships among theories, while the latter is ontological relationships among concepts. Knowledge structure of theories, therefore, boils down to structural and ontological relationships among those theories and elements (Fig. 3).

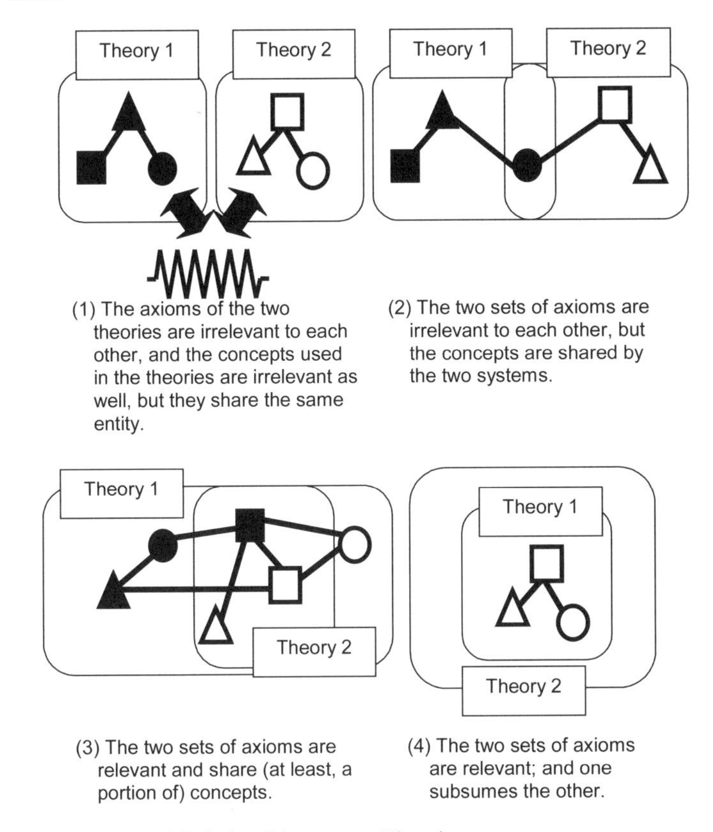

Fig. 3. Structural Relationships among Theories.

The structural relationships between these two theories can be categorized as follows.

1. The axioms of the two theories are irrelevant to each other, and the concepts used in the theories are irrelevant as well, but they share the same entity. Example: The same entity can be a spring in strength of materials as well as a coil in circuit theory.
2. The two sets of axioms are irrelevant to each other, but the concepts are shared by the two systems.
   Example: Strength of materials and vibration theory share the identical concept of spring.
3. The two sets of axioms are relevant and share (at least, a portion of) concepts. Example: Thermodynamics and statistical mechanics share a portion of concepts (such as temperature), but they simply provide two different views.
4. The two sets of axioms are relevant; and one subsumes the other. Example: The internal combustion engine is a special case of heat engines.

In addition, even if the two sets of theories are irrelevant and do not share any concepts, sometimes there can be analogical (or isomorphic) relationships among concepts. In Fig. 5, structural similarity can help analogy, for instance, because the same differential equation governs mechanical vibration and electrical vibration, thus allowing engineers to use the same mathematical solution principle.

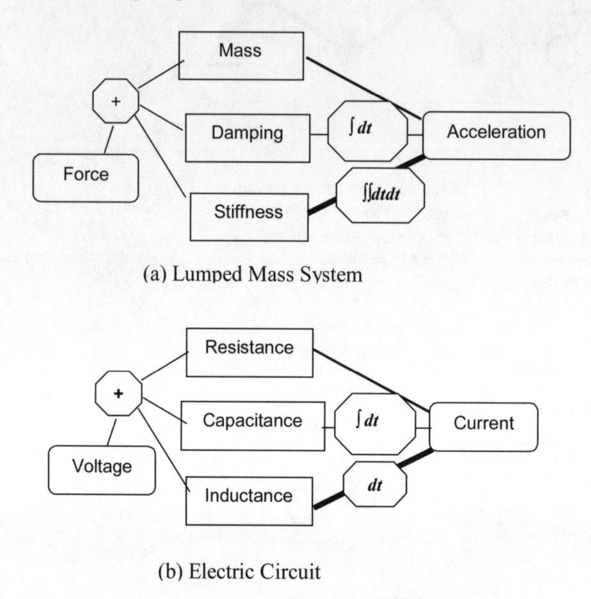

Fig. 4. Two Isomorphic Theories

The ontological relationships among concepts typically include such relationships as *part-of*, *super-sub*, and *is-an-instance-of*. Building a knowledge base about engineering design knowledge is a very demanding task, but such a knowledge base is an indispensable element of any advanced design support system (Tomiyama et al. 1992; Ishii et al. 1995; Sekiya and Tomiyama 1997; Sekiya et al. 1999).

These two type of relationships among theories, i.e., structural and ontological relationships, define knowledge structure that helps knowledge integration which is the next step in knowledge deployment.

Knowledge integration

As a consequence of knowledge structuring, knowledge will be organized in such a way that it can be easily integrated. With well-structured knowledge, now we are able to find a most relevant theory needed in one design situation. During design processes, as stated before, one of the key issues to arrive at good design is to integrate as many theories as possible. For this, abduction is considered to play a crucial role. The following discussion is a summary of our separate report (Tomiyama et al. 2003).

Abduction

Abduction is a reasoning method to arrive at "logically not deducible" conclusions (Hartshorne and Weiss 1931–1935; Burks 1958). It can be contrasted with deduction and induction (see formula (1)); abduction derives facts from axioms and theorems, deduction derives theorems from axioms and facts, and induction derives axioms from theorems and facts. Abduction plays a crucial role in synthesis (or design) to find an individual design solution that satisfies design requirements (i.e., properties that the solution should exhibit) using design knowledge. Here, we can give correspondences that design knowledge is considered axioms, design requirements are theorems, and the design solution is facts.

Schurz (2002) has compiled a (seemingly) complete model of abduction that classifies various types of abductive reasoning. Roughly speaking, he identified three major categories, factual abduction (first-order existential abduction), law abduction, and second-order existential abduction. Note that his classification does not imply clarification about all the necessary computational algorithms.

Abduction for (Factual) Creation

Within the design research community, it is often pointed out that synthesis is largely performed by abduction in the sense of factual abduction (Coyne 1988; Yoshikawa 1989). Indeed, first order existential abduction generates an entity that performs the given requirements. So, we name this abduction *abduction for (factual) cretion*. This can be explained as follows.

Given a theory in formula (1), i.e., given design knowledge (axiom) $A = \{\mathrm{O}(x) \rightarrow \mathrm{P}(x)\}$ and a design requirements (theorem) $\mathrm{P}(a)$ (an object a performs P), we obtain a design solution (fact) $F = \{\mathrm{O}(a)\}$ (a is an instance of object class O). This is simply retroduction or backward reasoning. *First order existential abduction* is a special form of factual abduction and generates a as a variable to be instantiated.

While philosophically this analogy seems valid, computationally (or from the design point of view), we can see that factual abduction does not really lead to creative and innovative design. First, it generates trivial facts from a known set of axioms and theorem (i.e., requirements) in a domain which is more or less covered by the axioms. In this sense, such a mode of abduction cannot go beyond what the axioms cover nor result in creative design.

Fig. 5 depicts this situation. First, we are given axioms as background knowledge and theorems as requirements. Factual abduction generates facts that describe a design solution. If given different requirements (i.e., a different enclosing domain), we arrive at different design solutions. However, these domains are already implicitly defined by axioms!

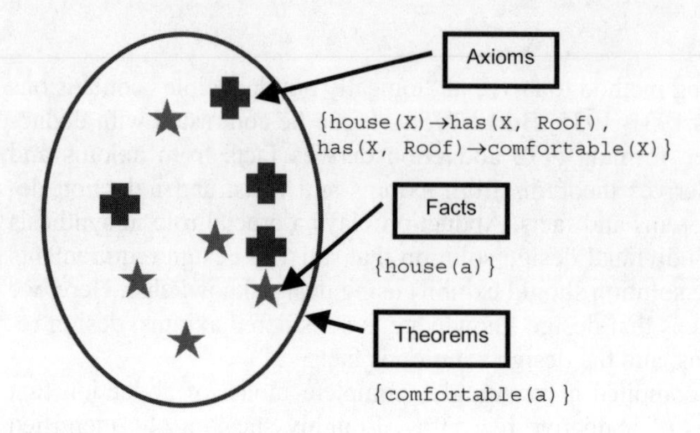

Fig. 5. Abduction for (Factual) Creation

Abduction for Integration

While abduction is a crucial concept as discussed in the previous section, abduction also plays another important role in integrating multiple theories (Takeda et al. 2001). Given a problem and a set of theories, if judged impossible to find a solution within the domain, abduction can introduce an appropriate set of relevant theories to form a new set of theories, so that solutions can be found with the new set of theories. For instance, as long as our knowledge is limited to the structural strength of materials of given shape, we will never reach such an innovative design as "drilling holes" for lighter structure while maintaining the strength. This is only possible when we have a piece of knowledge that removing material that does not contribute to strength does not make any harm but only makes the whole object lighter.

Fig. 6 depicts abduction for integrating theories. First, we are given axioms 1 as background knowledge and the combined domain of theorems 1 and 2 as requirements (Fig. 6 (a)). However, we may notice that there is no way to arrive at design solutions that can cover the domain designated by (potentially) theorems 2 with only axioms 1 (hence theorems 1). Computationally, this check can be performed by conducting all possible factual abduction to see if the results of abduction cover the entire domain of theorems 1 and 2.

Results of this check may request us to incorporate a new theory, i.e., axioms 2 that may be able to cover this domain (Fig. 6 (b)). After factual abduction using both of axioms 1 and 2, we may arrive at facts 1 and 2 that describe a design solution for these requirements (Fig. 6 (c)). Logically, this situation can be represented as follows.

$$\mathbf{A_1} \cup \mathbf{F} \not\models \sigma\ \mathbf{Th_1} \cup \mathbf{Th_2}$$

$$\mathbf{A_1} \cup \mathbf{A_2} \cup \mathrm{F} \vdash \sigma\ \mathbf{Th_1} \cup \mathbf{Th_2} \qquad (2)$$

However, notice that as a consequence of taking into consideration additional axioms 2 besides axioms 1, we effectively integrated axioms 1 and 2. This is an example of innovative design coming from innovative combination of knowledge. In Schurz's classification, this abduction for integrating theories seems to be carried out by combination of modes of second order existential abduction.

For instance, we can think about the following two-step algorithm to integrate multiple theories from different domains (that are superficially irrelevant to each other); first to identify the applicability and the domain of the theories to be introduced, and second to integrate the new set of theories. The first step identifies the structural or ontological relevance theories, and very much the same as analogical abduction. The second step actually does the integration based on, for instance, law abduction or second-order existential abduction.

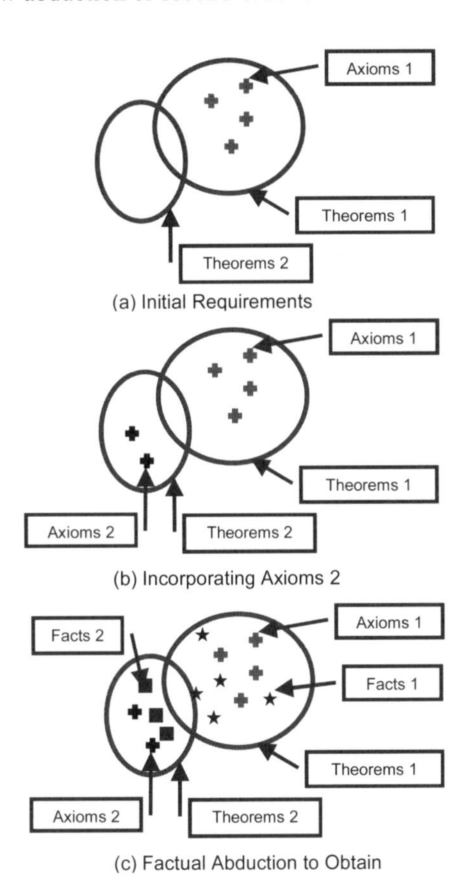

Fig. 6. Abduction for Integration

Conclusions

This paper proposed knowledge deployment that that is an activity from a meta-level point of view to understand and to develop a mechanism for systematizing, structuring, and integrating design knowledge. It summarized the research results of my group in element technologies of knowledge deployment consisting of knowledge systematization, knowledge structuring, and knowledge integration. However, there are many issues to be solved, including both theoretical and computational aspects. In the future, we then need to address more details of their mechanisms, algorithms, and tools.

As an application of this research, besides creative design, we can also address educational issues. Traditional engineering design education primarily focused on teaching design object knowledge and less so about design process knowledge and design methodology. However, it was not concerned about how to effectively deploy various pieces of knowledge in design. In this regard, research in knowledge deployment has implications what to teach in engineering design education. We should teach, for example, more about how to use various theories about engineering and its knowledge structures (i.e., relationships among different theories).

References

Coyne R (1988) Logic models of design, Pitman, London

Hartshorne C, Weiss P (eds) (1931–1935) The collected papers of Charles Sanders Peirce, vol. I–VI, Harvard University Press, Cambridge, MA

Burks A (ed) (1958) The Collected papers of Charles Sanders Peirce, vol. VII–VIII, Harvard University Press, Cambridge, MA

Ishii M, Sekiya T, Tomiyama T (1995) A very large-scale knowledge base for the knowledge intensive engineering framework. In: Mars NJI (ed) Towards very large knowledge bases, IOS Press, Amsterdam, Oxford, Tokyo, Washington, DC, pp 123–131

Schurz G (2002) Models of abductive reasoning, TPD preprints annual 2002 No. 1, Schurz G, Werning M (eds) Philosophical prepublication series of the chair of Theoretical Philosophy at the University of Düsseldorf, (http://service.phil-fak.uni-duesseldorf.de/ezpublish/index.php/article/articleview/70/1/14/). To appear in Synthese (Kluwer Academic Publishers) in 2003

Sekiya S, Tomiyama T (1997) Case studies of ontology for the knowledge intensive engineering framework. In: Mäntylä M, Finger S, Tomiyama T (eds) Knowledge intensive CAD, vol 2, Chapman & Hall, London, pp 139–156

Sekiya T, Tsumaya A, Tomiyama T (1999) Classification of knowledge for generating engineering models. In: Finger S, Tomiyama T, Mäntylä M (eds) Knowledge intensive computer aided design, Kluwer Academic Publishers, Boston, Dordrecht, London, pp 73–90

Takeda H, Yoshioka M, Tomiyama T (2001) A general framework for modelling of synthesis – Integration of theories of synthesis. In Proc ICED '01, pp 307–314

Tomiyama T, Kiriyama T, Takeda H, Xue D, Yoshikawa H (1990) Metamodel: A key to intelligent CAD systems. Res in Engineering Design, 1:19–34

Tomiyama T, Takeda H, Yoshioka M, Shimomura Y (2003) Abduction for creative design. To appear in Proc DETC 2003 Design Theory and Methodology, Paper No DETC2003/DTM-48650, ASME, New York

Tomiyama T, Tsumaya A, Hew KP, Kiriyama T, Murakami T, Washio T, Takeda H, Umeda Y, Yoshioka M (2000) A model of synthesis from the viewpoint of knowledge operations. In: Horváth I, Medland AJ, Vergeest JSM (eds) TMCE 2000, Delft University Press, Delft, pp 249–262

Tomiyama T, Umeda Y, Ishii M, Yoshioka M, Kiriyama T (1996) Knowledge systematization for a knowledge intensive engineering framework. In: Tomiyama T, Mäntylä M, Finger S (eds) Knowledge intensive CAD, vol 1, Chapman & Hall, London, pp 33–52

Tomiyama T, Xue D, Umeda Y, Takeda H, Kiriyama T, Yoshikawa H. (1992) Systematizing design knowledge for intelligent CAD systems. In: Olling GJ, Kimura F (eds) Human aspects in computer integrated manufacturing, IFIP Transactions B-3, North-Holland, Amsterdam, pp 237–248

Yoshikawa H (1989) Design philosophy: The state of the art. Annals of the CIRP, 38/2:579–586

Yoshikawa H, Tomiyama T, Kiriyama T, Umeda Y (1994) An integrated modeling environment using the metamodel, Annals of the CIRP, 43/1:121–124

Reconsidering the divergent thinking guidelines for design idea generation activity

Remko van der Lugt, Delft University of Technology
Faculty of Industrial Design Engineering, Netherlands

Introduction

When designers need to generate ideas, they tend to take paper and pencil, and start to produce idea sketches. Or, they may call an idea generation meeting. The available techniques for generating ideas in such idea generation meetings are mainly based on writing as a working medium. This research project started with the notion that idea generation techniques for design could be enhanced if sketching could be included in such idea generation meetings. For, many researchers regard sketching to be instrumental not only to the designer's creative process (e.g. Fish and Scrivener 1990; Goldschmidt 1991; Schon and Wiggins 1992; Goel 1995) but also to the design team's creative process (Bly 1988; Tang and Leifer 1988; Scrivener and Clark 1994).

In our empirical studies (Van der Lugt 2000, 2001), however, we found that these functions of sketching in design cannot simply be added as attributes to brainstorming meetings; including sketching appears to substantially change the idea generation process. In this paper we will briefly describe the findings from the empirical studies, followed by a theoretical reflection on the differences in idea generation processes, when sketching or writing is used as a primary working medium. We will then re-assess each of Osborn's (1963) four well-known guidelines for divergent thinking (defer judgment, strive for quantity, freewheel, build on each other's ideas) and provide a new set of guidelines that is specifically aimed at idea generation activity that supports the creative re-interpretation found in idea sketching.

Two empirical studies that explore the functioning of sketching in idea generation meetings

In our empirical work, we took a process perspective, focusing on the linking structure of the idea generation process as a means to assess the quality of the idea generation process (rather than focusing on the resulting ideas as dependent variable, which has been the norm in over fifty years of research on idea generation meetings). In our view, well-integratedness (Dorst 1997) is the primary process quality measure. A process is well-integrated if it shows a strong network of links between ideas, without early crystallization (Goel 1995) taking place. We adapted linkography (Goldschmidt 1996) as a research approach to analyze the structure of the idea generation processes. In linkography for each idea direct connections or 'links' with all earlier ideas are determined by gathering and evaluating evidence of such connections. The resulting system of links is then used as material for further analysis.

In the first study (Van der Lugt 2000) we explored different ways in which sketching could be included in brainstorming meetings. In total four meetings were held. In meeting A, the facilitator recorded ideas by sketching them on the flipchart. In meeting B the participants sketched ideas on large post-its notes, which were then handed to the facilitator to post on a flipchart. In meeting C, the facilitator recorded ideas by writing, but the participants were encouraged to add sketches on post-it notes to enrich the representations of the ideas. Meeting D functioned as a control condition. In this meeting, the facilitator recorded ideas by writing them down on a flipchart. In contrast to our initial expectations, we found that using sketching did not necessarily enhance the process. Instead, we encountered a breakdown of the process, especially with the fully graphic techniques. Compared to regular brainstorming, in the fully graphic meetings substantially less ideas were produced, while there were also considerably less links between ideas. Especially in meeting B, where the designers sketched their own ideas, there was also a very low amount of novel connections made, which we considered to be a sign of premature crystallization. Our interpretation of these results was that apparently, the process of idea sketching is not compatible with the guidelines of divergent thinking that are elementary to the brainstorming process (Osborn 1953): Defer judgment, strive for quantity, freewheel, and build on each other's ideas.

In the second empirical study (Van der Lugt 2001) we moved the search to gaining a further understanding of the structure of the differences in process between sketching and writing as a working medium. We applied an adaptation to brainsketching (Van der Lugt 2002) as a representative of idea generation techniques that use sketching. Brainsketching is a graphic variation of the more widely known brainwriting technique (Geschka, Schaude and Schlicksupp 1973). During brainsketching, participants sketch ideas individually in short rounds. After each round they briefly share their ideas and switch papers. In the next round they use the ideas already present on the worksheet as a source of inspiration. Usually this procedure is repeated about five times.

In contrast to the techniques used in the first study, brainsketching does not appear to deteriorate the process, because it allows for an individual cycle of re-interpretation within the group process (see Van der Lugt 2002). We compared brainsketching to brainstorming with post-its (Isaksen et al. 1994) as a representative of idea generation techniques that use writing as the primary working medium. See Fig. 1 for examples of the resulting flipcharts with ideas for the two techniques.

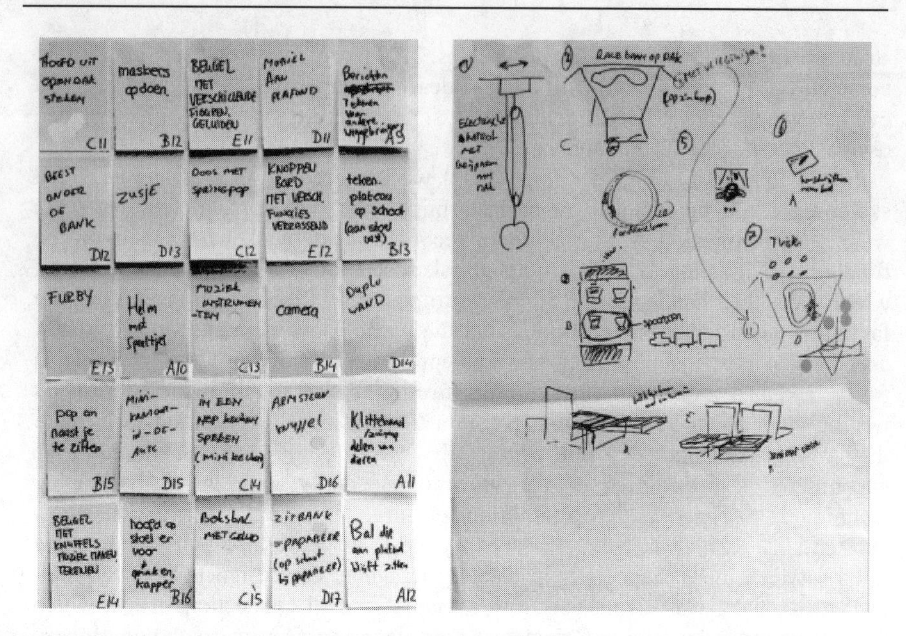

Fig. 1. Data from the second empirical study: A brainsketching flipchart and a brainstorming with post-its flipchart

In each of four experimental meetings both the brainsketching and the brainstorming were applied which allowed us to perform a paired comparison analysis of the results for the individual participants (n=20). Each meeting consisted of five advanced product design students who were involved in a course in facilitating creative problem solving meetings. The meetings were moderated by experienced professional creative problem solving facilitators.

When using brainsketching, we found that significantly (paired sample t-test, $p<0.005$) more connections with the earlier ideas were made, while the designers produced significantly ($p<0.005$) lower number of ideas when brainsketching, in comparison to brainstorming. This means that brainstorming with post-its complied better with the 'strive for quantity' guideline for generating ideas, and brainsketching complied better with the 'build on each other's ideas' guideline.

During brainsketching designers made significantly ($p<0.01$) more small alteration-type links, which were hardly present in the brainstorming segments. These incremental steps did not refer to early crystallization taking place, as we found no notable difference in novel connections made between the techniques. In that sense, brainsketching appears to entail a more balanced problem solving process.

The different characteristics of the brainsketching and brainstorming processes suggest that they may serve different purposes in design. Brainstorming-like techniques may better serve the traditional role of idea generation techniques, which is to generate a large number and variety of design ideas, of which some can be selected to further develop into design solutions. Brainsketching may be more suitable when, instead of a large number of ideas, a smaller but more refined collection of novel ideas are desired. For instance, brainsketching may be applied in a design project start-up meeting to provide a quick simulation of the design process to come. Such a simulation allows the members of a team to gain a shared understanding (Valkenburg 1998) of the design task by discussing possible pathways towards solutions that came up when generating ideas.

So, in the empirical work we found structural differences between the brainsketching and the brainstorming process that suggest that they may serve different purposes in design. How, then, can these differences be explained on a theoretical level? And, as the existing guidelines for divergent thinking are instrumental to sentential idea generation, can we develop an alternative set of guidelines to support the different structure and applicability of the graphic idea generation process? These questions will be addressed in the remainder of paper.

Differences in activities in graphic and sentential idea generation

Here we will examine the theoretical differences in the basic activities that take place in idea generation meetings that use sketching or writing as a primary working medium. To ease the discussion, we will refer to idea generation activity that use sketching as the primary working medium as *graphic* idea generation, and idea generation activity that uses written language as the primary working medium as *sentential* idea generation.

A representation of the current view on the activities in the creative problem solving stage of idea-finding is provided in the figure below. The stage consists of a divergent phase that focuses on idea production and a convergent phase of idea evaluation. According to Osborn (1963), the divergent phase is intended to to *"create a checklist of possible leads to solution" (p.124)*, from which one can select ideas in the adjacent convergent phase.

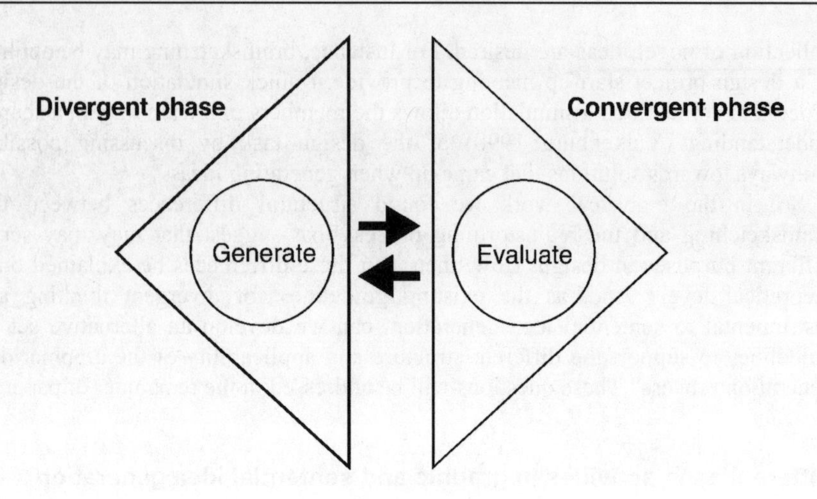

Fig. 2. The creative problem solving stage of idea finding

An alternative view is provided by Finke, Ward and Smith (1992), who propose, what they call, a 'heuristic model of creativity' called 'Geneplore' (see figure below) in their efforts to develop a creative cognition approach to understanding creativity. 'Geneplore' is a combination of the verbs 'generate' and 'explore'. Finke et al argue that creative cognition involves a repeating cycle that contains a generative phase, in which so-called pre-inventive structures are constructed, and an exploratory phase, in which the generated pre-inventive structures are interpreted. The results of these interpretations lead to insights that can be used to either focus on specific issues, or to expand conceptually, by modifying the pre-inventive structures. In the model, product constraints can be imposed at any time during the generative or exploratory phase.

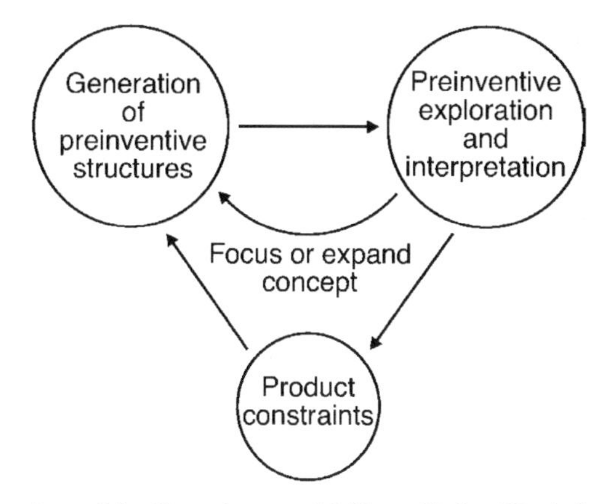

Fig. 3. The basic structure of the Geneplore model (From Finke, Ward, Smith 1992)

The pre-inventive structures are central in Finke et al's theory. They observe: "These (pre-inventive) structures can be thought of as internal precursors to the final, externalized products of a creative act. They can be generated with a particular goal in mind or simply as a vehicle of open-ended discovery. They can be complex and conceptually focused or simple and relatively ambiguous, depending on the situation and the requirement of the task." (Ward, Smith, Finke 1999)

According to Ward et al, both idea sketches and words can represent such pre-inventive structures. With the Geneplore model in mind, we can now start to infer differences in activities for the graphic and the sentential conditions. In the figure below the two phases of the Geneplore model are included in the divergent phase of the idea-finding stage.

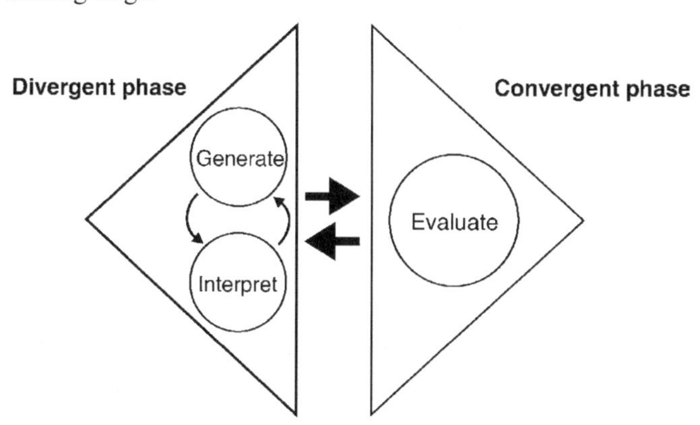

Fig. 4. A cyclical model of the creative problem solving stage of idea-finding

The depictive characteristics of sketches provide a rich basis for analysis by allowing the designers to envision the consequences of the ideas. Fish & Scrivener (1990) consider that: "The necessity to sketch arises from the need to foresee the results of the synthesis or manipulations of objects without actually executing such operations" (p. 117). In contrast, written language is descriptive in nature, and refers to classes of objects rather than specific objects. This does not allow the designers to envision the consequences of the specific ideas. So, written language stimulates the generation of ideas by allowing the designers to make free associations, but the inability to envision the consequences of the generated ideas, makes writing as a working medium less conducive to the interpret phase in the Geneplore cycle.

This corresponds with the findings in the second empirical study (Van der Lugt 2001). Where in the sentential condition (brainstorming with post-its) designers produced significantly more ideas, while making significantly fewer connections with earlier ideas compared to the graphic condition (brainsketching). Conceivably, there is little place for interpreting ideas during brainstorming. Participants do build on each other's ideas – this is supported by one of the guidelines – but the nature of such connections made can be regarded to be mostly concerned with brief inspections that spur further associations, rather than more detailed interpretation of the ideas. The guidelines for divergent thinking stimulate this, by dismissing any type of analytical thinking from the divergent phase.

The graphic idea generation process appears more closely connected to the generate-interpret cycle. In design research, various idea-sketching cycles have been proposed, which suggest that designers reflect on what they are sketching, while they are sketching it. For instance, Goldschmidt (1990) refers to design sketching activity as a 'dialectic of sketching', and in her research she showed how designers switch between making graphic propositions, followed by language-like interpretations of these graphic propositions. Schön and Wiggins (1992) considered the designer to be involved in a 'reflective conversation' with the drawing surface, meaning that the designer interprets the consequences of a drawing act while he or she is making the mark. idea generation through sketching can be considered to consist of two basic steps between which the designer shifts continuously. One step is related to producing (part of) a sketch of an idea on the paper and the other step consists of exploring and interpreting the sketch made in order to find directions for further idea sketching.

Reconsidering the divergent thinking guidelines when sketching is used

With the notion of the different area of application, and with the different activities in mind, we can now reconsider the appropriateness of Osborn's (1963) guidelines for divergent thinking when sketching is concerned. And, we can develop a new set of guidelines for divergent thinking that emphasizes the qualities of the graphic idea generation process.

Deferring judgment

Osborn (1963) and other Creative Problem Solving authors (e.g. Isaksen et al. 1994) use the words judgment, criticism, analysis, reflection, and evaluation interchangeably. The guideline suggests postponing the use of these kinds of analytical thinking towards the convergent phase. As mentioned before, the main purpose of divergent thinking is to create a catalogue or checklist of ideas, which can be inspected upon completion (Osborn 1963). There is no interpretation of earlier ideas when using brainstorming, beyond using earlier ideas as a springboard for further association.

We have noticed that total deferment of judgment does not appear possible, and is not desired, when engaged in idea sketching. The creative potential of sketching appears to be located in the interpretation phase of the Geneplore model (Finke et al. 1992). Dismissing this phase through enforcing the deferred judgment guideline is bound to deteriorate the idea sketching process. Besides that, it does not appear possible for the designers to avoid interpreting their sketches while producing them, which means that the designers cannot conform to the deferment of judgment rule, even if they want to.

Rather than dismissing any judgment in the divergent phase, for graphic idea generation techniques, efforts should be directed towards the participants engaging in an inquisitive type of reflection or interpretation of idea sketches. Analytical thinking within the divergent phase should be directed at finding new clues or directions for further idea generation, rather than being directed towards judging the quality of the idea at hand. The evaluation of ideas should then still be located in separate convergent phases within the creative problem solving model.

Striving for quantity

The 'striving for quantity' guideline also needs to be reconsidered. Fluency of idea generation is a relevant process characteristic for both sentential and graphic techniques. However, the underlying objective is different. In brainstorming, the main motive for asking for a large quantity of ideas is based on an almost statistical type of argument that the large quantity of ideas is bound to include more high quality ideas. Indeed, for the brainstorming technique a high correlation between quantity and quality of ideas was found by various researchers (e.g. Stroebe and Diehl 1994). Some researchers even state that, because of the high correlation between idea quality and the quantity of ideas generated, the quantity of ideas generated is the only relevant dependant variable to be used when investigating the effectiveness of brainstorming groups (e.g., Hocevar 1979).

For brainsketching, however, the constructive reflection within the idea sketching cycle makes for a more directed approach to generating ideas, which means that a high quantity of ideas does not necessarily imply a higher quality of ideas. It is more likely that the effectiveness of the reflective cycle strongly influences the quality of the generated ideas. This effectiveness is -in addition to the quality of the generative and the reflective phases- related to the swiftness in which cycles take place. A high number of cycles contribute to the idea sketching process by allowing for many opportunities for redirecting the process. Another reason for promoting fluency in the idea sketching process is to avoid over-development of specific ideas. When idea sketching, some participants may be inclined to produce advanced drawings of ideas, rather than quick and messy sketches. Such advanced drawings are likely to induce premature crystallization (Goel 1995) and are therefore undesirable in idea generation through sketching. Furthermore, a high fluency of generating ideas prevents participants from lingering in the interpretive phase, thereby encouraging constructive interpretation, rather than critical evaluation. Instead of the quantity of ideas generated, the swiftness of the idea generation process is important when idea sketching. This suggests a more process-oriented guideline of 'ideational fluency'.

Building on each other's ideas

The results from the experimental studies in this project suggest that 'building on each other's ideas' is an important guiding principle when using sketching in idea generation. For brainstorming, this guideline relates to using earlier ideas as a mere springboard for spurring the generation of more ideas. In the graphic idea generation process, the building on each other's ideas guideline is especially relevant. The primary objective for inviting designers to make connections is to incorporate information from ideas that were generated earlier on, into the current idea generation process. Building on other ideas stimulates a well-integrated idea generation process.

Freewheeling

Finally, the freewheeling guideline, which stimulates the participants to express their wilder ideas, is especially relevant for the graphic situation. It is an additional means of avoiding premature crystallization. As mentioned before, sketching tends to direct the idea generation process towards idea development, even when this is not yet desired. The freewheeling guideline assists the participants in opening up new directions for generating ideas, by stimulating the participants to generate wild or seemingly irrelevant ideas, which can then be interpreted towards more feasible ideas.

The analysis of the data from the second study indicates that, while brainsketching, especially connections made with one's own ideas open up novel directions through wild leap-type connections. Freewheeling appears most relevant for the individual interpretive idea generation cycle.

Four alternative guidelines for graphic idea generation techniques

The appraisal of the appropriateness of Osborn's four divergent thinking guidelines leads to suggesting the following four alternative divergent thinking guidelines for stimulating the idea generation process when sketching is involved:

- *Build on each other's ideas.* Inspect each other's ideas for information that can be useful for integrating into new ideas.
- *Interpret ideas constructively.* Interpret ideas by seeking suggestions or directions for further idea generation, rather than assessing and estimating the value of the idea as an artifact. This guideline stimulates a directed search into novel directions.
- *Strive for ideational fluency.* Stay away from over-developing or keeping interpreting sketches of single ideas. Try to accomplish a flow of idea generation, in which you swiftly alternate producing ideas and interpreting them to gather clues for further idea generation.
- *Look for wild connections, especially when interpreting your own ideas.* Make sure that you share and interpret your wildest ideas, because they can lead towards interesting novel directions.

Final remarks

The alternative set of guidelines is based on the assumption made that being well-integrated corresponds with quality of the idea generation process. Results of a small-scale additional study provide some support for this notion (Van der Lugt and Hemberg 2002), but a larger-scale study is needed to provide more conclusive results.

The alternative guidelines may open up new directions for idea generation techniques to make optimal use of the cycle of creative interpretation in idea sketching. And, as the underlying theoretical Geneplore model is not specifically addressed towards sketching, but to creative cognition as a whole, these alternative guidelines may even provide a starting point for furthering creative problem solving methodology in general. However, this is a first attempt at providing alternative guidelines for divergent thinking when sketching is involved, and they need to be further developed and evaluated by means of further research.

References

Bly, S. A. (1988). A use of drawing surfaces in different collaborative settings. In. Greif, I. (Ed.), *Second Conference on Computer-Supported Cooperative Work* (pp. 250-256). New York: association for Computing Machinery.

Dorst, K. (1997). *Describing design: A comparison of paradigms.* Doctoral dissertation, Delft University of Technology.

Finke, R. A., Ward, T. B., & Smith, S. M. (1992). *Creative cognition: Theory, research, and applications.* Cambridge, MA: MIT Press.

Fish, J. & Scrivener, S. (1990). Amplifying the mind's eye: Sketching and visual cognition. *Leonardo, 23* (1), 117-126

Geschka, H., Schaude, G.R, & Schlicksupp, H. (1973, August). Modern techniques for solving problems. *Chemical Engineering*, 91-97.

Goel, V. (1995). *Sketches of thought*. Cambridge, MA: MIT Press.

Goldschmidt, G. (1991) The dialectics of sketching. *Creativity Research Journal, 4* (2), 123-143.

Goldschmidt, G. (1996). The designer as a team of one. In. Cross, N., Christiaans, H. & Dorst, C. (Eds.), *Analysing design activity*. Chichester, UK: Wiley.

Hocevar, D. (1979). Ideational fluency as a confounding factor in the measurement of originality. *Journal of Educational Psychology, 71*, 2, 191-196.

Isaksen, S. G., Dorval, K. B., & Treffinger, D. J. (1994). *Creative approaches to problem solving*. Dubuque, IA: Kendall & Hunt.

Osborn, A. F. (1963). *Applied imagination (third edition)*. New York: Scribner's.

Schön, D. A. & Wiggins, G. (1992). Kinds of seeing and their functions in designing. *Design Studies, 13* (2), 135-156.

Scrivener, S. A. R. & Clark, S. M. (1994). Sketching in collaborative design. In. MacDonald, L. & Vince, J. (Eds.), *Interacting with Virtual Environments*. Chichester, U.K: Wiley.

Stroebe, W. & Diehl, M. (1994). Why groups are less effective than their members: On productivity losses in idea-generating groups. In. Stroebe, W. & Hewstone, M. (Eds.), *European Review of Social Psychology, volume 5* (pp. 271-303). London: Wiley.

Tang, J. C. & Leifer, L. J. (1988). A framework for understanding the workspace activity of design teams. In. Greif, I. (Ed.) *Second Conference on Computer-Supported Cooperative Work* (pp. 244-249). New York: association for Computing Machinery.

Valkenburg R. C. (1998) Shared understanding as a condition for team design. *The journal of automation in construction, 7* (2-3), 111-123.

Van der Lugt, R. (2000). Developing a graphic tool for creative problem solving in design groups. *Design Studies, 21* (5), 505-522.

Van der Lugt, R. (2001). *Sketching in design idea generation meetings*. Doctoral dissertation, Delft University of Technology.

Van der Lugt, R. (2002). Brainsketching and how it differs from brainstorming. *Creativity and innovation management, 11* (1), 43-54.

Van der Lugt, R. & Hemberg, E. (2002). *The missing link: connecting the quality of the idea generation process to the quality of the resulting ideas*. Paper presented at the International Conference on Creativity and Leadership in Entrepreneurship. Greenwich, July 8-10.

Ward, T. B., Smith, S. M., & Finke, R. A. (1999). Creative cognition. In. Sternberg, R. J. (Ed.), *Handbook of creativity* (pp.189-212). Cambridge, UK: Cambridge University Press

Designers and Users – an Unhappy Love Affair?

Rüdiger von der Weth, Stuttgart

Introduction

It is a common idea that designers do not take into account the people who are confronted with the their ideas and products. Many studies about designers' activities (e.g. in engineering and software design) show that especially in the early stages of the design process (clarification of the task, conceptual design) usability aspects are neglected (e.g. Dylla 1991). Especially novices concentrate mainly on technical aspects (Ahmed 2001). Even if designers work for other designers, this is a major problem. This is shown by the research of Andreasen and others about integrating the perspective of other designers while developing design tools (Araujo et al. 1996; Andreasen and Mc Alonee 2001; Araujo 2001). Often instructions for customers are based on a deficiant model of the users knowledge, motives and habits (Hacker 1991). This situation is criticized by many engineers. Concepts like „usability engineering" (e.g. Nielsen 1993) show the importance and the necessity for improvements in this field. This can happen in several ways. (a) Methods for designers should be developed and evaluated which integrate the user's perspective in designers thinking and work process. (b) Users should participate in the design process not only by the formulation of requirements (if customer and user are identical) and by the testing prototypes and products. He should also be involved in other stages of the design processes e.g. conceptualisation of first ideas. By that way designers should get solutions which fit in a better way to users needs.

There exist economical, organisational and technical constraints, which hinder the integration of the user perspective, but also several psychological problems. Many design tasks are complex problems in sense of psychology (Dylla 1991; von der Weth and Frankenberger 1995; Günther and Ehrlenspiel 1998; for psychological definitions of complexity: Dörner 1996; Frensch and Funke 1995): Designers have to develop a cognitive model, not only about form and function of their product („characteristics"), but also about economical, ecological and production-related aspects („properties"). They have to integrate all these aspects in an interrelated cognitive model of the products, their users and their surrounding. In this context usability is difficult to handle, because the designer has to oversee the possible changes in work or everyday life, which are produced by his technical innovation. But only in regard to these possible changes the designer can decide, which technical concepts and solutions will be accepted and estimated by the users. According to the model of the design process designers have also take into account properties, which cannot be influenced directly by the designer (Pahl and Beitz 1995). Empirical research on design processes provides evidence, that there are great problems to organise the individual design process (Fricke 1993; von der Weth and Frankenberger 1995) and the work of design teams (Badke-Schaub and Frankenberger 1999) on an appropriate and complex model of the user and the design process itself. Most difficulties exist for novices in design work (Ahmed 2001; Günther and Ehrlenspiel 1998). Therefore the usage of such design methods should strengthen usability aspects, which have the aim to take into account characteristics as well as properties.

Another way to get information about usability aspects is a close contact with users in all phases of the design process. Successful technical innovation depends on a careful analysis of the existing knowledge of the prospective users. The technical innovation will be accepted if it allows to use and improve existing action patterns (Luhn 1999). In manufacturing technology and business administration models exist, in which way technical innovation influences the resulting activites of the user. Many authors take emphasis on the relevance of implicit knowledge in organisations and propose special innovation strategies (Nonaka and Takeuchi 1997). In most of these publications implicit knowledge is seen as knowledge which has no „official" character in the organisation. In psychology the concepts of implicit and tacit knowledge are used for the level of individuals (e.g. Broadbent 1986). But in any case it is necessary to use special strategies to know the user habits: Experts′ knowledge has for the most part implicit character and can be made explicit partially and only with special methods (Neubert and Tomczyk 1986; Hacker 1991; Luhn and von der Weth 1999). Basing on such investigations it should be possible to link new technical concepts with existing problem solving strategies and routines instead of eliminating them.

Two studies shall be introduced, following the two strategies described above. (a) The first study decribes the evaluation of a method which shall broaden the designers focus on more aspects then form and function (b) the second study describes a method of „participative" software design which involves users in early stages of the design process and should enable them to make knowledge explicit which is useful for designers to improve usability of their products.

Study 1: von der Weth and Weinert, 2001*

Ahmed (2001) found out, that expert designers have a different representation of design problems than novices. Their problem space is more differentiated and complex. Expert designers take into account more aspects such as customers, users, money and also the own work process. She found typical questions of experts, which help them to organise the design process. She developed a training to improve novices′ attention for these themes. Novices learn to improve their design process by the usage of questions concerning „experts′ aspects" (Ahmed, Wallace and Blessing 2000). The method was called C-QuARK-method, which was deviated from the first letters of the questions. In our study this method was used by 19 trainees in aviation industry (training group). 18 designers worked without using the method (control group). In both groups these designers worked together, four design teams in each group. All these teams had to cope with real work problems. The result of their work had an influence on their further carreer in the company. Several times during the design process both groups were observed and had to fill out questionnaires. The study lasted more than ten weeks. The research plan one can find in *table 1.*

The questionnaire data show, in which way the attitude to the C-QuARK-method changed during the design process. The experimental group should estimate the relevance and whether the method was used more or less intensively. The *Figs. 1 and 2* show the results comparing the questionnaires from the beginning and the end of the study.

Table 1. Evaluation plan, showing the process and used data collection methods

	Experimental group	Control group
Week 1:	• Observation of 2 teams during the customer meeting for approximately 1 hour • Introduction to the "C-QuARK-method" • Filling out of the Feedback form	• Observation of 2 teams during the customer meeting for approximately 1 hour • Filling out part I of the Feedback form
Week 2:	• Observation of the teams during group discussions, customer/ supplier meetings for ca. 1 hour • Filling out of the Progress form	• Observation of the teams during group discussion or customer meetings for ca. 1 hour • Filling out of the Progress form
Week 3:		
Week 4:	• Observation of each participant while working on the design project for half an hour • Filling out of the Progress form	• Observation of each participant while working on the design project for half an hour • Filling out of the Progress form
Week 5:		
Week 6:	• Observation of each participant while working on the design project for half an hour • Filling out of the Progress form	• Observation of each participant while working on the design project for half an hour • Filling out of the Progress form
Week 7:		
Week 8:		

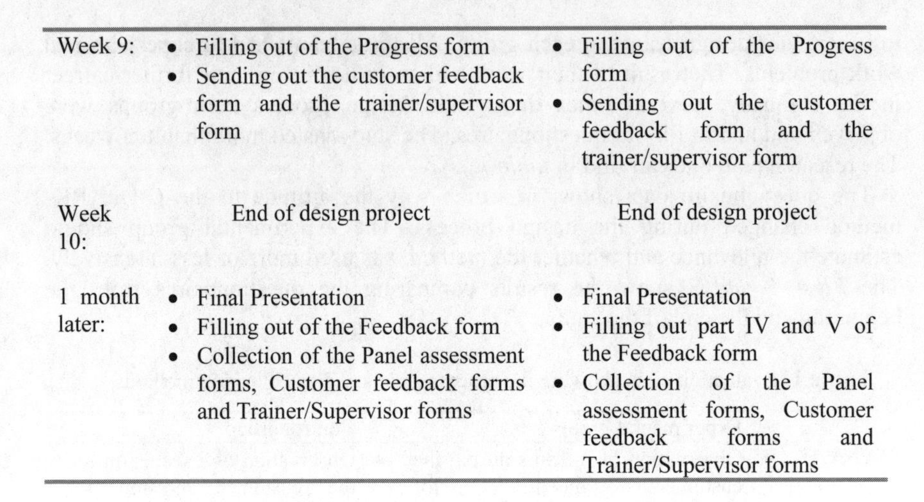

Week 9:	• Filling out of the Progress form • Sending out the customer feedback form and the trainer/supervisor form	• Filling out of the Progress form • Sending out the customer feedback form and the trainer/supervisor form
Week 10:	End of design project	End of design project
1 month later:	• Final Presentation • Filling out of the Feedback form • Collection of the Panel assessment forms, Customer feedback forms and Trainer/Supervisor forms	• Final Presentation • Filling out part IV and V of the Feedback form • Collection of the Panel assessment forms, Customer feedback forms and Trainer/Supervisor forms

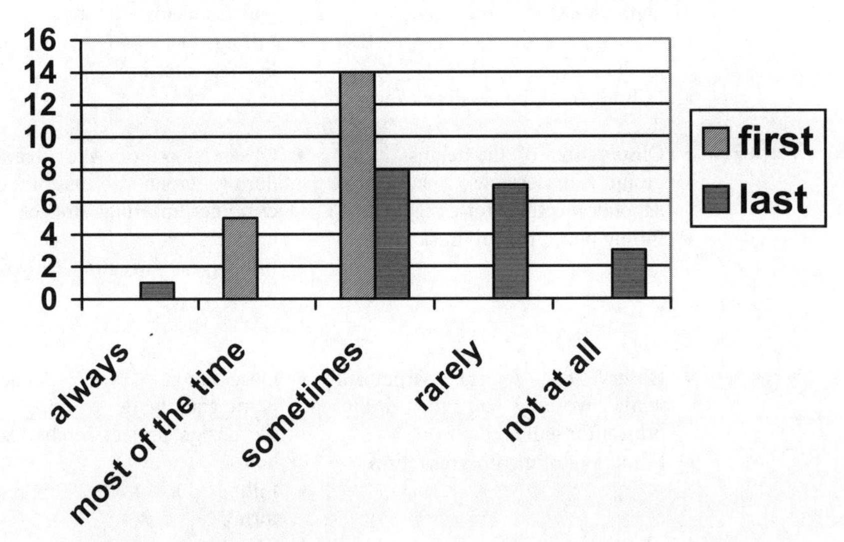

Fig. 1. Expected frequency of usage before the design processes and estimated frequency after the process (U-test: Z=-2.244*)

These results show a more negative attitude to the method and the opinion, that the method is used very seldom. According to the opinion of the memebrs of the experimental group expert questions did not play a major role for the design process.

Comparing these questionnaire data with the result of the obervations one can see a completely different result. Here we see a significant effect of the C-QuARK method on the experimental groups behaviour. They used more experts´questions than the control group and less typical novice questions. The statistical data are to be found on *table 2*.

Table 2. Observed usage of expert and novice questions. E: Mean of the experimental group C: Mean of the control group

	1st Observation	2nd Observation	3rd Observation
Expert questions	E: 10.05	E: 16.22	E: 13.11
	C: 4.83	C: 9.5	C: 9.62
	t= -2.834	t=-2.675	t=-1.221
	p<.01**	p<.05*	p=.23
Novice questions	E: 2.89	E: 2.50	E: 2.62
	C: 5.36	C: 4.00	C: 4.27
	t=2.875	t=1.115	t=1.824
	p<.01**	p=.28	p=.09

The expert questions were used and so the focus of the novice designers work apparantely has changed. But the users of this method didn´t notice that. One possible explaination might be, that novice designers are not able to supervise their own design process, because they have to concentrate on the technical aspects of their work. This corresponds with the results of Ahmed (2000). In that case apparantely many of them are not yet able to assign their actual own activities to methodical steps of the design process or to assess their own usage of a certain method. Further studies should examine the question, whether the effect of methods to improve usability engineering would be greater, if the „methodological self awareness“ could be improved.

Study 2: Hacker, von der Weth, Ishig and Luhn (in press)

Our second study deals with the problem to improve communication between designers and users. Work experience should be made explicit and made fruitful for the process of software design. Workers in microprocessor industry participated in the design of an information system for their work places. This system should support them in finding information about the work place quicker than before ...

- in the case of open questions and problems
- during the training of new processes
- giving instructions to new colleagues.

Therefore the design of the user interface should follow strictly the users´ work process. Because work procedures are often difficult to describe (e.g. try to describe how to make a paper for a scientific workshop), the cooperative software design process had several steps *(Fig. 3)*. (1) The cooperation process started with a mixed form of observation and interview on one work place with two different operators. Starting from this a first description of the work process and usage of information was made, which was controlled by the observed workers. (2) The result was used by the designer of the information system who developed the information system together with a mixed group of users and engineers working continiously in a TOID-team (Task oriented information exchange, Neubert and Tomzcyk 1986; Jahn et al. 2002), a special form of quality management groups. In this group also the goals of the design process were developed. They are described in table 3, as N° 1 to 11. (3) The results of this group were also used for the continuous optimisation of the automation process on this work place. So operators participated as well in the design of work information as in the design of the work process.

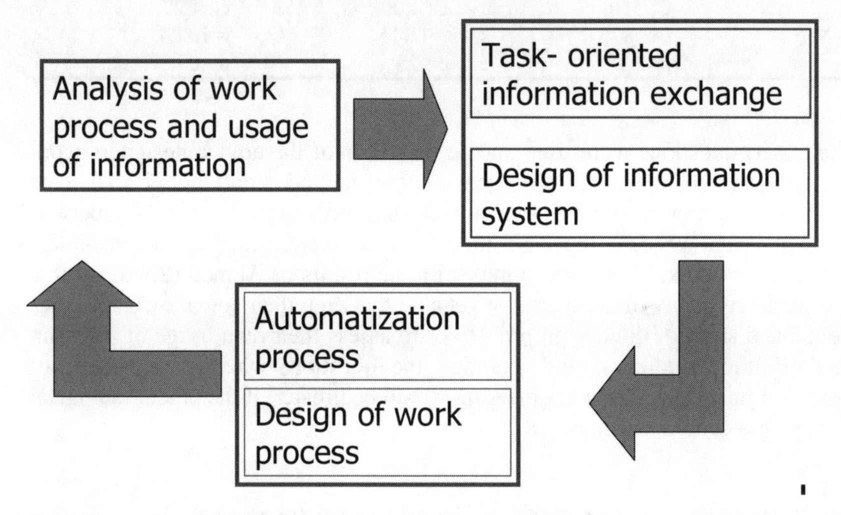

Fig. 2. Steps of user integration

Three months after the installation the users assessed the improvements by the information system. The results in many cases were significantly positive *(table 3)*. Additionally an analysis of the design process was carried out by two expert engineers working in that microprocessor plant. These expert ratings were reliable (κ=.85). The quality of the design process described above was compared with the common way to generate work information in that plant. The comparison was made with a checklist basing on ISO 13407. The cooperative design process was found more participative than the way, work information was generated before. Detailed results are described in Hacker et al. (submitted).

Table 3. Questionnaire results; Assessment of the workplace information system. Statistical analysis was carried out by comparing positive and negative results with the sign test (** $p < .01$; * $p<.05$).

	Information System Design and Actuality	++	+	0	-	--	p
1	Quicker access to information	2	6	1	0	0	**
2	Clearer information	2	5	2	0	0	**

3	Better structured information	3	2	4	0	0	*
4	More actual information	1	4	3	1	0	n.s.
5	Information is easier to find	2	6	1	0	0	**
6	Better form of representation	0	6	2	0	0	*
	Improvement of work process						
7	Quicker solutions for problems	2	1	4	1	0	n.s.
8	Quicker training of new colleagues	1	4	4	0	0	*
9	Quicker own training for new work processes	1	7	1	0	0	**
10	Less mistakes of new colleagues	0	4	3	2	0	n.s.
11	Less own mistakes	0	1	5	3	0	n.s.
	Development of Knowledge						
12	Laerning useful things from colleagues	1	1	5	2	0	n.s.
13	Learning from other´s mistakes	1	4	2	2	0	n.s.
14	Help for colleagues possible	0	7	1	0	0	**
15	Learning to explain own work in a better way	1	4	4	0	0	*
16	Stimulation new and useful ideas	0	4	4	1	0	n.s.

Conclusions from both studies

The C-QuARK-Method is a very simple tool to improve the designers attention for properties and the design process itself. In that direction it has significant effect on designers behaviour. But it seems to be, that especially for novices such changes are difficult to percept and to assess. This hinders a major access of such methods. Therefore question-based methods should be supplied by strategies which allow to control the own design process, whether usability and other properties played an adequate role. In any case the designer needs information about the user, especially about his knowledge and his habits. Participation of users was an effective tool to make knowledge and action patterns of the users explicit and to connect it with innovative design concepts. For this purposes it is necessary to start with a combination of interviews with systematical observation of user behaviour. In this case the best results in improving usability should be obtained, but this has to be assessed by further research. Both studies show that there are promising possibilities, but they also show what still hat to be done.

Designers have to learn much more about the following aspects:

- Models of human behavior, esp. the role of existing knowledge and action patterns for the use of new technical products.
- The effects of technical innovation on knowledge and action patterns of the users.
- Methods to predict these aspects especially in the early steps of the design process.
- These predictions have also to consider developments of aspects like market, politics and culture especially for long time product development processes.

References

Ahmed, S. (2001). *Understanding the Use and Reuse of Experience in Engineering Design.* PhD thesis, Engineering Department, University of Cambridge.

Ahmed, S., Wallace, K.M., Blessing, L.T.M. (2000). *Training Document - C-QuARK Method - The Experienced Designer's Approach to Design.* Unpublished, Engineering Design Centre, Cambridge University, Cambridge.

Anderson, J.R. (1983). *The architecture of cognition.* Cambridge, MA: Harvard University Press.

Andreasen, M.M. and Mc Alonee, T. (2001). Joining three heads – Experiences from mechatronic projects. In H. Meerkam (Ed.). *12 Symposium Design for X, Neukirchen 11-12 October.* Lehrstuhl für Konstruktionstechnik Friedrich Alexander Universität: Erlangen.

Araujo, C.S. (2001). *Akquisition of Product Development Tools in Industry – A Theoretical Contribution.* PhD Thesis. Technical University of Denmark.

Araujo, C.S., Benedetto-Netto, H. Campello, A.C., Segre F.M. and Wright, I.C. (1996). The Utilization of Product Development Methods. A Survey of UK Industry. *Journal of Engineering Design*, 73, 265- 277.

Badke-Schaub,P. and Frankenberger, E. (1999). Analysis of Design Projects. *Design Studies, 20*, 465-480.

Broadbent, D.E., Fitzgerald, P. and Broadbent, M.H.P. (1986). Implicit and explicit knowledge in the control of complex systems. *British Journal of Psychology, 77,* 33 - 50.

Dörner, D. (1996). *The logic of failure*. New York: Metropolitan Books.

Dörner, D. (1999). *Bauplan einer Seele.* Reinbek b.H.: Rowohlt.

Dylla, N. (1991). *Denk- und Handlungsabläufe beim Konstruieren.* München: Hanser.

Flick, U. (1991). Stationen des qualitiativen Forschungsprozesses. In: U. Flick, E. v. Kardorff, L. v. Rosenstiel, S. Wolff (Hrsg.). Handbuch Qualitative Sozialforschung. Grundlagen, Konzepte, Methoden und Anwendungen. München: PVU, p.148-173.

Frensch, P.A. and Funke, J. (1995). *Complex problem solving: The European Perspective*. Hillsdale, NJ: Lawrence Earlbaum Associates.

Fricke, G. (1993). Konstruieren als flexibler Problemlöseprozess. VDI Verlag: Düsseldorf.

Günther, J. and Ehrlenspiel, K. (1998). How Do Designers from Practice Design? In: P. Badke-Schaub, E. Frankenberger and H. Birkhofer. *Designers - The Key to Successful Product Development* (pp. 85-97). London: Springer.

Hacker, Winfried 1990: Arbeitstätigkeitsleitende Texte: Zu arbeitspsychologischen Grundlagen der Bewertung von Tätigkeits- und Bedienungsanleitungen. In: T. Becker, L. Jäger, W. Michaeli, H. Schmalen (Hrsg.). *Erfolgreich kommunizieren, verständlich schreiben: Technik-Wissen leserfreundlich vermittelt. Leitfaden für die Praxis mit Formulierungstraining.* Köln: TÜV Rheinland.

Hacker, W. (1992). *Expertenkönnen. Erkennen und Vermitteln.* Göttingen: Verlag für Angewandte Psychologie.

ISO 13407 „Benutzerorientierte Gestaltung interaktiver Systeme“ (1997, Entwurf).

Jahn, F., Wetzstein, A., Ishig, A. and Hacker, W. (2002). Der aufgabenbezogene Informationsaustausch (AI). Weiterentwicklung einer Methode zur Gestaltung und Optimierung von Arbeitsprozessen. *Projektberichte Heft 6, Institut für Psychologie I,* Arbeitsgruppe Wissen-Denken-Handeln, TU Dresden.

Luhn, G. (1999).*Implizites Wissen und technisches Handeln am Beispiel der Elektronikproduktion.* Bamberg: Meisenbach.

Luhn, G. and von der Weth, R. (1999). Abstraction and experience: engineering design in new contexts of cognitive and philosophical science. *Proceedings of the ICED 99,* 947-952.

Neubert, J. and Tomczyk, R. (1986). *Gruppenverfahren der Arbeitsanalyse und Arbeitsgestaltung.* Berlin: Deutscher Verlag der Wissenschaften.

Newell, A. (1991). *Unified Theories of Cognition.* Cambridge, MA: Harvard University Press.

Nielsen, J. 1993. *Usability Engineering.* Cambridge, MA: AP PROFESSIONAL.

Pahl, G. and Beitz, W. (1995[2]). Engineering Design. London: Springer.

Schaub, H. (2002). *Virtuelle Akteure als Testfeld psychologischer Theorien.* Positionsreferat, gehalten auf dem 43. Kongress der Deutschen Gesellschaft für Psychologie. Berlin: Humboldt Universität.

Nonaka, I. and Takeuchi, H. (1997). *Die Organisation des Wissens. Wie japanische Unternehmen eine brachliegende Ressource nutzbar machen.* Frankfurt a.M., New York: Campus.

Von der Weth, R. (2001). *Management der Komplexität.* Bern: Huber.

von der Weth, R. and Frankenberger, E. (1995). Strategies, competence and style - problem solving in engineering design. Learning and Instruction, 5, 357-383.

von der Weth, R. and Weinert, S. (2002). Richtige Fragen sind die halbe Antwort - Fragensysteme als Hilfsmittel beim Konstruieren. In W. Hacker (Hrsg.): *Denken in der Produktentwicklung.* Zürich: vdf, 36-48.

Von der Weth, R., Hacker, W and Ishig, A. (in Vorbereitung). *Wissensbasierte Arbeitsgestaltung in der Mikrochipfertigung.* Unveröffentlichtes Manuskript. Stuttgart: Hochschule für Technik.

Acknowledgements

* The research was carried out by the Technical University Dresden, Department of General Psychology, in collaboration with the University of Cambridge, Engineering Department and the Rolls-Royce Aerospace Plc. Funding was provided by a grant of the Prof. Dr-Ing. Erich Müller Stiftung.

Methods, tools and prerequisites: Summary of Discussion

Günter Höhne and Torsten Brix, Ilmenau, Technische Universität

1 Preliminary remarks

Session III worked primarily in the area of the following three main topics:
- Cognitive outsourcing (e.g. allocation of functions between humans and computers) and mental load,
- Possibilities and limitations of computer aided creativity and
- Pros and cons of training design methods.

In conjunction with the poster session (q.v. http://www.pe.mw.tu-muenchen.de/humanbehaviour) and the written contributions the subsequent discussion was focussed on key questions which are listed in the Table 1.

Table 1. Key Questions to "Methods and Tools for Design"

"Cognitive outsourcing"	Computer aided creativity	Training design methods
Is sketching (2D and 3D) the future of design? What is the future of sketching?	How can Virtual Reality/ Augmented Reality etc. stimulate creativity in design? For what problems that´s useful?	Lot of ideas/existing tools: how to use them in designers environment?
Are there methods for understanding design space, situation and success?	How shall we really understand internal and external visualisation?	How to support the designers with knowledge they don't have?
Should our "cognitive outsourcing" be represented by a design language or ontology shared by user and computer?		Shall the toolmaker be better or the user be more trained?

This questions initiated detailed discussions whose results were further questions, findings and proposals for activities in research, development and training in Engineering design. The following three sections give a summary of the results.

2 Further Questions

All activities concerning methods and tools for Engineering Design have to consider their use in the industrial practice. This aspect is very important and was expressed in the first question:

Q1: What does industry see as the future of tools, sketching, etc.?

The other questions are formulated in relation to some main aspects of the discussion based on the aforementioned key questions. Fig. 1 shows the main aspects and their relations.

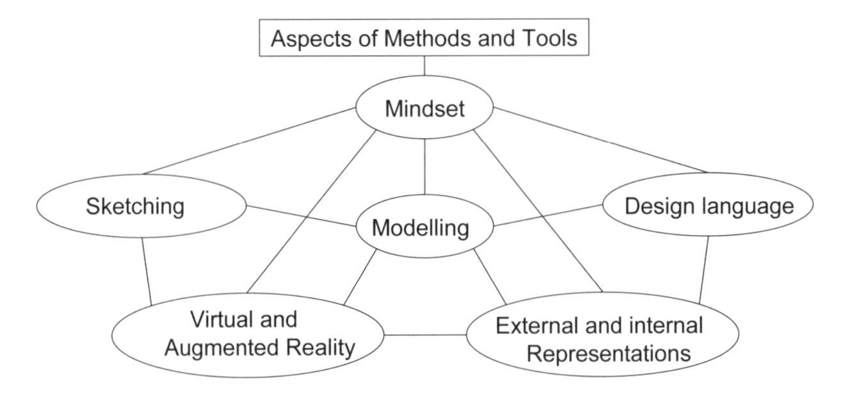

Fig. 1. Network of main aspects for developing and application of methods and tools

Mindset as a mental construct determines the operators procedural mode and the efficiency of the application of methods and tools in the design process. A number of questions result from it:

Q2: How much has the toolmaker to consider the mindset of the user and how much must the user learn about the tool?
Q3: How do prospective users of a tool work?
Q4: What information do they need?
Q5: What do they want to work quicker, better, cheaper, etc.?
Q6: Is it possible to have a set of (sub)methods that are individually adapted to the users mindset once and then can be used and combined for more „complexicated" methods?
Q7: What mixture of representations (with what level of interaction) can be used to efficiently communicate a mindset/tool to another person? Could virtual reality or augmented reality help here?
Q8: The designer knows about the knowledge he needs, but does not have. Does there exist knowledge that he doesn't know exists? How do we find it? How can we support the designer in this regard?
Q9: Should be tools and methods used in phase of creativity question oriented?

Q10: Is it possible to have a set of (sub)methods that are individually adapted to the users mindset once and then can be used and combined for more „complexicated" methods?

Q11: What mixture of representations (with what level of interaction) can be used to efficiently communicate a mindset/tool to another person? Could VR/AR help here?

Considering the sketching in 2D and 3D it came to further questions:

Q12: Use of different methods and tools for sketching depends on the task, but in which way?

Q13: Is the future of sketching depends upon the level of the sketch (formal or informal)?

Q14: Sketching is important for designing physical objects. What about non-physical objects?

Q15: What will people do when they are not with a computer? Will new generations learn with pen and paper?

Q16: Should be evaluated 3D sketching? Should be worked out scenarios?

Q17: Are we going to change from old style working (paper & ink leading to the real thing) to a new way (from the outset to the end working with a (semi)real thing)?

Q18: Can sketching, modelling, simulating, magic meeting, etc. be integrated in one (Virtual Reality) environment? What would this workspace be like?

The Modelling of the design objects on different levels of abstraction is essential for a successful problem solving. Following questions are related to this issues:

Q19: What is the optimum mix of verbal and graphical representation?

Q20: At which abstraction level should we use which type of external representation?

Q21: How do we quickly find good and comparable solutions?

Q22: Is externalisation truly possible?

Q23: Problem is not one of ontology, etc., but of modelling and representation?

A design language could be helpful for improving design work. Two questions result from this:

Q24: Do we need a common language or not?

Q25: Do we need a better way to communicate with computers?

Research tasks arise from the following questions:

Q26: Should we develop new methods to measure the success and/or quality of design?

Q27: Are the currently used prescriptive methodical design processes satisfactory? Do we need new ones?

Q28: Is there an acceptance amongst designers to use tools for early design stages? What are the requirements of such tools?

Q29: What is the effort of describing the design space for specific tasks?

3 Findings

Cognitive outsourcing

F1: While the power of sketching for ideation and communication is obvious. When it is used in a group context, we need to explore the additional requirements it imposes on group interaction.

F2: We should apply a useful distinction between passive memory aids (external representations), tool, assistive agents (machines). Outsourcing aids can help with figurative, textual or mixed information; however, conceptual/verbal and graphical/pictorial are strongly cross linked.

F3: Graphical representations can be too concrete, while verbal representation can easier support abstractions. As a result, verbal descriptions can sometimes support generation of new solutions better than graphical representations can.

F4: Useful domain independent ontology is emerging for design, encompassing both design methods (plan refinement, e.g.) as well as artefact specification (device, component, structure, behaviour, function, causal process, etc). Such ontology is essential for proper human-machine communication, and even human-human and machine-machine communication.

The following findings aim at sketching as one important activity for cognitive outsourcing:

F5: Sketching is and will be a powerful tool for the optimisation of ideas and solution. Sketch → Analyses → creating/developing/modifying ideas → Sketch. Creating/developing/modifying is mediated by the sketch-analysis cycle. 3D-Sketching is a good example of computer supported tool for the design process.

F6: Sketching is useful to represent ideas even in non spatial situations. Sketches aid in "scare folding", connecting little bits of implicit knowledge to make a stable structure of knowledge. Thus in group discussions this form of sketching may also be very useful. In many areas of discourse diagrams are extensively used to indicate various kinds of connections between entities, for example graph representations, semantic networks.

F7: Knowledge systemisation should be based on a good understanding of how to modularise knowledge into units and how to compose the units into larger structures of knowledge. Ontology research is necessary to support this form of modularisation and composition.

Computer-aided creativity and imagination

F8: While much has been written about creativity, there is still no consensus on its nature and mechanisms. It is not even clear that we are not engaged in false reification when we talk about creativity as a separable process.

F9: External visualisation (or more generally, perceptual and kinaesthetic simulation) can adopt more powerful techniques than is possible for internal imagination. Thus, for example, automated morphing of shapes over specified ranges may aid in imagining possibilities. VR can help here.

F10: Virtual environments can help a designer "dream away" to find new ideas; the environment stimulates internal visualisation.

F11: Supporting the designer with tools that provide insight (raise questions, focus, find or suggest new information, etc.) is very important. Some tools exist for this purpose such as design space explorers.

F12: Supporting creativity requires better ways of retrieving from graphical memories, such as sketches and other visual representations, and displaying this information.

Training, tools and methods

F13: Usability of tools/methods has to be improved – the "mindset" of the designer has to be kept in mind by the toolmaker. The development of all tools, whether for the designer or anyone else, should involve the user in all stages.

F14: A process of considering the design space involving trying the sources of underdetermination in the design space and exploration of extreme situations in the space. Such a process can help in learning about the relationship between underdetermined parameters and design performance measures.

F15: During the phase of integration of a new tool or method the task of actually using that tool and method should go further and further into the background. When the training is over the user should be able to focus on the task without needing to constantly focus on the tool or method.

F16: Historically tools comes about in two distinct ways; need-driven and technology-driven; in the former tools are built to satisfy needs; in the latter the technological possibility (3D sketching, say, VR) opens up, but at first it's not clear how best to deploy it, if at all. The latter requires a period of co-evolution between the user community (in this case, designers) and the tool-builders, during which the possibility-space is explored and the appropriate mode of deployment is discovered together. This co-evolution may also result in radical changes in the practice of design; for example: concurrent design, made possible by advances in computing and communication, has radically changed the practice of design in big companies.

These findings form a basis for investigations concerning the questions in section 1 as well as for proposals of the further work.

4 Proposals

Very important for the next period of investigations is to define a strategic roadmap which should include:

P1 We should achieve a common understanding: Build an ontology of human behaviour in designing.
P2 Whole concept of tools must be revisited.
P3 We should determine essential elements of early design

Following proposals have importance for developing tools and methods:

P4 We should investigate how and when given tools can be used – virtual reality (VR) and augmented reality for conceptual design (synthesis) or for embodiment design (analysis).
P5 We should develop requirements for VR-systems ("VR-systems should be accessible").
P6 We need essential revisions to our approach to toolmaking:
- treat tool- and method-making as a design process and
- "Fundamentals of knowledge need to be delivered to toolmakers".
P7 There is more to design than cognition and recognition.
P8 We need a clearer understanding of the types of modeling (2D-sketching, 3D-sketching, CAD, mathematical models, spreadsheets, ...) and their function in the design process.
P9 We should carefully distinguish between scientific results, hypotheses and beliefs.
P10 What are reliable scientific results anyway?

Future Issues in Design Research

Lucienne Blessing

Technical University Berlin

Introduction

While making notes during the conference and trying to summarise what had come out during two days of intensive discussion, it became more and more clear that the three issues had started to merge. The rather distinct questions that were formulated before the conference, were reformulated and refined during the first discussions into a set of questions that guided the subsequent discussions. Although this was done seperately in each of the thematically different streams, some overlap started to emerge. The subsequent discussions, formal as well as informal, showed that from which point of view this research area was considered - individual, group or computer support – common issues emerged that turned out to be concerned with the effectiveness and efficiency of the process of doing design research, rather than its research questions.

When presenting these in the summary session of the conference, some additions and refinements were made, but a general agreement seemed to exist that these were the issues that needed resolving and the need was expressed to finalise the conference by setting priorities and suggesting ways of resolving the issues.

This chapter contains in its first part the core issues and some of the discussion as well as personal reflections. The second part presents the most urgent issues and ways to proceed.

Issues

Terminology and common understanding

Although the participants formed a specialised part of the design resarch community as a whole, the issue of differences in terminology became very clear. Terms borrowed from other disciplines had been interpreted in different ways, the disciplines represented by the participants had their own terminology unfamiliar to the others and the same terms coming from participants from different cultures had a different meaning. This resulted in confusion, discussion and a clearly expressed need to have a common terminology. A common terminology, was seen as essential for a research area. Only in this way a common understanding and building on existing work can be realised. Both are nevessary for progress. One major stumbling block are the links between the terms: if one term changes, others may have to change too. When defining conceptual design, the definition of the preceeding and following phase have to in line with this definition, at least within one school of thought. These interrelationships hinder the formulation of standard definitions.

The issue of terminology has been adressed in the past, e.g. Hubka wrote a dictionary in three languages [Hubka, 81], at the Engineering Design Centre in Cambridge an attempt was made to develop a glossary in one language, accepting multiple definitions to allow for different definitions in different schools. [Chakrabarti, 94]. Unfortunately these and other attempts did not lead to a generally accepted common source.

Closer collaboration between disciplines on theories, methodologies and research methods.

This issue is related to the issue of terminology. Terms and research methods have been borrowed from other disciplines, but not always correctly because the paradigms and theories were not known. A closer collaboration with other disciplines could not only prevent terminology confusion but provide theories, such as action theory and group dynamic theory, that could be very relevant starting points for design research. Other disciplines can further provide interesting research methods to study design, as the research in this conference showed. The conference enabled the exchange between researchers, several of which had not known eachother, despite the fact that many of the participants have been working in this area for many years. An overview of those involved in this area of research is obviously missing.

Common model

The wish for a common model or a set of (partially shared) models came up frequently. Few models exist, most research results remain unconnected. Such a model would indicate a better understanding of design and would provide a shared understanding, a basis on which to do research. It remained open as to whether the development process, the product, the thinking process or a combination of aspects should be modelled. Formulating an all-encompassing model seemed, at least for the time being, unrealistic. A far better understanding of all the aspects is needed before such a common model could be developed. Model, rather than theory was used as term, indicating the fact that common models would be a first step to be taken.

Consolidation

The two previous issues lead to the wish for consolidation. An increasing number of researchers has been involved in empirical design resarch, focusing on different aspects of design, using different methods. Referencing islands have emerged and few attempts exist to brings it all together. A consolidation is necessary in the form of an overview and subsequent analysis of the research and its findings. This will allow resarchers to build on eachother's work, to gradually come to common models and to a more fundamental understanding of design. This would in turn help the development of methods and tools for design, by providing the input necessary to develop support that adresses core issues and has a higher likelihood of being accepted.

Classification of research area and resarch findings

Design research is one of the few, if not only, resaerch area that, despite the wide variety of topics that it covers, does not have a commonly accepted classification (ontology, taxonomy) of its research topics and findings. This is not only visible in the different and broad listings of topics covered by design conferences, but also in the problems authors sometimes have to assign their paper to a particular topic. A classification would help to structure the research topics and findings and to identify their interrelationships. It furthermore would provide an overview in this exponentially grown research area and allow the identification of areas in which research is still lacking. Care has to be taken that this classification will not lead to larger, but more disconnected islands of research.

Development of a research methodology

Design research is multi-facetted, focusing on process, product, organisation, tools, etc. and covers a large number of strongly interconnected aspects. This results in a large variety of research questions and the need for a large variety of research methods to be able to answer these questions. Design research does not have its own research methods (yet) but uses those from other disciplines. The problems are that many possibly useful methods are not known, that the paradigms on which these methods are based are often not considered (emphasising the need for collaboration), but more importantly, many of the existing methods have to be adapted to the special characteristics of design. Not even the research approach proposed in other disciplines can be applied directly. Design research could use its own research methodology, based on existing methodologies and the typical characteristics of design as field of study.

Research objective and goals

In design research, resaerch objectives and goals are often stated at a very holistic level (improving design) or remain fuzzy. A study of the proceedings of two ICED conferences showed that in 47% of the 331 papers on tools and methods, motivations are absent: only in 33% of papers were they defined precisely. The issues of implementation in industrial settings is only dealt with in 37% of the papers [Cantamessa, 01]. The latter is interesting, as the goals are often formulated as an industrial need, such as reducing lead time. A validation as to whether the goals have been achieved, does not take place. Validation may not be possible, because of the available research time (it may be years before the effects of research influences an aspect such as lead time) or because the goal is formulated at too high a level (a large number of interrelated factors may have influenced the goal, not just the one we are adressing) A possible consequence of the lack of validation or at least evaluation may be the fact that many research results are not used in practice.

Common set of criteria for good research

The preceeding issues give rise to the need for a commonly accepte set of criteria for good research. This will give guidance to researchers and help the identification of research results that form a solid basis. This set should address which of the many methods are suitable for use in design research, which modifications are allowed, and what validity and objectivity in the area of design research mean.

Communalities and specificities between research findings and domains

The conference showed differences in findings in the different domains that had been investigated (machine design, architecture, etc.). The need was expressed to clarify the communalities and specifities in the various domains to obtain a better understanding of design in general and of design in the different domains.

Changing models, dynamics of models

The need was expressed for models that can change to take into account the dynamics of models. Most models are static descriptions.

Activities

During the discussion of the identified core issues, the need was expressed to come up with concrete ideas on how to adress at least some of these issues. The discussion revealed that several initiatives existed of which many participants were unaware, confirming the observed islands of research. A more collaborative effort and overviews of existing research findings and experiences was considered to be of utmost importance for the research community to become more established. The Design Society (www.designsociety.org) was considered as a worldwide platform through which to address some of the issues. The following is a listing of the proposed actions and some of the initiatives that exist.

A Special Interest Group on Human Behaviour in Design

A Special Interest Group within the Design Society will be set up (Lindemann), which will act as a forum to start common activities and become at least a repository for events and work resulting from the area of empirical design research.

Map of research areas and researchers

This map supports communication and provides a basis for consolidation. The Design Society could help establishing at least the network. As the list of researchers and research areas grows, the earlier described classification of research will become increasingly important.

Courses in research methods applicable in design research

Some of those courses already exists, but are not known. These courses should be announced more widely, e.g. through the pages of the Design Society.

Summerschool to learn about design research

The International Summer school on Engineering Design Research, is an example in which PhD students are taught about existing research and about how to do design research (Andreasen and Blessing). An overview of such summer schools should be published.

A web-based database with empirical design research methods

The unique feature of the database would be the inclusion of the purpose for which specific research methods were used and the experiences - both positive and negative - with these methods. The discussions this information will give rise to, will improve the methods, give guidance to new researchers and gradually form an established set of methods for doing design research.

Database of empirical studies in design, their setup and findings

This database is under development in Berlin (Blessing) based on an earlier database which was not publicly available. Taking up other relevant findings from other disciplines in this database seemed a useful addition.

A set of test problems / test tasks

Building on the idea of providing overviews, the development of a repository of test problems and tasks for use by other researchers was proposed. Developing a design (or other) task for an empirical study requires several iterations. Using those developed and tested by others will not only save much time but also make research results more comparable. An example is the wall-mounted swivel mechanism used by several research groups (see Ehrlenspiel in this book).

A set of protocols for common use

Not only the developmnt of test problems is time-consuming, data collection and in particular transcription occupies a large amount of time. Making available transcribed protocols, including all the details of the empirical study itself, could reduce research time and make results more comparable. Those protocols could be particularly useful in explorative studies for new researchers to obtain a feeling for the type of data particular methods provide and to help them find a focus. An example of the use of one protocol analysed by various researchers using different methods is [Frankenberger et al, 1998]

Database with teaching material

(as above but for teaching design)

Contact/communication with other societies

The conference showed that the researchers collectively knew a large number of other societies, global as well as regional, but that this information had not been shared. The Design Society and its members were asked to start listing the interesting societies and make this list available.

Conclusions

Many of the issues listed in the first part of this chapter will not be addressed by these activities, at least not directly. General agreement existed, however, that a basis of shared knowledge, information and data is necessary before the issues can be addressed in an effective and efficient way.

The conference confirmed my feeling that human behaviour in design is not only a necessary, but also an interesting and exciting area of research with a richness that offers an enormous potential for further exploration but with a lack of interconnection. Collaborative efforts are required to put this research area on a firm basis. The agreed activities will help form this basis.

References

Cantamessa M (2001) Design research in perspective: a meta-resaerch on ICED97 and ICED99. In: S Culley et al (eds) Design Resarch – Theories, methodologies and product modelling, ICED01. Professional Engineering, London, pp 29-36

Chakrabarti A (1994) A Framework for a Glossary of Terms in Engineering Design. Cambridge University Engineering Department, Technical Report CUED/C-EDC/TR19

Frankenberger E. et al (eds) (1998) Designers - the Key to Successful Product Development. Springer Verlag, Heidelberg

Hubka V (1981) Terminology of the Science of Design Engineering in 6 languages. WDK 3 Heurista Publisher, Zurich

Printing: Saladruck Berlin
Binding: Stürtz AG, Würzburg